Neil F. Cramer

The Physics of Alfvén Waves

Neil F. Cramer

The Physics of Alfvén Waves

Berlin · Weinheim · New York · Chichester · Brisbane · Singapore · Toronto

Dr Neil F. Cramer
School of Physics,
University of Sydney, Australia
e-mail: cramer@physics.usyd.edu.au

1st edition

Library of Congress Card No: applied for

British Library Cataloguing-in-Publication Data: A catalogue record for this book is available from the British Library.

Die Deutsche Bibliothek – CIP Cataloguing-in-Publication-Data
A catalogue record for this publication is available from Die Deutsche Bibliothek

© WILEY-VCH Verlag Berlin GmbH, Berlin (Federal Republic of Germany), 2001

ISBN 3-527-40293-4

Printed on non-acid paper.

Printing: Strauss Offsetdruck GmbH, D-69509 Mörlenbach
Bookbinding: Wilhelm Osswald & Co., D-67433 Neustadt (Weinstraße)

Printed in the Federal Republic of Germany.

WILEY-VCH Verlag Berlin GmbH
Bühringstrasse 10
D-13086 Berlin

Dedicated to
Christine

Preface

This book was written with the aim of providing a unified treatment, suitable for beginning researchers, of the properties and applications of Alfvén waves and related waves in magnetized plasmas, that is, ionized gases in a magnetic field strong enough to affect the behaviour of the ions and electrons. These waves are among the most fundamental features of magnetized plasmas. The term "Alfvén wave" is used with a number of different meanings in the literature, and it is intended in this book to provide a guide to these different usages.

The book covers the basic properties of the low–frequency wave modes in magnetized plasmas, the Alfvén waves and magnetoacoustic waves. (Henceforth in this book the general term "Alfvén waves" will be understood to encompass magnetoacoustic waves as well). In addition, it covers results of the latest research in applications of the waves in the contexts of laboratory, space and astrophysical plasmas, in particular some of the results achieved since the publication over a decade ago of the two "standard" references on Alfvén waves, *The Alfvén Wave* by Hasegawa & Uberoi (1982), and *An Introduction to Alfvén Waves* by Cross (1988). This book also covers a somewhat broader range of topics than the two earlier books.

There is much potential for cross–fertilization between the different areas of application of plasma physics. This book aims to facilitate this cross–fertilization by showing the common features of the physics of Alfvén waves across the various plasma environments. There is, of course, an enormous volume of work on Alfvén waves in the literature, so I have primarily selected topics in which I have research experience. Hopefully this selection will give starting researchers a flavour of this interesting field of physics, and at least help to point them in the right direction for their specific interests.

The book deals especially with nonideal effects, such as multi–species, collisional and kinetic effects. These physical processes have consequences for the dispersion relations and absorption properties of linear and nonlinear Alfvén waves that are not predicted by ideal or pure magnetohydromagnetic (MHD) theory. It has been common in the literature to use ideal MHD theory to treat the behaviour of the waves, particularly when treating problems of propagation in nonuniform plasmas. This approach has the virtue that the basic equations are relatively simple differential equations, which may often be solved analytically. They lead to many interesting mathematical problems, such as the treatment of the absorption of wave energy at the "Alfvén resonance" by analysing the behaviour of the solutions at the singularities of the differential equations, as well as nonlinear properties of the waves.

However, the emphasis of this book is the analysis of the effects of nonideal corrections to the MHD theory on the properties of the waves. These corrections are intended to reflect the physics of the realistic applications of the theory more closely.

Some of the corrections which have been treated in the literature are: finite frequency effects (i.e. allowance for the wave frequency to approach the ion cyclotron frequency), inclusion of minority ion species and charged dust grains, resistivity, viscosity, friction with neutral particles, and kinetic theory effects. All of these modifications of the plasma change the dispersion relation of the waves, and most cause damping of the waves. The disadvantage of including nonideal effects compared to the MHD approach is that the starting equations are higher-order fluid differential equations, in the case of finite frequency, resistivity and viscosity effects, or are coupled integro-differential equations in the case of kinetic theory effects. The greater complexity of the equations reflects the greater number of wave modes allowable in the plasma when nonideal effects are taken into account, and they can often only be solved numerically.

A unique feature of the "shear" Alfvén wave in ideal MHD is the fact that wave energy propagates along the magnetic field, regardless of the angle of the wave front with the magnetic field. This feature leads to several fascinating phenomena, the discussion of which in the literature has at times been controversial, such as localized propagation, the absence of discrete eigenmodes in nonuniform plasmas, and resonance absorption. This book endeavours to provide a coherent and unified explanation of such phenomena in terms of the above–mentioned nonideal effects.

The plan of the book is as follows. In the first chapter we establish the basic models used to describe the plasma, and give the equations used in the subsequent analysis of the waves. The multi–fluid, ideal MHD and kinetic theory models are discussed. Chapter 2 deals with the waves in a uniform plasma, employing the different plasma models, and then Chapter 3 discusses the waves in nonuniform plasmas, in particular the case of stratified plasmas. Chapter 4 follows on from Chapter 3 with a treatment of surface waves in strongly nonuniform plasmas. The instabilities of nonequilibrium plasmas that produce Alfvén and magnetoacoustic waves, and the theory of nonlinear Alfvén waves, and the instabilities of the waves themselves, are reviewed in Chapter 5. The applications of Alfvén waves in laboratory plasmas are discussed in Chapter 6, including the use of the waves in devices for controlled nuclear fusion. Chapter 7 investigates the natural occurrence of the waves in space and solar plasmas, ranging from the Earth's magnetosphere to the interplanetary plasma and the Sun's atmosphere. Finally, in Chapter 8 we look at two problems in astrophysical plasmas: waves in partially ionized and dusty interstellar clouds, and in the relativistic and very strongly magnetized plasma of pulsar magnetospheres.

I owe a great deal to my close research collaborators on the physics of Alfvén waves in Sydney, initially Ian Donnelly and, more recently, Sergey Vladimirov, for their stimulation. Amongst my overseas collaborators, I would like to particularly thank Frank Verheest and Jun-Ichi Sakai. I have also drawn much inspiration over the years from the group of Australian Alfvén wave enthusiasts, including the late Frank Paoloni, Rod Cross, Bob Dewar and Robin Storer. Don Melrose has inspired me through his vast knowledge of plasma physics, and has provided a stimulating research environment in Sydney. Dave Galloway has been my valued local expert on solar Alfvén waves. I thank Bob May and Les Woods for introducing me to Alfvén waves at the beginning of my career. Lastly, my students George Rowe, Robert Winglee, Ken Wessen and Lap Yeung, have made invaluable contributions to our research in Sydney on Alfvén waves.

Part of the writing of this book was accomplished during a study leave spent at Toyama University, Japan, the University of Gent, Belgium, and the Max Planck Institute for Extraterrestrial Physics, Garching, Germany, and I thank Professors Sakai, Verheest and Morfill for their hospitality. The work encompassed in this book has also been supported by grants from the Australian Research Council.

Sydney
June 2001 *Neil Cramer*

Contents

1 Descriptions of Magnetized Plasmas

1.1 Introduction

It was discovered, some five decades ago, that low–frequency electromagnetic waves are able to propagate in conducting fluids, such as plasmas, even though they cannot propagate in rigid conductors. Hannes Alfvén, in 1942, investigated the properties of plasmas, assuming the plasma medium to be a highly conducting, magnetized and incompressible fluid. He found that a distinctive wave mode arises in the fluid, propagating along the magnetic field direction (Alfvén 1942). This wave is now called the shear or torsional Alfvén wave. The existence of the wave, in the conducting fluid mercury, was experimentally verified by Lundquist (1949). The importance of the waves discovered by Alfvén for space and astrophysical plasmas was soon realized, and the compressible plasma case, which leads to the fast and slow magneto-acoustic waves in addition to the shear Alfvén wave, was treated by Herlofsen (1950).

The Alfvén and magnetoacoustic waves, which are the basic low–frequency wave modes of magnetized plasmas, have been the subject of intense study in the succeeding decades. The main reason for the great interest in these waves is that they play important roles in the heating of, and the transport of energy in, laboratory, space and astrophysical plasmas. The "Alfvén wave heating" scheme has been investigated theoretically and experimentally as a supplementary heating scheme for fusion plasma devices, and it has been invoked as a model of the heating of the solar and stellar coronae. The waves are believed to underlie the transport of magnetic energy in the solar and stellar winds, transfer angular momentum in interstellar molecular clouds during star formation, play roles in magnetic pulsations in the Earth's magnetosphere, and provide scattering mechanisms for the acceleration of cosmic rays in astrophysical shock waves. These and other applications of Alfvén and magnetoacoustic waves in the fusion, space physics and astrophysics fields are the subject of this book.

In realistic physical problems in all plasma environments, Alfvén and magnetoacoustic waves propagate in nonuniform plasmas. As a result, the waves may be reflected, transmitted or absorbed. The practical question of the heating to high temperatures of laboratory fusion plasmas that are contained in a vessel, and are therefore necessarily nonuniform, involves such processes. The space and astrophysical environments where the waves are found are also inevitably nonuniform.

Wave energy can be concentrated in plasma regions of nonuniform density and/or magnetic field, and in the limiting case of density or magnetic field discontinuities, when a well-defined surface is present, wave eigenmodes exist whose amplitudes decay approximately exponentially in each direction away from the surface. These are the Alfvén surface wave eigenmodes, which have been shown by theory and experiment to play an important role in Alfvén wave heating, because they can be easily excited by an antenna in a laboratory plasma. Alfvén surface waves are also expected to exist in astrophysical plasmas where jumps in density or magnetic field occur, such as the surfaces of magnetic flux tubes in the solar and stellar atmospheres, or the boundaries between plasmas of different properties in the Earth's magnetosphere. The properties of the waves when they propagate in nonuniform plasmas, including the phenomenon of Alfvén surface waves, are treated here in some depth.

A number of different models of the plasma, namely the multi–fluid, ideal MHD, Hall-MHD and kinetic theory models, are presented in this chapter, as well as a brief summary of Fourier transform theory. The results are used in later chapters to describe the self–consistent response of the plasma medium to the presence of the waves. All the models use Maxwell's equations for the electric field $\boldsymbol{E}$ and magnetic field $\boldsymbol{B}$:

$$\nabla \cdot \boldsymbol{E} = \frac{\rho_e}{\epsilon_0} \tag{1.1}$$

$$\nabla \cdot \boldsymbol{B} = 0 \tag{1.2}$$

$$\nabla \times \boldsymbol{E} = -\frac{\partial \boldsymbol{B}}{\partial t} \tag{1.3}$$

$$\nabla \times \boldsymbol{B} = \mu_0 \boldsymbol{J} + \frac{1}{c^2} \frac{\partial \boldsymbol{E}}{\partial t}. \tag{1.4}$$

Here ρ_e is the electric charge density, $\boldsymbol{J}$ is the current density, and c is the speed of light *in vacuo*. SI units are used throughout.

1.2   The Multi–Fluid Equations

We consider first the multi–fluid model of a plasma, in which each distinct species of particle is specified by the index α, with mass m_α and charge $Z_\alpha e$, where e is the fundamental unit of electric charge. Each collection of particles of a specific type is supposed to act as a fluid, with its own velocity $\boldsymbol{v}_\alpha$, mass density ρ_α, number density n_α and pressure p_α. Each fluid is "collision dominated". This means that the time for relaxation of each type of species to a Maxwellian velocity distribution with a unique temperature T_α, through collisions of like particles, is short compared with the other time–scales of interest. Each fluid may be acted on by the electric and magnetic fields, and may act on the other fluids via collisions, which may have characteristic times of the order of the time–scales of interest.

The equation of motion for the fluid corresponding to the species α is

$$\frac{d\boldsymbol{v}_\alpha}{dt} = -\frac{1}{\rho_\alpha}\nabla p_\alpha + \frac{Z_\alpha e}{m_\alpha}(\boldsymbol{E} + \boldsymbol{v}_\alpha \times \boldsymbol{B}) - \sum_{\alpha'} \nu_{\alpha\alpha'}(\boldsymbol{v}_\alpha - \boldsymbol{v}_{\alpha'}) \tag{1.5}$$

where $\nu_{\alpha\alpha'}$ is the collision frequency of a particle of species α with particles of species α'. The continuity equation for each fluid is, if sources and sinks for the particles are neglected,

$$\frac{\partial \rho_\alpha}{\partial t} + \nabla \cdot (\rho_\alpha \boldsymbol{v}_\alpha) = 0. \tag{1.6}$$

A partial pressure $p_\alpha = n_\alpha k_B T_\alpha$, where k_B is Boltzmann's constant, may be associated with each species. The thermal speed for species α is then defined as

$$V_\alpha = \sqrt{k_B T_\alpha / m_\alpha}. \tag{1.7}$$

A strong magnetic field B may enable the pressure parallel to B to be different to the pressure perpendicular to B, leading to distinct parallel and perpendicular temperatures for each species. Such concepts are useful if the relaxation times for the pressures in the two directions are longer than the wave periods being considered.

The multi–fluid equations (1.5) and (1.6) may be combined and simplified under various assumptions, as is covered in elementary plasma physics texts such as the books by Schmidt (1979) and Tanenbaum (1967). A careful discussion of the basis of multi–fluid models in terms of relative time–scales is to be found in the book by Woods (1987).

The simplest approximation of the multi–fluid model is that of magnetohydrodynamics (MHD), where the inertia of the electron fluid is neglected, and the motions of the different ion and neutral species are combined such that the plasma is assumed to act like a single fluid. If the collisions between electrons and ions are allowed for, the electron momentum equation reduces to Ohm's law, which relates the electric field E' in the rest frame of the fluid to the current density J.

The form of Ohm's law used in "collisional" or "resistive" MHD is

$$E' = E + v \times B = \eta J \tag{1.8}$$

where v is the single fluid velocity, and η is the electrical resistivity, related to the electron-ion collision frequency ν_{ei} by

$$\eta = \nu_{ei} / \epsilon_0 \omega_{pe}^2 \tag{1.9}$$

where

$$\omega_{pe} = \left(\frac{n_e e^2}{\epsilon_0 m_e} \right)^{1/2} \tag{1.10}$$

is the electron plasma frequency. Here n_e is the electron density. The "ideal" MHD model, in which the resistivity is neglected in Eq. (1.8), is discussed further in the next section.

1.3   The Magnetohydrodynamic Model

Magnetohydrodynamics, or hydromagnetics, is a fluid model which describes a magnetized plasma in which both the ions and the electrons are said to be strongly magnetized or tied to the magnetic field lines, which is to say that the magnetic field is strong enough that the cyclotron periods of all the charged species are well below all other time–scales of interest. The entire plasma acts like a single normal fluid with a single well–defined temperature and pressure. The equations derived from the multi–fluid equations (1.5) and (1.6) correspond to the plasma being treated as a single fluid of density ρ and velocity $\boldsymbol{v}$, and include the continuity equation:

$$\frac{\partial \rho}{\partial t} + \nabla \cdot (\rho \boldsymbol{v}) = 0 \tag{1.11}$$

and the equation of fluid motion:

$$\rho \frac{\mathrm{d}\boldsymbol{v}}{\mathrm{d}t} = -\nabla p + \boldsymbol{J} \times \boldsymbol{B} \tag{1.12}$$

where p is the thermal pressure of the plasma particles, that is, the sum of the partial pressures due to each species of plasma particle.

The third of the MHD equations, in the presence of resistivity, is Ohm's law (Eq. 1.8). In the derivation of Eqs. (1.8) and (1.11) (for example by Woods, 1987), the plasma is assumed to be charge neutral, and the electron mass is assumed negligible compared to the ion mass. In the absence of resistivity or other collisional processes, the magnetic lines of force are said to be frozen–in to the plasma. This concept is valid provided there is no electric field along the magnetic field direction.

We also need an equation of state linking the pressure to the other state variables, such as the adiabatic equation of state

$$\frac{\mathrm{d}}{\mathrm{d}t}(p\rho^{-\gamma}) = 0 \tag{1.13}$$

with γ the adiabatic index. On the other hand, for an incompressible plasma, the equation of state is

$$\frac{\mathrm{d}\rho}{\mathrm{d}t} = 0 \tag{1.14}$$

or equivalently, from Eq. (1.11),

$$\nabla \cdot \boldsymbol{v} = 0. \tag{1.15}$$

In the case of an adiabatic equation of state for a nonionized fluid, the speed of sound is given by

$$c_{\mathrm{s}} = (\gamma p/\rho)^{1/2}. \tag{1.16}$$

We thus see from Eqs. (1.13) and (1.14) that the incompressible equation of state (1.14) corresponds to an infinite adiabatic index and infinite speed of sound.

In Ampère's law Eq. (1.4) in the MHD model we usually neglect the displacement current term, the reason being that the characteristic speeds are normally much less than the speed of light *in vacuo* (an exception is considered in Chapter 8). In this case we may write, for the magnetic force in Eq. (1.12),

$$\boldsymbol{J} \times \boldsymbol{B} = -\frac{1}{2\mu_0}\nabla B^2 + \frac{1}{\mu_0}(\boldsymbol{B} \cdot \nabla)\boldsymbol{B}. \tag{1.17}$$

Table 1.1: Representative values of the electron density n_e, temperature T, magnetic field B, Alfvén speed v_A, sound speed c_s and plasma β, in different physical regimes.

	$n_e(\mathrm{m}^{-3})$	$T(\mathrm{K})$	$B(\mathrm{T})$	$v_A(\mathrm{ms}^{-1})$	$c_s(\mathrm{ms}^{-1})$	β
Laboratory plasma	$10^{18}-10^{24}$	10^6	1	10^7-10^4	10^5	$10^{-4}-10^2$
Ionosphere	10^8-10^{12}	10^3	$10^{-4.5}$	10^7-10^5	10^3	$< 10^{-4}$
Solar corona	10^{13}	10^6	10^{-4}	10^6	10^5	10^{-2}
Solar atmosphere	10^{18}	10^4	10^{-1}	10^6	10^4	10^{-4}
Gaseous nebula	10^9	10^2	10^{-9}	10^3	10^3	1
Interstellar gas	10^6	10^2	10^{-9}	10^4	10^3	10^{-2}

The first term on the right–hand side of Eq. (1.17) corresponds to minus the gradient of an effective *magnetic pressure*,

$$p_B = B^2/2\mu_0 \tag{1.18}$$

which may be combined with the particle pressure p in Eq. (1.12) to give an effective total pressure

$$p_T = p + p_B. \tag{1.19}$$

The strength of the particle pressure compared with the magnetic pressure is measured by the plasma beta,

$$\beta = \frac{p}{(B^2/2\mu_0)}. \tag{1.20}$$

β is proportional (with the constant of proportionality depending on the equation of state) to the ratio of the square of the sound speed c_s to the square of the Alfvén speed v_A, which as we shall see is the characteristic speed of low–frequency shear Alfvén waves , and is given by

$$v_A = B/(\mu_0\rho)^{1/2}. \tag{1.21}$$

Thus for an adiabatic equation of state, we have

$$\beta = \frac{2}{\gamma}\frac{c_s^2}{v_A^2}. \tag{1.22}$$

Typical values of v_A, c_s and β, for the particle densities, temperatures and magnetic fields applicable to the various physical regimes to be covered in this book, are shown in Table 1.1.
  It is also useful to distinguish the ion and electron betas separately:

$$\beta_i = \frac{p_i}{(B^2/2\mu_0)}, \qquad \beta_e = \frac{p_e}{(B^2/2\mu_0)} \tag{1.23}$$

where p_i and p_e are the partial pressures of the ions and the electrons respectively. We note that in an incompressible plasma, β can be finite, even though c_s becomes infinite.

The second term on the right–hand side of Eq. (1.17), $(\boldsymbol{B} \cdot \nabla)\boldsymbol{B}/\mu_0$, can be decomposed into two components. Defining $\boldsymbol{b}$ as the unit vector in the direction of the magnetic field, the component aligned with the magnetic field may be written as

$$\boldsymbol{bb} \cdot \nabla B^2/2\mu_0.$$

This component cancels the field–aligned component of the magnetic pressure gradient. Thus only the components of the magnetic pressure gradient perpendicular to the magnetic field exert force on the plasma. If this magnetic pressure gradient force exists, there is said to be *magnetic compression*.

The component of $(\boldsymbol{B} \cdot \nabla)\boldsymbol{B}/\mu_0$ that is perpendicular to the field may be written as (e.g. Kivelson 1995a)

$$-\boldsymbol{n}B^2/\mu_0 R_c \tag{1.24}$$

where $\boldsymbol{n}$ is the outward normal vector and R_c is the local radius of curvature of the magnetic field. This component is antiparallel to the radius of curvature of the field lines, and is called the *magnetic tension* or the *curvature force*. It is present only for curved field lines, and is analogous to the perpendicular force exerted by tension in a curved string. It acts to reduce the curvature of the field line. These concepts of magnetic compression and tension are useful for gaining an intuitive idea of the behaviour of Alfvén waves, as we will find in the next chapter.

A generalization of the MHD model, sometimes used to describe waves in collisionless plasmas with anisotropic pressures and a strong magnetic field, is the CGL or double–adiabatic approximation (Chew, Goldberger & Low 1956, Schmidt 1979). Replacing Eq. (1.13) are two adiabatic equations of state for the pressures parallel and perpendicular to the magnetic field:

$$\frac{\mathrm{d}}{\mathrm{d}t}\left(\frac{p_\perp^2 p_{||}}{\rho^5}\right) = 0, \qquad \frac{\mathrm{d}}{\mathrm{d}t}\left(\frac{p_\perp}{\rho B}\right) = 0. \tag{1.25}$$

The first relation follows from parallel motion of the particles being independent of the perpendicular motion, while the second expresses constancy of magnetic moments (Schmidt 1979). Instead of Eq. (1.12) with Eq. (1.17), we have two equations of motion for the parallel and perpendicular components of $\boldsymbol{v}$:

$$\rho\left(\frac{\mathrm{d}\boldsymbol{v}}{\mathrm{d}t}\right)_{||} = -\nabla_{||}p_{||} - (p_\perp - p_{||})\left(\frac{\nabla B}{B}\right)_{||} \tag{1.26}$$

$$\rho\left(\frac{\mathrm{d}\boldsymbol{v}}{\mathrm{d}t}\right)_\perp = -\nabla_\perp\left(p_\perp + \frac{B^2}{2\mu_0}\right) + \frac{1}{\mu_0}\left((\boldsymbol{B} \cdot \nabla)\boldsymbol{B}\right)_\perp \left(\frac{p_\perp - p_{||}}{B^2/\mu_0} + 1\right). \tag{1.27}$$

1.4   The Hall-MHD Model

The "two–fluid" model is the next fluid approximation, and assumes the plasma to consist of an electron fluid and a single ion species fluid. The electrons are considered to be magnetized (electron cyclotron period much shorter than time–scales of interest), while the ions are not completely magnetized (ion cyclotron period comparable with the other time–scales). In its simplest form this model is often referred to as Hall MHD, since the Hall term is present on the right–hand side of Ohm's law (assuming resistivity to be negligible):

$$E + v \times B = \frac{1}{n_i e} J \times B,$$ (1.28)

where n_i is the ion number density. The absence of the Hall term in Ohm's law Eq. (1.8) used in MHD means that ion cyclotron effects are absent in the MHD model, in other words the ion cyclotron frequency

$$\Omega_i = \frac{Be}{m_i}$$ (1.29)

is assumed much higher than the wave frequency, whereas in the Hall-MHD model the wave frequency can be comparable with the ion cyclotron frequency.

In the static Hall effect, for a conductor of finite extension across the current flow and a static magnetic field applied perpendicular to the current, an electric field across the conductor is observed. In dynamic situations such as waves, alternating Hall currents are produced. The Hall-MHD model allows wave frequencies up to and beyond the ion cyclotron frequency, but below the electron cyclotron frequency, to be considered.

We can employ the single–particle point of view of a plasma to provide a physical explanation of the effect of the Hall term. If the motion of a particle is followed over several gyration orbits, that is if the time–scale of interest, such as the period of a wave, is larger than the gyration period about the magnetic field, the particle drifts in a direction perpendicular to both the electric and magnetic field with the velocity

$$u_D = E \times B / B^2.$$ (1.30)

This drift does not introduce currents into the plasma, since u_D is independent of both the charge q and the mass m of the particle. However, the electrons and ions will move in different directions if the wave frequency becomes comparable with the ion cyclotron frequency. Such differential motions of charges constitute a current, the Hall current. Another way of expressing this is to say that the individual electron and ion Hall currents no longer cancel each other.

The Hall-MHD model is used to study the wave heating of laboratory plasmas at frequencies approaching the ion cyclotron frequency (see Chapter 6). It has also been found to capture important macroscopic effects in the simulation of the interaction of the solar wind with the Earth's magnetosphere (Winglee 1994, Huba 1996) and of magnetic field line reconnection (Lottermoser & Scholer 1997). Ion cyclotron effects may also play a role in solar coronal heating (Cranmer, Field & Kohl 1999) (see Chapter 7).

As an extension of the Hall-MHD model of the plasma, the nonideal effects of resistivity, electron inertia and electron pressure that arise from the species equations of motion can be included in the following generalized Ohm's law:

$$E + v \times B = \eta J + \frac{m_e}{n_i e^2} \frac{\partial J}{\partial t} + \frac{1}{n_i e} J \times B - \frac{1}{n_i e} \nabla p_e.$$ (1.31)

The first term on the right–hand side of Eq. (1.31) is the usual resistive term and the second term is the electron-inertia term, which allows the effect of a finite plasma frequency to be included. The third term is the Hall term and the fourth is an electron pressure gradient term, with p_e the electron gas pressure. This generalized Ohm's law allows the description of wave modes with very short wavelength in a direction perpendicular to the magnetic field, an important aspect of the process of Alfvén resonance absorption as we will see in Chapters 2 and 3.

1.5   Fourier Transforms

Both the fluid and kinetic descriptions of a plasma employ the theory of Fourier transforms , which we summarize here. If the equilibrium state of the plasma is assumed uniform, and the fluid equations are linearized, the coefficients in the resulting wave equations are constants. A Fourier transform in space and time will then yield an algebraic equation in the Fourier amplitude of, for example, the perturbation velocity v_1, which is a function of time t and space x. The Fourier transform of v_1 is defined as

$$v(\omega, k) = \int dt d^3x \exp(i(\omega t - k \cdot x)) v_1(t, x) \tag{1.32}$$

with the inverse transform

$$v_1(t, x) = \int \frac{d\omega d^3k}{(2\pi)^4} \exp(-i(\omega t - k \cdot x)) v(\omega, k). \tag{1.33}$$

Taking the Fourier transform in the uniform plasma case is simply equivalent to seeking plane wave solutions, that is, to assuming the form

$$\exp[i(k_x x + k_y y + k_z z - \omega t)] \tag{1.34}$$

for the wave fields, with k_x, k_y and k_z the constant wavenumbers in a Cartesian coordinate system. However, this is not the case for a nonuniform plasma or for nonlinear wave fields, where Fourier methods are not so useful.

1.6   The Kinetic Theory

The disadvantage of a fluid description of a plasma is that some effects, such as Landau and cyclotron damping, caused by a resonance of the wave with particles, cannot be modelled. The description of such effects requires a kinetic theory of the plasma. The theory of collisionless plasmas is well developed, so we can simply quote the relevant results from the theory, for example from the book of Melrose (1986).

The kinetic theory proceeds from the Vlasov theory of the collisionless plasma. After Fourier transforming Maxwell's equations and the Vlasov equations for the ions and electrons, we use expansions in terms of Bessel functions to derive the frequency and wavenumber dependent dielectric tensor K_{ij}, which is defined in terms of the conductivity tensor σ_{ij}:

$$K_{ij}(\omega, k) = \delta_{ij} + \frac{i}{\varepsilon_0 \omega} \sigma_{ij}(\omega, k). \tag{1.35}$$

The results of the calculation of the dielectric tensor from Vlasov theory are given in the Appendix. The plasma is said to be spatially dispersive if K_{ij} depends on $\boldsymbol{k}$.

The Fourier transform of the current density induced by the electric field imposed on the plasma is given in component form by

$$J_i(\omega, \boldsymbol{k}) = \sigma_{ij}(\omega, \boldsymbol{k})E_j(\omega, \boldsymbol{k}). \tag{1.36}$$

The wave equation for the electric field derived from Maxwell's equations is

$$\nabla \times \nabla \times \boldsymbol{E} + \frac{1}{c^2}\frac{\partial^2 \boldsymbol{E}}{\partial t^2} = -\mu_0 \frac{\partial \boldsymbol{J}}{\partial t}. \tag{1.37}$$

Fourier transforming Eq. (1.37) and using Eq. (1.35) and Eq. (1.36) yields

$$\Lambda_{ij}(\omega, \boldsymbol{k})E_j(\omega, \boldsymbol{k}) = 0 \tag{1.38}$$

with the wave tensor

$$\Lambda_{ij}(\omega, \boldsymbol{k}) = \frac{c^2}{\omega^2}(k_i k_j - k^2 \delta_{ij}) + K_{ij}(\omega, \boldsymbol{k}) \tag{1.39}$$

where $k^2 = k_x^2 + k_y^2 + k_z^2$. The dispersion equation is then obtained by setting the determinant, $\Lambda(\omega, \boldsymbol{k})$, of the 3×3 matrix $\Lambda_{ij}(\omega, \boldsymbol{k})$ to zero.

For example, in the next chapter we will assume the wavevector to lie in the x-z plane, so $k_y = 0$, in which case the determinant is

$$\Lambda(\omega, \boldsymbol{k}) = \begin{vmatrix} -c^2 k_z^2/\omega^2 + K_{11} & K_{12} & c^2 k_x k_z/\omega^2 + K_{13} \\ K_{21} & -c^2 k^2/\omega^2 + K_{22} & K_{23} \\ c^2 k_x k_z/\omega^2 + K_{31} & K_{32} & -c^2 k_x^2/\omega^2 + K_{33} \end{vmatrix}. \tag{1.40}$$

The dispersion relation for some mode M is given by a solution, $\omega = \omega_M(\boldsymbol{k})$, of the dispersion equation $\Lambda = 0$. The direction of the electric field $\boldsymbol{E}(\omega_M(\boldsymbol{k}), \boldsymbol{k})$ for waves in the mode M is described by the unimodular polarization vector $\boldsymbol{e}_M(\boldsymbol{k})$, which satisfies the relation

$$\boldsymbol{e}_M(\boldsymbol{k}) \cdot \boldsymbol{e}_M^*(\boldsymbol{k}) = 1. \tag{1.41}$$

An important quantity that characterizes the magnitude and direction of the flow of energy in waves is the group velocity vector, given by

$$\boldsymbol{v}_g = \frac{\partial \omega}{\partial \boldsymbol{k}}. \tag{1.42}$$

The energy flux in the waves is equal to the group velocity times the energy density in the waves. Provided the plasma is not spatially dispersive (i.e., there is no thermal energy in the wave (Stix 1992)), the energy flux is purely due to the electromagnetic energy flux, given by the Poynting vector

$$\boldsymbol{S} = \boldsymbol{E} \times \boldsymbol{B}/\mu_0. \tag{1.43}$$

In that case, the group velocity vector has the same direction as the Poynting vector. Further discussion of energy flux and the group velocity is to be found in the books by Cross (1988) and Melrose (1986).

2 Waves in Uniform Plasmas

2.1 Introduction

In this chapter we consider the properties of small amplitude linear waves in a spatially uniform plasma, using the models of the plasma introduced in Chapter 1. The properties of small amplitude waves in the uniform plasma form a basis for the discussion in following chapters of the waves in nonuniform plasmas, and of nonlinear waves. The waves will be assumed to have frequencies below or of the order of the ion cyclotron frequency, and we shall concentrate on waves that are predominantly electromagnetic. The classification of these waves is not as difficult a task as in the general case for the full range of frequencies covering the electron plasma frequency and the electron cyclotron frequency. In that case recourse may be made to techniques such as the Clemmow-Mullaly-Allis diagram to classify the waves (Stix 1992).

We discuss the waves first with the ideal MHD or hydromagnetic model, suitable for low–frequency waves in thermal plasmas with no interspecies collisions. Then the frequency range is extended to encompass the ion cyclotron frequency, using the Hall-MHD model for both cold and warm plasmas. A description of the cold plasma modes in terms of the dielectric tensor approach follows. The nonideal effects of interspecies collisions and multiple ions are then considered. Next, kinetic effects using the Vlasov theory are included, for both low and high plasma beta. Finally, the "kinetic" and "inertial" Alfvén waves are discussed, employing both fluid and kinetic theory.

2.2 Waves with the MHD Model

Let us assume an equilibrium with the plasma at rest and with no zero–order electric field. The plasma will be modelled in this section by the MHD equations (1.8), (1.11) and (1.12), with the adiabatic equation of state (1.13). We assume initially that the plasma has zero resistivity (the ideal MHD model). If subscripts 0 denote the equilibrium state, and subscripts 1 denote the first–order perturbations associated with the wave motion, the equilibrium satisfies the force balance equation obtained from Eq. (1.12),

$$\nabla p_0 = \boldsymbol{J}_0 \times \boldsymbol{B}_0. \tag{2.1}$$

From Eq. (1.11) and Eq. (1.13), the perturbed density and pressure satisfy

$$\frac{\partial \rho_1}{\partial t} + \nabla \cdot (\rho_0 \boldsymbol{v}_1) = 0 \tag{2.2}$$

and

$$p_1 = \frac{\gamma p_0}{\rho_0}\rho_1.$$
(2.3)

From the equation of fluid motion (1.12), the perturbed fluid velocity satisfies

$$\rho_0 \frac{\partial \boldsymbol{v}_1}{\partial t} = -\nabla p_1 + \boldsymbol{J}_0 \times \boldsymbol{B}_1 + \boldsymbol{J}_1 \times \boldsymbol{B}_0.$$
(2.4)

The perturbed electric and magnetic fields satisfy the equations

$$\frac{\partial \boldsymbol{B}_1}{\partial t} = -\nabla \times \boldsymbol{E}_1$$
(2.5)

and, from Eq. (1.8) with $\eta = 0$,

$$\boldsymbol{E}_1 = -\boldsymbol{v}_1 \times \boldsymbol{B}_0.$$
(2.6)

If the wavelengths are much shorter than the scale–lengths over which the equilibrium quantities ρ_0, p_0 and B_0 change, these quantities can be assumed to be constants, and the plasma is effectively uniform. The equilibrium current density $\boldsymbol{J}_0$ can therefore be neglected in Eq. (2.4). If we also neglect the displacement current in Eq. (1.3), assuming the characteristic speeds are much less than the speed of light *in vacuo* (we remove this assumption in Chapter 8), we have from Eq. (1.4)

$$\mu_0 \boldsymbol{J}_1 = \nabla \times \boldsymbol{B}_1.$$
(2.7)

The uniform equilibrium magnetic field is chosen to lie along the z-axis. It is then convenient to use the following perturbation variables to describe the wave fields:

$$\nabla \cdot \boldsymbol{v}_1, \quad v_{1z}, \quad B_{1z}, \quad J_{1z}, \quad \rho_1, \quad \zeta_{1z}$$
(2.8)

where

$$\zeta_{1z} = (\nabla \times \boldsymbol{v}_1)_z$$
(2.9)

is the fluid vorticity in the magnetic field direction.

Equations (2.2)-(2.7) can then be manipulated to yield the following set of six differential equations:

$$\rho_0 \frac{\partial \zeta_{1z}}{\partial t} - B_0 \frac{\partial J_{1z}}{\partial z} = 0$$
(2.10)

$$\mu_0 \frac{\partial J_{1z}}{\partial t} - B_0 \frac{\partial \zeta_{1z}}{\partial z} = 0$$
(2.11)

$$\rho_0 \frac{\partial}{\partial t}\nabla \cdot \boldsymbol{v}_1 + \frac{B_0}{\mu_0}\nabla^2 B_{1z} + c_s^2 \nabla^2 \rho_1 = 0$$
(2.12)

$$\frac{\partial B_{1z}}{\partial t} + B_0 \left(\nabla \cdot \boldsymbol{v}_1 - \frac{\partial v_{1z}}{\partial z} \right) = 0 \tag{2.13}$$

$$\rho_0 \frac{\partial v_{1z}}{\partial t} + c_s^2 \frac{\partial \rho_1}{\partial z} = 0 \tag{2.14}$$

$$\frac{\partial \rho_1}{\partial t} + \rho_0 \nabla \cdot \boldsymbol{v}_1 = 0. \tag{2.15}$$

It is seen that the two differential equations (2.10) and (2.11) for ζ_{1z} and J_{1z} are uncoupled from the four differential equations (2.12)-(2.15) for $\nabla \cdot \boldsymbol{v}_1$, B_{1z}, v_{1z} and ρ_1. We should also note that in the equations for ζ_{1z} and J_{1z}, the spatial derivatives are only in the equilibrium magnetic field direction. Taking the Fourier transforms, defined in Eq. (1.32), of Eq. (2.10) and Eq. (2.11) (or simply substituting the plane wave solution Eq. (1.34) into the differential equations), we obtain a consistency equation for a nontrivial solution which relates the frequency to the wavenumber. This is the dispersion equation for waves described by the variables ζ_{1z} and J_{1z}:

$$\omega^2 - v_A^2 k_z^2 = 0 \tag{2.16}$$

where the Alfvén speed v_A in the equilibrium plasma is given by

$$v_A = B_0 / (\mu_0 \rho_0)^{1/2}. \tag{2.17}$$

The dispersion equation (2.16) is independent of the components of the wavevector perpendicular to the equilibrium magnetic field, and is also independent of the sound speed c_s.

Taking the Fourier transforms of Eqs. (2.12)-(2.15) yields a separate dispersion equation for waves described by the variables $\nabla \cdot \boldsymbol{v}_1$, B_{1z}, v_{1z} and ρ_1:

$$\omega^4 - \omega^2 (v_A^2 + c_s^2) k^2 + v_A^2 c_s^2 k^2 k_z^2 = 0 \tag{2.18}$$

where $k = |\boldsymbol{k}|$. This dispersion equation does involve the perpendicular components of the wavevector, and the sound speed, in contrast to Eq. (2.16).

It is evident that the two dispersion equations (2.16) and (2.18), together with their corresponding sets of characteristic wave field variables, correspond to two distinct types of wave mode. The waves described by Eq. (2.16) are called *Alfvén* waves, and the waves described by Eq. (2.18) are called *magnetoacoustic* (or *magnetosonic*) waves (see Table 2.1). The magnetoacoustic mode may be further split up into two distinct modes, the fast and slow magnetoacoustic waves. An arbitrary low–frequency disturbance can be represented as a superposition of the Alfvén wave and the fast and slow magnetoacoustic waves.

Let us define the angle θ between the wavevector and the magnetic field $\boldsymbol{B}_0$, so that $k_z = k \cos \theta$. The first dispersion equation (2.16) then gives the positive frequency solution

$$\omega_A = v_A |k_z| = v_A k |\cos \theta| \tag{2.19}$$

of the Alfvén mode. The second dispersion equation (2.18) gives two positive frequency solutions: the fast magnetoacoustic mode, with

$$\omega_F^2 = \frac{k^2}{2} \left(v_A^2 + c_s^2 + \left((v_A^2 + c_s^2)^2 - 4 v_A^2 c_s^2 \cos^2 \theta \right)^{1/2} \right) \tag{2.20}$$

Table 2.1: The dispersion equations and characteristic variables for the Alfvén and magnetoacoustic modes in the ideal MHD model.

	Dispersion equation	Characteristic variables
Alfvén wave	$\omega^2 - v_A^2 k_z^2 = 0$	$J_{1z}, \ \zeta_{1z}$
Magnetoacoustic waves	$\omega^4 - \omega^2 (v_A^2 + c_s^2) k^2$	$\nabla \cdot \boldsymbol{v}_1, \ \ v_{1z}, \ \ B_{1z}, \ \ \rho_1$
	$+ v_A^2 c_s^2 k^2 k_z^2 = 0$	

and the slow magnetoacoustic mode, with

$$\omega_S^2 = \frac{k^2}{2} \left(v_A^2 + c_s^2 - \left((v_A^2 + c_s^2)^2 - 4 v_A^2 c_s^2 \cos^2 \theta \right)^{1/2} \right). \tag{2.21}$$

We note that the phase velocity $v_{\rm ph} = \omega/k$ is independent of k for all three modes, so all the modes are nondispersive, although they are anisotropic because $v_{\rm ph}$ depends on the angle of propagation θ. The characteristic phase velocity surfaces, that is, polar plots of the phase velocities of the three modes against the angle θ, have often been presented in texts on MHD (see for example Shercliff (1965)).

Defining $\boldsymbol{b}$ as the unit vector in the direction of $\boldsymbol{B}_0$, and $\boldsymbol{\kappa}$ as the unit vector along the wavevector $\boldsymbol{k}$, we have $\boldsymbol{\kappa} \cdot \boldsymbol{b} = \cos \theta$. To discuss the polarization properties of the three modes, it is convenient to define two mutually orthogonal unit vectors, each orthogonal to the $\boldsymbol{\kappa}$ vector:

$$\boldsymbol{a} = -\frac{\boldsymbol{k} \times \boldsymbol{B}_0}{|\boldsymbol{k} \times \boldsymbol{B}_0|} \tag{2.22}$$

and

$$\boldsymbol{t} = \boldsymbol{a} \times \boldsymbol{\kappa}. \tag{2.23}$$

Without loss of generality, for a uniform plasma we can choose the $\boldsymbol{k}$ vector to lie in the x-z plane. We then have

$$\boldsymbol{b} = (0, 0, 1), \quad \boldsymbol{\kappa} = (\sin \theta, 0, \cos \theta) \tag{2.24}$$

and the $\boldsymbol{a}$ and $\boldsymbol{t}$ vectors become

$$\boldsymbol{a} = (0, 1, 0), \quad \boldsymbol{t} = (\cos \theta, 0, -\sin \theta). \tag{2.25}$$

These vectors are shown in Figure 2.1. We now proceed to discuss the three modes in some detail.

2.2.1   The Alfvén Mode

If the wave is purely in the Alfvén mode, we can assume the characteristic variables listed in Table 2.1 for the magnetoacoustic mode to be zero. Thus we have (with $k_y = 0$),

$$v_{1z} = 0 \quad \text{and} \quad \nabla \cdot \boldsymbol{v}_1 = \mathrm{i} k_x v_{1x} = 0. \tag{2.26}$$

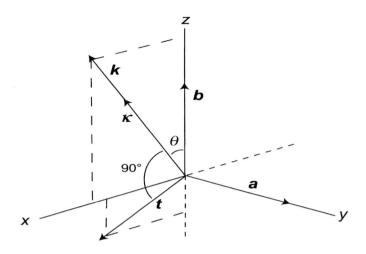

Figure 2.1: The basic unit vectors κ, b, t and a defined in relation to the equilibrium magnetic field (along the z-axis) and the wavevector k.

Provided that $k_x \neq 0$, (i.e. $\sin\theta \neq 0$), we conclude that $v_{1x} = 0$ (the case of $k_x = 0$, i.e. propagation parallel or antiparallel to the magnetic field, is discussed in the next section). Thus the velocity perturbation $v_1 = v_A$ of the Alfvén mode is in the y–direction only, that is, it is parallel to the vector a, perpendicular to both the equilibrium magnetic field and the wavevector directions.

Since $B_{1z} = 0$ and $\nabla \cdot B_1 = 0$, it follows that $B_{1x} = 0$, so that this mode only has a magnetic field perturbation in the y–direction. Using Eq. (2.6), we find that the electric field perturbation is in the x–direction. We therefore have

$$v_1 \propto B_1 \propto a \quad \text{and} \quad E_1 \propto \cos\theta\, t + \sin\theta\, \kappa. \tag{2.27}$$

Thus the wave electric field is not purely transverse to the direction of propagation, although it is perpendicular to the magnetic field direction.

Since $\nabla \cdot v_1 = 0$, there is no density or particle pressure perturbation for this mode. There is also no magnetic field perturbation in the equilibrium magnetic field direction, which implies, since we have $B_1 \cdot B_0 = 0$ and $B^2 \simeq B_0^2 + 2B_1 \cdot B_0$ (to first order), that the magnetic field strength is constant to first order. The strengths of the magnetic and particle pressures in the plasma are thus each conserved; there is no compressive stress due to this wave.

However, the field lines are bent, giving rise to a magnetic tension given by Eq. (1.24). There is no compression of the magnetic field, and the Alfvén wave in this low–frequency limit is said to be a noncompressional *shear* wave . The phase speed of the wave lies between the speeds of the slow and fast magnetoacoustic waves, so the Alfvén wave is also sometimes referred to as the *intermediate* wave. Plasma flow across the magnetic field can increase the bending of the field; the shear wave acts to reduce the additional curvature of the field line. The currents that are set up to reduce the bending are closed partly along the field, so the shear wave introduces the field–aligned current J_{1z} (Kivelson 1995b). In a cylindrical geometry, the shear Alfvén wave is referred to as a *torsional* wave , with adjacent magnetic surfaces able to shear past each other without coupling to each other.

An important point to note is that the Alfvén wave dispersion relation is independent of the wavevector component perpendicular to the equilibrium magnetic field, and the wave cannot propagate perpendicular to the field (i.e. the phase velocity drops to zero). The group velocity

$$v_g = \frac{\partial \omega}{\partial \boldsymbol{k}} = v_A \cos \theta \, \boldsymbol{b} \tag{2.28}$$

is aligned with the background magnetic field, that is, the wave energy is always carried along the direction of the background field, regardless of the direction of the wavevector. We shall see later that when nonideal effects, such as collisions and thermal effects, are taken into account, the energy can propagate obliquely to the magnetic field direction.

2.2.2 The Fast and Slow Magnetoacoustic Modes

For a wave in the fast or slow magnetoacoustic mode, the characteristic variables for the shear Alfvén mode are zero. Thus, with the choice (Eq. 2.24) for the direction of $\boldsymbol{k}$, we have

$$\zeta_{1z} = \mathrm{i}\boldsymbol{k} \times \boldsymbol{v}_1|_z = \mathrm{i}k_x v_{1y} = 0 \tag{2.29}$$

so $v_{1y} = 0$. Using Eqs. (2.14) and (2.15), we find for the velocity perturbation $\boldsymbol{v}_{F,S}$ of the fast and slow modes,

$$\boldsymbol{v}_1 = \boldsymbol{v}_{F,S} \propto (\omega^2 - c_s^2 k_z^2, \; 0, \; c_s^2 k_x k_z) \tag{2.30}$$

which can be written as

$$\boldsymbol{v}_{F,S} \propto \cos \theta \, (\omega^2 - c_s^2 k^2) \, \boldsymbol{t} + \sin \theta \, \omega^2 \, \boldsymbol{\kappa}. \tag{2.31}$$

Using

$$\boldsymbol{B}_1 = -\frac{k B_0}{\omega} (\boldsymbol{\kappa} \times (\boldsymbol{v}_1 \times \boldsymbol{b})) \tag{2.32}$$

and Eq. (2.6), we find that the directions of the perturbation magnetic and electric fields are given by

$$\boldsymbol{B}_1 \propto \boldsymbol{t} \quad \text{and} \quad \boldsymbol{E}_1 \propto \boldsymbol{a}. \tag{2.33}$$

There are density perturbations in the wave (since $\nabla \cdot \boldsymbol{v}_1 \neq 0$), and perturbations of the magnetic field parallel to $\boldsymbol{B}_0$ (and thus of the field strength). Thus these modes are compressive in nature, even if the sound speed is zero. In contrast to the shear Alfvén wave, the fast and slow magnetoacoustic waves act to reduce magnetic or particle pressure gradients in the plasma.

The fast wave is sometimes called a *compressional Alfvén* wave. In the case of a low–β plasma, that is with $v_A^2 \gg c_s^2$, the fast wave dispersion relation is

$$\omega = v_A k \tag{2.34}$$

and from Eq. (2.30), v_F is perpendicular to $\boldsymbol{B}_0$. The fast wave can propagate and transport energy in any direction. The particle pressure and magnetic pressure in the fast wave increase and decrease in phase. In the slow wave, the particle and magnetic pressures vary out of phase. In an incompressible plasma ($c_s \to \infty$), the magnetic pressure is balanced by the particle pressure in the slow wave, so that only the magnetic tension remains effective. In that case the slow wave has the same dispersion relation (2.19) as the shear Alfvén wave, although its magnetic field is polarized in the direction of the t vector, that is, transverse to the direction of propagation but with a component along the equilibrium magnetic field direction.

For the fast wave in a low–β plasma, the group velocity and Poynting vector are parallel to the wavevector. On the other hand, for the slow wave, the group velocity and the Poynting vector are respectively almost parallel and perpendicular to the equilibrium magnetic field. An extensive discussion of the phase and group velocities of the waves can be found in the book by Cross (1988).

For parallel or antiparallel propagation ($\sin\theta = 0$), the fast mode loses its compressive character and becomes degenerate with the Alfvén mode, in that both modes have a phase speed v_A, with a perturbation magnetic field perpendicular to the equilibrium field. The slow mode in this case is a pure sound mode with phase speed c_s.

2.3 The Hall-MHD Model

In this section the basic properties of the relevant wave modes in a uniform plasma with a single ion species are surveyed using a fluid theory, in which the electrons are strongly magnetized, but the ions are not completely magnetized. This model is sometimes referred to as Hall MHD, since the Hall term is retained in Ohm's law (Eq. 1.28). The model allows frequencies up to and beyond the ion cyclotron frequency, but well below the electron cyclotron frequency, to be considered. It is convenient to consider separately the cases of a cold plasma ($\beta = 0$) and a warm plasma ($\beta \neq 0$).

2.3.1 Cold Plasma

In the MHD model, if the plasma is cold, with $c_s = 0$ and $\beta = 0$, the slow magnetoacoustic mode disappears and, from Eq. (2.34), the fast magnetoacoustic wave is isotropic, that is, it has the same phase velocity, v_A, in all directions. If the Ohm's law (Eq. 1.28) retaining the Hall term is now used, and c_s is again assumed zero, the equations replacing Eqs. (2.10)-(2.13) and Eq. (2.15) are:

$$\rho_0 \frac{\partial \zeta_{1z}}{\partial t} - B_0 \frac{\partial J_{1z}}{\partial z} = 0 \tag{2.35}$$

$$\mu_0 \frac{\partial J_{1z}}{\partial t} - B_0 \frac{\partial \zeta_{1z}}{\partial z} + \frac{v_A^2}{\Omega_i} \frac{\partial}{\partial z} \nabla^2 B_{1z} = 0 \tag{2.36}$$

$$\rho_0 \frac{\partial}{\partial t} \nabla \cdot \boldsymbol{v}_1 + \frac{B_0}{\mu_0} \nabla^2 B_{1z} = 0 \tag{2.37}$$

$$\frac{\partial B_{1z}}{\partial t} + B_0 \nabla \cdot \boldsymbol{v}_1 + \frac{v_A^2}{\Omega_i} \mu_0 \frac{\partial J_{1z}}{\partial z} = 0 \tag{2.38}$$

$$\frac{\partial \rho_1}{\partial t} + \rho_0 \nabla \cdot \boldsymbol{v}_1 = 0. \tag{2.39}$$

In addition, we have for the cold plasma the result $v_{1z} = 0$.

The variables $\nabla \cdot \boldsymbol{v}_1$ and ζ_{1z} can be eliminated from Eqs. (2.35)-(2.38) to yield two coupled second-order differential equations in J_{1z} and B_{1z}:

$$\left(\frac{\partial^2}{\partial t^2} - v_A^2 \frac{\partial^2}{\partial z^2} \right) J_{1z} + \frac{v_A^2}{\mu_0 \Omega_i} \frac{\partial^2}{\partial z \partial t} \nabla^2 B_{1z} = 0 \tag{2.40}$$

$$\left(\frac{\partial^2}{\partial t^2} - v_A^2 \frac{\partial^2}{\partial z^2} \right) B_{1z} + \frac{\mu_0 v_A^2}{\Omega_i} \frac{\partial^2}{\partial z \partial t} J_{1z} = 0. \tag{2.41}$$

It should be noted that the characteristic variables given in Table 2.1 for the MHD versions of the Alfvén mode and the fast mode, are now coupled together because of the ion cyclotron terms in Eqs. (2.36) and (2.38), and in Eqs. (2.40) and (2.41). Thus the two sets of characteristic variables can no longer be separately associated with the two modes.

It is straightforward to obtain the following dispersion equation from Eqs. (2.40) and (2.41):

$$(\omega^2 - v_A^2 k_z^2)(\omega^2 - v_A^2 k^2) = \left(\frac{\omega}{\Omega_i} \right)^2 v_A^4 k_z^2 k^2. \tag{2.42}$$

We note that the two factors on the LHS of Eq. (2.42), which, individually set to zero, give the Alfvén and fast magnetoacoustic wave dispersion equations (2.16) and (2.18), are coupled together by the RHS of Eq. (2.42). The factors are uncoupled at low frequency, $\omega \ll \Omega_i$, yielding the dispersion relations (2.19) and (2.34), when the effect of the Hall term may be neglected and the ideal MHD model (with $c_s = 0$) is applicable.

Let us define the dimensionless wavenumber α:

$$\alpha = \frac{v_A k}{\Omega_i}. \tag{2.43}$$

The frequencies of the two modes may then be derived from Eq. (2.42) to be (again assuming the wavevector direction to be given by Eq. (2.24)):

$$\omega_\pm^2 = \frac{v_A^2 k^2}{2} \left(1 + (1 + \alpha^2) \cos^2 \theta \pm \left[1 - 2(1 - \alpha^2) \cos^2 \theta + (1 + \alpha^2)^2 \cos^4 \theta \right]^{1/2} \right). \tag{2.44}$$

The mode with the − sign gives the shear Alfvén wave in the limit $\alpha \to 0$, and is sometimes referred to as the *ion cyclotron mode*. The *ion cyclotron resonance*, where $k \to \infty$, occurs for this mode as $\omega \to \Omega_i$, provided $\cos \theta \neq 0$. At a cyclotron resonance point in a plasma with a varying magnetic field, where the cyclotron frequency becomes equal to the wave frequency, the group velocity goes to zero, the wave energy accumulates and the wave fields grow to values that are limited by dissipative or nonlinear mode conversion processes.

The solution with the + sign is still called the *fast magnetoacoustic wave*, since it reduces to the MHD version of that wave if $\omega \ll \Omega_i$. The fast magnetoacoustic wave is not greatly

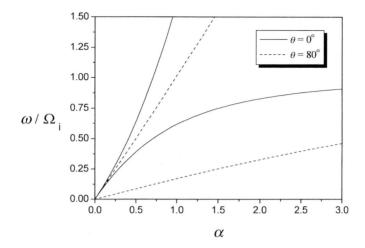

Figure 2.2: The dispersion relations of the two modes in the cold Hall-MHD model, for parallel propagation and oblique propagation ($\theta = 80°$). Here the normalized wavenumber is $\alpha = v_A k/\Omega_i$. The upper curve in each case corresponds to the fast Alfvén wave, and the lower to the slow Alfvén wave.

affected by the ion cyclotron resonance, and becomes the electron whistler wave at frequencies $\gg \Omega_i$ where the electron inertia must be taken into account.

For parallel propagation, $\sin\theta = 0$, the two frequencies are

$$\omega_\pm = \frac{v_A k_z}{2}\left((4+\alpha^2)^{1/2} \pm \alpha\right).\qquad(2.45)$$

The ion cyclotron mode, given by the $-$ sign in Eq. (2.45), in this case has left–hand circular polarization of the electric field, and the magnetoacoustic mode, corresponding to the $+$ sign, has right–hand circular polarization. The ion cyclotron mode has phase speed $< v_A$, experiences the ion cyclotron resonance and is referred to here as the *slow Alfvén wave*. The magnetoacoustic mode has phase speed $> v_A$, and is sometimes referred to as the *fast Alfvén wave*. Thus the inclusion of the Hall term has led to the removal of the low–frequency degeneracy of the parallel propagating Alfvén and magnetoacoustic waves. These features of the dispersion relations for parallel and oblique propagation are shown in Figure 2.2. For $\alpha \ll 1$, the two parallel propagating modes have the same phase speed v_A, while for oblique propagation the fast and slow waves have phase speeds v_A and $v_A \cos\theta$ respectively.

2.3.2 Warm Plasma

We now include the effect of a nonzero sound speed in the Hall-MHD model. The Hall term is included in the generalized Ohm's law Eq. (1.28), and the overall fluid pressure gradient term is retained in Eqs. (2.12) and (2.14). However, the electron inertia and electron pressure gradient terms in Eq. (1.31) are neglected, and resistivity is also neglected. The differential equations for the characteristic variables $\nabla \cdot \boldsymbol{v}_1$ and so forth are now

$$\rho_0 \frac{\partial \zeta_{1z}}{\partial t} - B_0 \frac{\partial J_{1z}}{\partial z} = 0 \tag{2.46}$$

$$\mu_0 \frac{\partial J_{1z}}{\partial t} - B_0 \frac{\partial \zeta_{1z}}{\partial z} + \frac{v_A^2}{\Omega_i} \frac{\partial}{\partial z} \nabla^2 B_{1z} = 0 \tag{2.47}$$

$$\rho_0 \frac{\partial}{\partial t} \nabla \cdot \boldsymbol{v}_1 + \frac{B_0}{\mu_0} \nabla^2 B_{1z} + c_s^2 \nabla^2 \rho_1 = 0 \tag{2.48}$$

$$\frac{\partial B_{1z}}{\partial t} + B_0 \left(\nabla \cdot \boldsymbol{v}_1 - \frac{\partial v_{1z}}{\partial z} \right) + \frac{v_A^2}{\Omega_i} \frac{\partial J_{1z}}{\partial z} = 0 \tag{2.49}$$

$$\rho_0 \frac{\partial v_{1z}}{\partial t} + c_s^2 \frac{\partial \rho_1}{\partial z} = 0 \tag{2.50}$$

$$\frac{\partial \rho_1}{\partial t} + \rho_0 \nabla \cdot \boldsymbol{v}_1 = 0. \tag{2.51}$$

Upon Fourier transforming Eqs. (2.46)-(2.51), the following dispersion equation results (Stringer 1963):

$$(\omega^2 - v_A^2 k_z^2)(\omega^2(\omega^2 - c_s^2 k^2) - v_A^2 k^2(\omega^2 - c_s^2 k_z^2)) = \left(\frac{\omega}{\Omega_i}\right)^2 v_A^4 k_z^2 k^2 (\omega^2 - c_s^2 k^2). \tag{2.52}$$

The term on the right–hand side of Eq. (2.52), which arises from the Hall term in Ohm's law, couples the Alfvén dispersion equation (arising from the first factor on the left–hand side of Eq. (2.52)) to the fast and slow magnetoacoustic dispersion equation (the second factor). Defining

$$\beta' = \frac{c_s^2}{v_A^2} \tag{2.53}$$

Eq. (2.52) may be written

$$(f^2 - \alpha^2 \cos^2 \theta)(f^2(f^2 - \alpha^2 \beta') - \alpha^2(f^2 - \alpha^2 \beta' \cos^2 \theta))$$
$$= f^2 \alpha^4 \cos^2 \theta (f^2 - \alpha^2 \beta'). \tag{2.54}$$

The Hall term reduces the speed of the slow magnetoacoustic wave and increases that of the fast magnetoacoustic wave, so the terms *slow*, *intermediate* and *fast* may still be used to identify the waves, at least for oblique angles of propagation. Figure 2.3 shows the dispersion relations of the three modes obtained from Eq. (2.54), for $\beta' = 0.5$. For parallel propagation the intermediate (Alfvén) wave is uncoupled from the magnetoacoustic waves, and experiences the ion cyclotron resonance. However, for $\theta \neq 0$ the slow magnetoacoustic and intermediate modes reconnect, and the new "intermediate" mode indeed has a phase speed intermediate between the fast and slow speeds at all frequencies. Moreover, the intermediate mode no longer experiences a resonance, but the slow magnetoacoustic mode experiences a resonance at $\omega = v_A k \cos \theta$.

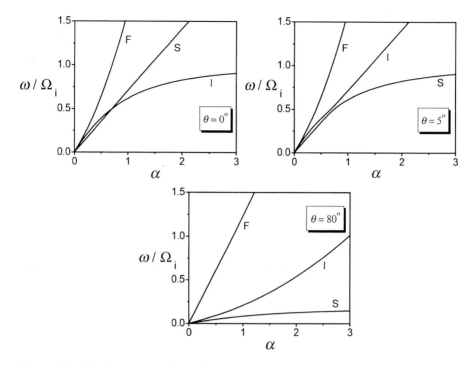

Figure 2.3: The dispersion relations of the three modes in the warm Hall-MHD model, for parallel propagation, and two examples of oblique propagation ($\theta = 5°$ and $\theta = 80°$). The normalized wavenumber is $\alpha = v_A k/\Omega_i$. The fast magnetoacoustic (F), slow magnetoacoustic (S) and intermediate (I) modes are indicated. Here $\beta' = 0.5$.

2.4   Cold Collisionless Plasmas

A derivation of the properties of Alfvén and magnetoacoustic waves, which is an alternative to the derivation from the fluid equations considered so far, proceeds from the kinetic theory describing collisionless plasmas. This approach starts from the assumption that the particle velocity distribution functions are arbitrary, that is, that collisions amongst particles driving the distributions to thermal Maxwellians are ineffective. However it is usual to make the further assumption that the distributions are indeed Maxwellians, which may however correspond to different temperatures for motions parallel and perpendicular to the background magnetic field, or which may be drifting relative to each other for different species. The theory is more naturally cast in terms of the dielectric tensor of the plasma introduced in Chapter 1 (see, e.g., the book by Melrose (1986) for more details).

At the low frequencies of primary interest in this book, in general there are three wave modes in a collisionless magnetized plasma, which are analogous to the modes found in the MHD treatment. Two of these modes exist in the cold plasma (zero β) case: the Alfvén mode and the fast magnetoacoustic mode. The third mode for $\beta \neq 0$ is the collisionless analogue of the slow magnetoacoustic mode. In this section we shall assume that the temperatures of all the species go to zero, and so obtain the analogues of the MHD Alfvén and fast magnetoacoustic modes for a cold plasma. The kinetic theory for hot plasmas is described in Section 2.7. We also use the dielectric tensor formulation to discuss collisional effects in the Hall-MHD model (Section 2.5.2), and waves in cold multi–ion plasmas (Section 2.6).

The dielectric tensor derived from the Vlasov theory for a uniform plasma may be written down from the general dielectric tensor presented in the Appendix, Eq. (A.2), using Eqs. (A.5)-(A.7), in the small gyroradius limit for a cold plasma. It may also be derived directly using the particle equations of motion for the ions and the electrons, that is, the fluid equations of motion (1.5) neglecting the pressure gradient of each species (Stix 1992). The result for the dielectric tensor in either case is

$$K_{ij} = \begin{bmatrix} S & -iD & 0 \\ iD & S & 0 \\ 0 & 0 & P \end{bmatrix} \tag{2.55}$$

with

$$S = \frac{1}{2}(R_+ + R_-), \quad D = \frac{1}{2}(R_+ - R_-) \tag{2.56}$$

and

$$R_\pm = 1 - \sum_\alpha \frac{\omega_{p\alpha}^2}{\omega^2} \frac{\omega}{\omega \pm \epsilon_\alpha \Omega_\alpha}, \quad P = 1 - \sum_\alpha \frac{\omega_{p\alpha}^2}{\omega^2} \tag{2.57}$$

where we have allowed for an arbitrary number of charged species, denoted by the index α, with ϵ_α the sign of the charge, $\omega_{p\alpha}$ the plasma frequency, and Ω_α the (positive) cyclotron frequency, of each of the species.

If the equilibrium magnetic field lies along the z–axis and the wavenumber has a direction specified by Eq. (2.24), the tensor Λ_{ij} appearing in the Fourier transformed wave equation (1.38) is given by

$$\Lambda_{ij} = \begin{bmatrix} S - n^2 \cos^2\theta & -iD & n^2 \sin\theta\cos\theta \\ iD & S - n^2 & 0 \\ n^2 \sin\theta\cos\theta & 0 & P - n^2 \sin^2\theta \end{bmatrix} \tag{2.58}$$

where

$$n = ck/\omega \tag{2.59}$$

is the refractive index for the wave.

The determinant Λ of the matrix Λ_{ij} may be written as a quadratic function of n^2:

$$\Lambda = An^4 - Bn^2 + C \tag{2.60}$$

with

$$
\begin{aligned}
A &= P\cos^2\theta + S\sin^2\theta \\
B &= (S^2 - D^2)\sin^2\theta + PS(1 + \cos^2\theta) \\
C &= P(S^2 - D^2).
\end{aligned}
\tag{2.61} \tag{2.62} \tag{2.63}
$$

Setting $\Lambda = 0$ yields the solutions

$$n^2 = n_\pm^2(\omega, \theta) = \frac{B \pm F}{2A} \tag{2.64}$$

where

$$F^2 = (PS - S^2 + D^2)^2 \sin^4 \theta + 4P^2 D^2 \cos^2 \theta. \tag{2.65}$$

In the case of parallel propagation ($\theta = 0$), it is straightforward to show from Eq. (2.57) and Eq. (2.60) that

$$n_\pm^2 = R_\pm. \tag{2.66}$$

The unimodular electric polarization vector for a general mode M, introduced in Section 1.6, may be written in terms of the unit vectors defined in Eqs. (2.23) and (2.24) as (Melrose 1986)

$$e_M = \frac{K_M \boldsymbol{\kappa} + T_M \boldsymbol{t} + i\boldsymbol{a}}{(K_M^2 + T_M^2 + 1)^{1/2}}. \tag{2.67}$$

The quantity K_M describes the longitudinal part of the polarization, and the transverse part of the polarization is determined by the quantity T_M. In general the transverse polarization is elliptical, and T_M is the axial ratio of the polarization ellipse. The values $T_M = \pm 1$ correspond to right and left–hand circular polarizations respectively, $T_M = 0$ to linear polarization along $\boldsymbol{a}$, and $T_M = \infty$ to linear polarization along $\boldsymbol{t}$.

The polarization vector may be calculated from the tensor wave equation (1.38), using the refractive index solutions n_M given by Eq. (2.64) (Melrose 1986). The results for T_M and K_M are

$$T_M = \frac{DP \cos \theta}{An_M^2 - PS} \tag{2.68}$$

and

$$K_M = \frac{(P - n_M^2)D \sin \theta}{An_M^2 - PS}. \tag{2.69}$$

In the limit where the wave frequency is much less than the ion or electron plasma frequencies, P defined in Eq. (2.57) is large and negative, so letting $1/P \to 0$, we obtain for the refractive index and the transverse and longitudinal parts of the polarization vector,

$$n_M^2 = \frac{S(1 + \cos^2 \theta) \pm [S^2 \sin^4 \theta + 4D^2 \cos^2 \theta]^{1/2}}{2 \cos^2 \theta} \tag{2.70}$$

$$T_M = \frac{S \sin^2 \theta \pm [S^2 \sin^4 \theta + 4D^2 \cos^2 \theta]^{1/2}}{2D \cos \theta} \tag{2.71}$$

and

$$K_M = T_M \tan \theta. \tag{2.72}$$

Let us define a new scaling of the dielectric tensor components, with dimensions of wavenumber squared:

$$u_1 = \left(\frac{\omega^2}{c^2}\right) S, \quad u_2 = -\left(\frac{\omega^2}{c^2}\right) D \tag{2.73}$$

and

$$u_3 = \left(\frac{\omega^2}{c^2}\right) P. \tag{2.74}$$

The dielectric tensor can be written in terms of these components as

$$K_{ij} = \frac{c^2}{\omega^2} \begin{bmatrix} u_1 & iu_2 & 0 \\ -iu_2 & u_1 & 0 \\ 0 & 0 & u_3 \end{bmatrix}. \tag{2.75}$$

The dispersion equation, gained by setting Eq. (2.60) to zero, can be written in terms of the total wavenumber k and the wavevector components k_z in the direction of, and k_x perpendicular to, the equilibrium magnetic field, rather than in terms of the refractive index and the angle of propagation. The resulting dispersion equation is (for $k_y = 0$):

$$u_3[k^2 k_z^2 - u_1(k^2 + k_z^2) + u_1^2 - u_2^2] - (u_1^2 - u_2^2 - u_1 k^2)k_x^2 = 0. \tag{2.76}$$

In the large $|P|$ or $|u_3|$ limit, the dispersion equation is:

$$k^2 k_z^2 - u_1(k^2 + k_z^2) + u_1^2 - u_2^2 = 0. \tag{2.77}$$

Another form of the dispersion equation (2.77) we find useful is

$$k^2 - k_z^2 = k_x^2 = \frac{G^2 - H^2}{G} \tag{2.78}$$

where

$$G = u_1 - k_z^2, \quad H = u_2. \tag{2.79}$$

In the limit $\omega \ll |\Omega_e|$, that is, for frequencies of interest such that the electrons are completely magnetized, the quantities S and D may be written

$$S = 1 + \sum_i \frac{\omega_{pi}^2}{(\Omega_i^2 - \omega^2)} \tag{2.80}$$

and

$$D = -\sum_i \frac{\epsilon_i \omega_{pi}^2 \omega}{\Omega_i(\Omega_i^2 - \omega^2)} \tag{2.81}$$

where the sums are over the ion species, that is, excluding the electrons, and overall charge neutrality is assumed.

If we define the Alfvén speed v_A to be based on the total mass density of all the ion species, the following identity holds:

$$\sum_i \frac{\omega_{\mathrm{pi}}^2}{\Omega_i^2} = \frac{c^2}{v_\mathrm{A}^2}. \tag{2.82}$$

Then in the limit $\omega \ll \Omega_i$ for all i, $D \to 0$ and

$$S = 1 + \frac{c^2}{v_\mathrm{A}^2}. \tag{2.83}$$

Assuming the angle of propagation to the magnetic field, θ, is not too small, and the condition

$$4D^2 \cos^2\theta \ll S^2 \sin^4\theta \tag{2.84}$$

holds, the term in D in the square brackets of Eqs. (2.70) and (2.71) can be neglected. The Alfvén mode (A) is then identified with the $+$ solution of Eq. (2.70), with

$$n_\mathrm{A}^2 = \frac{S}{\cos^2\theta} = \frac{c^2}{v_\mathrm{A}^2 \cos^2\theta} \tag{2.85}$$

assuming $v_\mathrm{A} \ll c$, and

$$T_\mathrm{A} = \infty. \tag{2.86}$$

The electric polarization vector of the Alfvén mode is then obtained from Eq. (2.67) as

$$e_\mathrm{A} \propto \tan\theta\, \boldsymbol{\kappa} + \boldsymbol{t}$$

that is,

$$e_\mathrm{A} = (1, 0, 0), \tag{2.87}$$

and the dispersion relation is

$$\omega = v_\mathrm{A} k |\cos\theta| = v_\mathrm{A} |k_z| \tag{2.88}$$

in agreement with the MHD results in Eqs. (2.19) and (2.27).

The fast magnetoacoustic mode (F) is identified with the $-$ solution of Eq. (2.70), with

$$n_\mathrm{F}^2 = S = \frac{c^2}{v_\mathrm{A}^2} \tag{2.89}$$

assuming $v_\mathrm{A} \ll c$, and

$$T_\mathrm{F} = 0. \tag{2.90}$$

The polarization vector of the F mode is then in the direction of the unit vector $\boldsymbol{a}$ and the dispersion relation is $\omega^2 = v_\mathrm{A}^2 k^2$. Thus the fast magnetoacoustic wave is isotropic, that is, it has the same phase velocity, v_A, in all directions. These results were also obtained with the MHD theory in Section 2.2.2 in the limit $c_s^2 \ll v_A^2$.

We can also use the results in Eqs. (2.70) and (2.71), and in Eqs. (2.80) and (2.81) with a single ion species, to describe the waves at frequencies approaching the ion cyclotron frequency but much less than the electron cyclotron frequency. Thus in the nonrelativistic limit $v_A \ll c$, we have

$$S = -\frac{c^2}{v_A^2} \frac{\Omega_i^2}{\omega^2 - \Omega_i^2} \tag{2.91}$$

$$D = \frac{c^2}{v_A^2} \frac{\omega \Omega_i}{\omega^2 - \Omega_i^2}. \tag{2.92}$$

Inserting these relations into Eqs. (2.73) and (2.77) yields the same dispersion equation (2.42) as derived from the Hall-MHD model.

In the case of propagation of the waves almost parallel to the equilibrium magnetic field, such that the condition Eq. (2.84) is reversed, we have $K_M = 0$ and $T_M = \pm 1$ from Eqs. (2.71) and (2.72), so the modes are oppositely circularly polarized. The general dispersion equations for a cold plasma are given in this case by

$$n_A^2 = R_- = S - D, \quad T_A = \frac{D \cos \theta}{|D \cos \theta|} \quad \text{(left hand)} \tag{2.93}$$

and

$$n_F^2 = R_+ = S + D, \quad T_F = -\frac{D \cos \theta}{|D \cos \theta|} \quad \text{(right hand)}. \tag{2.94}$$

In the nonrelativistic limit, and for $\omega \ll \Omega_e$ and a single ion species, we have

$$R_\pm = -\frac{c^2}{v_A^2} \frac{\Omega_i}{\Omega_i \pm \omega}. \tag{2.95}$$

The dispersion equations (2.93) and (2.94) then give rise to the same dispersion relations (2.45) for parallel propagation of the two oppositely circularly polarized modes as derived from the Hall fluid model discussed in Section 2.3.1.

Note that the left–hand polarized (A) mode in this case is referred to as the *slow Alfvén wave*, while the right–hand polarized (F) mode is called the *fast magnetoacoustic* or *fast Alfvén wave*. In the low–frequency limit the two modes have the same dispersion relation, $\omega^2 = v_A^2 k_z^2$, and their fields can be combined together to give a linearly polarized wave with phase velocity v_A, which is the shear Alfvén wave derived from the ideal MHD theory.

2.5 Collisional Damping

Appreciable damping of the Alfvén and magnetoacoustic waves can occur if collisional processes between and amongst the species have a similar time–scale to the wave period. The energy in the wave motions is dissipated to ultimately produce thermal equilibrium in the plasma. The dissipation can include mutual ion collisions that act to reduce any velocity gradients in the plasma (viscosity), ion-neutral and electron-neutral collisions that act to reduce the streaming of ions and electrons through a gas of neutral particles, or electron-ion collisions that tend to reduce the electric current in the plasma, that is, the streaming of electrons relative to the ions, giving rise to resistivity of the plasma.

We treat here the effects of several of these dissipative processes on Alfvén and magnetoacoustic waves in a uniform plasma, employing the MHD and Hall fluid models. It is interesting to note that some collisional processes, such as ion-neutral collisions, just modify the existing wave modes (for a given real frequency), while others, such as resistivity, introduce new modes. We discuss first the case of collisionally damped waves with frequencies well below the ion cyclotron frequency, and then consider ion cyclotron effects with the Hall model.

2.5.1   Low Frequency

Ion-Neutral Collisions

We consider first the effects of collisions of the ions with a gas of neutral particles. Damping of a wave occurs because the neutrals are dragged along with the ion motions in the wave. If the ions and the neutrals consist of the same atomic or molecular species, a strong resonant charge exchange interaction can occur, with resultant strong damping. The MHD model can be modified by including the equation of motion of the fluid of neutrals, and including collisional momentum exchange between the plasma and the neutral fluid, as in the fluid model equations (1.5).

For simplicity we assume both the plasma and the neutrals are cold, so we can neglect the particle pressure gradient terms in the equations of motion. The Hall term in Ohm's law is initially neglected. The conservation of momentum equation (2.4) for the uniform plasma fluid of equilibrium mass density ρ_0, in the perturbed plasma velocity $\boldsymbol{v}_1$ and perturbed current density $\boldsymbol{J}_1$, is replaced by

$$\rho_0 \frac{\partial \boldsymbol{v}_1}{\partial t} = \boldsymbol{J}_1 \times \boldsymbol{B}_0 - \rho_0 \nu_{\mathrm{in}}(\boldsymbol{v}_1 - \boldsymbol{v}_{1\mathrm{n}}) \tag{2.96}$$

where $\boldsymbol{v}_{1\mathrm{n}}$ is the perturbed velocity of the fluid of neutrals, and ν_{in} is the collision frequency of an ion with the neutrals. The corresponding equation of motion for the neutral fluid, of mass density ρ_{n}, is

$$\rho_{\mathrm{n}} \frac{\partial \boldsymbol{v}_{1\mathrm{n}}}{\partial t} = -\rho_{\mathrm{n}} \nu_{\mathrm{ni}}(\boldsymbol{v}_{1\mathrm{n}} - \boldsymbol{v}_1) \tag{2.97}$$

with ν_{ni} the collision frequency of a neutral particle with the ions. The collision frequencies satisfy the relation $\rho_0 \nu_{\mathrm{in}} = \rho_{\mathrm{n}} \nu_{\mathrm{ni}}$.

Taking the curl of Eqs. (2.96) and (2.97) yields equations in the vorticity of the neutral fluid

$$\zeta_{1\mathrm{n}z} = (\nabla \times \boldsymbol{v}_{1\mathrm{n}})_z \tag{2.98}$$

and the plasma vorticity ζ_{1z} introduced in Section 2.2:

$$\rho_0 \frac{\partial \zeta_{1z}}{\partial t} + \rho_0 \nu_{\text{in}} \zeta_{1z} - \rho_0 \nu_{\text{in}} \zeta_{1nz} - B_0 \frac{\partial J_{1z}}{\partial z} = 0 \tag{2.99}$$

$$\frac{\partial \zeta_{1nz}}{\partial t} + \nu_{\text{ni}} \zeta_{1nz} - \nu_{\text{ni}} \zeta_{1z} = 0. \tag{2.100}$$

Noting that the other equation (2.11) linking ζ_{1z} and J_{1z} is unchanged, we see that the Alfvén mode is now described by the characteristic variables ζ_{1z}, ζ_{1nz} and J_{1z}, satisfying Eqs. (2.11), (2.99) and (2.100).

Similarly, taking the divergence of Eqs. (2.96) and (2.97), we find that the magnetoacoustic mode is described by the characteristic variables $\nabla \cdot \boldsymbol{v}_1$, $\nabla \cdot \boldsymbol{v}_{1n}$ and B_{1z}, satisfying Eq. (2.13) (with $v_{1z} = 0$ since $c_{\text{s}} = 0$), plus the two following equations,

$$\rho_0 \frac{\partial \nabla \cdot \boldsymbol{v}_1}{\partial t} + \rho_0 \nu_{\text{in}} \nabla \cdot \boldsymbol{v}_1 - \rho_0 \nu_{\text{in}} \nabla \cdot \boldsymbol{v}_{1n} - B_0 \frac{\partial J_{1z}}{\partial z} = 0 \tag{2.101}$$

$$\frac{\partial \nabla \cdot \boldsymbol{v}_{1n}}{\partial t} + \nu_{\text{ni}} \nabla \cdot \boldsymbol{v}_{1n} - \nu_{\text{ni}} \nabla \cdot \boldsymbol{v}_1 = 0. \tag{2.102}$$

For the Alfvén mode, the resulting dispersion equation reads

$$s\omega^2 = v_{\text{A}}^2 k_z^2 \tag{2.103}$$

where v_{A} is the Alfvén speed based on the plasma density ρ_0, and s is a complex mass loading factor:

$$s = 1 + \frac{\rho_{\text{n}}}{\rho_0} \frac{1 + i\tau}{1 + \tau^2} \tag{2.104}$$

with

$$\tau = \frac{\omega}{\nu_{\text{ni}}}. \tag{2.105}$$

We note that the Alfvén wave dispersion relation is still independent of the wavevector component perpendicular to the equilibrium magnetic field.

The ion-neutral collisions enter the magnetoacoustic mode dispersion equation in a similar way:

$$s\omega^2 = v_{\text{A}}^2 k^2. \tag{2.106}$$

The factor s in Eqs. (2.103) and (2.106) can be said to produce a complex density $\rho' = s\rho_0$ or complex Alfvén speed $v_{\text{A}}' = v_{\text{A}}/\sqrt{s}$ (Hasegawa & Uberoi 1982). For a real wavenumber, there are three solutions of the dispersion equation (2.103). As shown in Figure 2.4(a), one solution is the Alfvén wave of positive frequency, damped in time, and another is an overdamped mode. There is also a damped Alfvén wave of negative frequency. For large wavenumber, the limiting damping rate of the Alfvén wave is $\nu_{\text{in}}/2$, while the overdamped mode has the limiting damping rate of ν_{ni}. If we assume the frequency to be a real number, the two solutions for the wavenumber resulting from Eq. (2.103) correspond to the right and left–hand propagating spatially damped Alfvén mode, as shown in Figure 2.4(b). The limiting spatial damping rate for large frequency is $\nu_{\text{in}}/2v_{\text{A}}$. The fast magnetoacoustic wave described by Eq. (2.106) has the same damping properties as the Alfvén wave, if k_z is replaced by k. Further discussion of ion–neutral damping effects can be found in the book by Woods (1987).

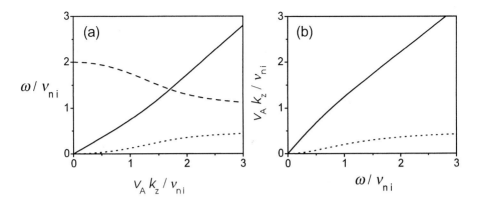

Figure 2.4: (a) The normalized frequency (solid curve) and damping rate (dotted curve) of the Alfvén wave damped by ion-neutral collisions, and the damping rate (dashed curve) of the overdamped mode, against (real) normalized wavenumber. (b) The real part of the normalized wavenumber (solid curve) and spatial damping rate (dotted curve) of the Alfvén wave against (real) normalized frequency. Here $\rho_n/\rho_0 = 1$.

Resistivity

Turning now to the effects of resistivity η on the waves in a cold, fully ionized plasma, we can use the form Eq. (1.8) of Ohm's law, again assuming the frequency is low so that the Hall term can be neglected. The Alfvén mode equations are now Eq. (2.10), and a new version of Eq. (2.11):

$$\mu_0 \frac{\partial J_{1z}}{\partial t} - B_0 \frac{\partial \zeta_{1z}}{\partial z} - \eta \nabla^2 J_{1z} = 0. \tag{2.107}$$

The resulting dispersion equation is

$$\omega \left(\omega + i \frac{\eta k^2}{\mu_0} \right) = v_A^2 k_z^2. \tag{2.108}$$

We note that the electric field in the equilibrium magnetic field direction is no longer zero for the Alfvén mode: $E_{1z} = \eta J_{1z}$. Another important point to note from Eq. (2.108) is that the dispersion relation is no longer independent of the wavevector component perpendicular to the equilibrium magnetic field. This has consequences for the discussion of propagation in nonuniform plasmas in Chapter 3.

Rewriting Eq. (2.108) as

$$k^2 = \frac{\omega^2}{(v_A^2 \cos^2 \theta + i \eta \omega / \mu_0)} \tag{2.109}$$

we see that for a given real frequency, the wavenumber acquires an imaginary part, and spatial damping of the wave takes place. The least damping of the wave occurs for parallel propagation. Maximum damping occurs for perpendicular propagation, when

$$k = \left(\frac{\omega \mu_0}{2\eta} \right)^{1/2} (1 - i). \tag{2.110}$$

In the latter case the wave acts like an electromagnetic wave in a good conductor, with the damping length or skin depth equalling the wavelength divided by 2π (Griffths 1989).

Equation (2.108) can be written in nondimensional form as

$$f \left(f + i \frac{1}{R_m} \alpha^2 \right) = \alpha^2 \tag{2.111}$$

where $f = \omega/\omega_0$, $\omega_0 = v_A/L$, $\alpha = k_z L$, L is a reference length, and the magnetic Reynolds number is

$$R_m = \frac{\mu_0 L v_A}{\eta}. \tag{2.112}$$

The dispersion relation of the resistively damped Alfvén wave obtained from Eq. (2.111) is shown in Figure 2.5. Figure 2.5(a) shows the frequency and damping rate plotted against a real wavenumber, while Figure 2.5(b) shows the real part of the wavenumber and the spatial damping rate plotted against real frequency. When the wavenumber becomes greater than $2R_m/L$, the real part of the frequency is zero and there are two purely damped modes.

The fast magnetoacoustic mode equations are now Eq. (2.12), and a new version of Eq. (2.13):

$$\frac{\partial B_{1z}}{\partial t} + B_0 \nabla \cdot \boldsymbol{v}_1 - \frac{\eta}{\mu_0} \nabla^2 B_{1z} = 0 \tag{2.113}$$

with the resulting dispersion equation

$$\omega \left(\omega + i \frac{\eta k^2}{\mu_0} \right) = v_A^2 k^2. \tag{2.114}$$

It follows from Eq. (2.114) that the magnetoacoustic wave is still isotropic, that is, for a given frequency the wavenumber and phase velocity are still independent of the angle of propagation θ, although damping of the wave occurs.

2.5.2   Hall Effects

If we now consider a plasma with ion-neutral collisions and resistivity, the inclusion of the Hall term in the Ohm's law again induces a coupling between the characteristic variables of Table 2.1 for the Alfvén and magnetoacoustic modes. We note that the inclusion of the Hall term as well as collisional processes introduces a number of characteristic time–scales, with which the period of the wave must be compared. For example, if the only collisions are those of ions with neutrals, the characteristic frequencies are the ion cyclotron frequency Ω_i and the ion-neutral collision frequency ν_{in}, and the relative importance of the two effects is measured by the ratios of the characteristic frequencies with the wave frequency.

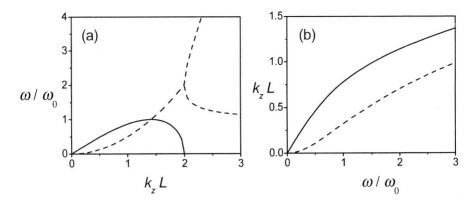

Figure 2.5: (a) The frequency (solid curve) and damping rate (dotted curve) of the Alfvén wave damped by resistivity, and the damping rates (dashed curves) of the overdamped modes, against (real) wavenumber. (b) The real part of the wavenumber (solid curve) and the spatial resistive damping rate (dotted curve) of the Alfvén wave against (real) frequency. Here the complex frequency is normalized by $\omega_0 = v_A/L$ and the complex wavenumber by L^{-1}. The magnetic Reynolds number R_m is 1.

It is instructive first to consider the case where the ions collide with a stationary background of neutral particles, neglecting resistivity, and to use the dielectric tensor approach of Section 2.4. The ion and electron equations of motion (1.5) may be used to derive the following form of the electric field-current density relationship:

$$J = \sigma_{||}E_{||} + \sigma_P E_\perp + \sigma_H B \times E_\perp/B = \begin{bmatrix} \sigma_P & -\sigma_H & 0 \\ \sigma_H & \sigma_P & 0 \\ 0 & 0 & \sigma_{||} \end{bmatrix} \cdot E \qquad (2.115)$$

where $E_{||}$ and $E_\perp$ are the components of the electric field aligned respectively parallel and perpendicular to the magnetic field. The frequency–dependent conductivity tensor σ_{ij} appearing in Eq. (2.115), which is to be used to calculate the dielectric tensor Eq. (2.55) and thence the dispersion equation, has the components $\sigma_{||}$ (the parallel conductivity), σ_P (called the *Pedersen conductivity*), and σ_H (called the *Hall conductivity*) (Luhmann 1995).

In the case of ion collisions with stationary neutrals, the conductivities are

$$\sigma_P = \frac{n_0 e}{B_0} \frac{\Omega_i(\nu_{in} - i\omega)}{\Omega_i^2 + (\nu_{in} - i\omega)^2} \qquad (2.116)$$

$$\sigma_H = \frac{n_0 e}{B_0} \frac{(\nu_{in} - i\omega)^2}{\Omega_i^2 + (\nu_{in} - i\omega)^2} \qquad (2.117)$$

$$\sigma_{||} = \frac{n_0 e^2}{m_i(\nu_i - i\omega)} \qquad (2.118)$$

where n_0 and B_0 are respectively the equilibrium electron number density and magnetic field. Here we have assumed the wave frequency and the electron-neutral collision frequency to be much less than the electron cyclotron frequency. The corresponding expressions with electron effects included are to be found in Luhmann (1995).

The *Cowling conductivity*, defined as

$$\sigma_C = \sigma_P + \frac{\sigma_H^2}{\sigma_P} \tag{2.119}$$

is a useful quantity because the rate of work done on the plasma is, if $E_z = 0$,

$$\boldsymbol{J} \cdot \boldsymbol{E} = J^2/\sigma_C. \tag{2.120}$$

Using Eqs. (2.116)-(2.118), noting the relationship Eq. (1.35) between the dielectric tensor K_{ij} and the conductivity tensor σ_{ij}, and assuming the nonrelativistic limit $v_A \ll c$, we find the following expressions for the quantities S and D entering the dielectric tensor Eq. (2.55):

$$S = -\frac{c^2}{v_A^2} \frac{(\omega + i\nu_{in})\Omega_i^2}{\omega\left((\omega + i\nu_{in})^2 - \Omega_i^2\right)} \tag{2.121}$$

$$D = \frac{c^2}{v_A^2} \frac{(\omega + i\nu_{in})^2\Omega_i}{\omega\left((\omega + i\nu_{in})^2 - \Omega_i^2\right)}. \tag{2.122}$$

These expressions should be compared with the corresponding noncollisional quantities of Eqs. (2.91) and (2.92). Use of these expressions for S and D in Eqs. (2.73) and (2.77) yields the following collisional version of the cold plasma Hall model dispersion equation (2.42):

$$\left(\omega(\omega + i\nu_{in}) - v_A^2 k_z^2\right)\left(\omega(\omega + i\nu_{in}) - v_A^2 k^2\right) = \left(\frac{\omega + i\nu_{in}}{\Omega_i}\right)^2 v_A^4 k_z^2 k^2. \tag{2.123}$$

Defining the Hall parameter β_i;

$$\beta_i = \frac{\Omega_i}{\nu_{in}} \tag{2.124}$$

and using normalized frequency $f = \omega/\Omega_i$ and wavenumber $\alpha = v_A k/\Omega_i$, the dispersion equation (2.123) becomes

$$\left(f(f + i/\beta_i) - \alpha^2 \cos^2\theta\right)\left(f(f + i/\beta_i) - \alpha^2\right) = (f + i/\beta_i)^2 \alpha^4 \cos^2\theta. \tag{2.125}$$

The resulting damped fast Alfvén wave and slow Alfvén (ion cyclotron) wave dispersion relations are shown in Figure 2.6, for the case of parallel propagation. Note that there is strong dispersion of both modes at low α, and the ion cyclotron wave is strongly damped, with the limiting damping rate ν_{in} as $\alpha \to 0$.

We can allow for the motion of the neutral gas, and both anisotropic resistivity and ion-neutral collisions, and use either the fluid model or the dielectric tensor approach to derive the following generalization of the Hall model dispersion equation (2.42) for a cold plasma and a cold gas of neutral species (Woods 1962, 1987):

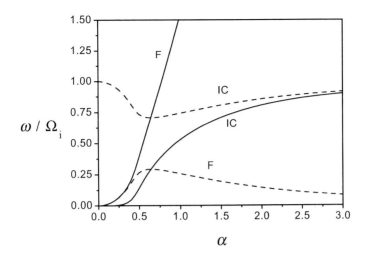

Figure 2.6: The frequency (solid curves) and damping rate (dashed curves) of the fast Alfvén wave (F) and the slow Alfvén (ion cyclotron) wave (IC), damped by ion-neutral collisions, plotted against the normalized real wavenumber $\alpha = v_A k/\Omega_i$. Here $\theta = 0°$ and the Hall parameter $\beta_i = 1$.

$$\left(s\omega^2\left(1 + i\delta_\perp k_z^2 + i\delta_{||}k_x^2\right) - v_A^2 k_z^2\right)\left(s\omega^2\left(1 + i\delta_\perp k^2\right) - v_A^2 k^2\right)$$
$$= s^2 \left(\frac{\omega}{\Omega_i}\right)^2 v_A^4 k_z^2 k^2 \tag{2.126}$$

where

$$\delta_{||,\perp} = \frac{\eta_{||,\perp}}{\mu_0\omega} \tag{2.127}$$

with $\eta_{||,\perp}$ the resistivities, respectively parallel and perpendicular to the equilibrium magnetic field. If $\tau \gg 1$, in which case

$$s \simeq 1 + i\frac{\nu_{in}}{\omega} \tag{2.128}$$

the stationary neutrals result in Eq. (2.123) follows from Eq. (2.126).

2.6   Multiple Ion Species

Several different ion species commonly exist in fusion and space plasmas, and charged dust grains (usually negatively charged) can occur in laboratory and space plasmas. These additional charged species can be taken into account by using a multi–fluid model, or by the dielectric tensor approach. We consider for simplicity the case of a cold plasma made up of two charged ion species, plus electrons, with overall charge neutrality in the equilibrium state, that is,

$$Z_1 n_1 + Z_2 n_2 = n_e \qquad (2.129)$$

where Z_1 and Z_2 are the (signed) charge numbers on each type of ion, and n_1 and n_2 are the number densities of each ion. The magnitudes of the cyclotron frequencies of the two types of ions are Ω_1 and Ω_2.

We define the dimensionless parameters

$$b = \rho_2/\rho_1, \quad g = \epsilon_2 \Omega_2/\epsilon_1 \Omega_1 \qquad (2.130)$$

where ρ_1 and ρ_2 are the respective mass densities of the ions. We use a frequency normalized to the first ion cyclotron frequency, $f = \omega/\Omega_1$. Assuming the first ion species has a positive charge and is the lighter ion (such as a proton), the parameter g has the sign of the charge of the second species and $|g| < 1$. The case of positive g applies to fusion plasmas with heavy positive secondary ions and contaminants, while the case of negative g is relevant to space plasmas with negatively–charged dust grains to be discussed in Chapters 7 and 8.

If we first consider parallel propagation, $k_x = 0$, the dispersion equation in the large $|P|$ and $v_A \ll c$ limit may be derived from Eq. (2.77) or Eq. (2.78) as

$$k_z^2 = u_1 \pm u_2 \qquad (2.131)$$

where, using Eqs. (2.73), (2.80) and (2.81), we find

$$u_1 = \frac{\omega^2}{v_A^2}\left(\frac{1}{1-f^2} + \frac{b}{1-f^2/g^2}\right) \qquad (2.132)$$

$$u_2 = \frac{\omega^3}{\Omega_1 v_A^2}\left(\frac{1}{1-f^2} + \frac{b/g}{1-f^2/g^2}\right). \qquad (2.133)$$

Here the Alfvén speed v_A is based on the mass density of the first (lighter) ion species. In terms of the dimensionless wavenumber $\alpha = v_A k_z/\Omega_1$, Eq. (2.131) reduces to

$$\alpha^2 = f^2\left[\frac{g(1+b) \pm f(1+bg)}{(1 \pm f)(g \pm f)}\right]. \qquad (2.134)$$

Let us consider positive g first. There are three positive frequency solutions of Eq. (2.134) for the parallel propagating modes for given α, as shown in the plot of frequency against α in Figure 2.7(a) in the vicinity of the cyclotron frequency of the more massive ion. The lowest frequency mode is left–hand circularly polarized (L–mode), and has a resonance in k_z at $\omega = \Omega_2$, that is, it is an ion cyclotron wave (in the heavier ion). It has the following dispersion relation at low k_z:

$$f \simeq \frac{1}{(1+b)^{1/2}}\alpha - \frac{|b/g+1|}{2(1+b)^2}\alpha^2. \qquad (2.135)$$

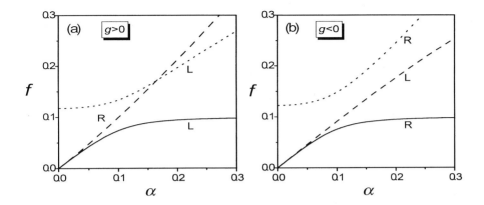

Figure 2.7: The normalized frequency $f = \omega/\Omega_1$ of the parallel propagating left–hand (L) and right–hand (R) circularly polarized waves in a two-ion-species plasma, plotted against the normalized wavenumber $\alpha = v_A k_z/\Omega_1$. The density ratio b is 0.2. (a) The secondary species is positively charged, and $g = 0.1$. (b) The secondary species is negatively charged, and $g = -0.1$.

Noting that the Alfvén speed based on the total mass density of both ion species is given by

$$v_{AT} = v_A/(1+b)^{1/2} \tag{2.136}$$

the dispersion relation at very low k_z and $\omega \ll \Omega_2$ is

$$\omega \simeq v_{AT} k_z. \tag{2.137}$$

The next highest frequency mode (at low wavenumber) is a right–hand circularly polarized mode (R–mode), with a dispersion relation at low k_z given by

$$f \simeq \frac{1}{(1+b)^{1/2}}\alpha + \frac{|b/g+1|}{2(1+b)^2}\alpha^2. \tag{2.138}$$

This mode is essentially the fast Alfvén wave, with small modifications because of the ion cyclotron effects. For $\omega \ll \Omega_2$ this mode has the same dispersion relation (2.137) as the left–hand mode, and the two modes together can form the linearly polarized Alfvén wave with phase velocity v_{AT}. It is interesting to note that if $|g|$ is small and α is large enough, the term quadratic in α in Eq. (2.138) can dominate, and we have a quadratic dispersion relation characteristic of a whistler wave. Further discussion of this point for plasmas with charged dust grains can be found in Chapter 7.

The highest frequency mode (at low wavenumber) is an L–mode, with a dispersion relation for low k_z given by

$$f \simeq \frac{g(1+b)}{1+bg} + \frac{|b/g+1|}{2(1+bg)^2}\alpha^2 \tag{2.139}$$

so there exists a k_z cutoff (where $k_z = 0$) at the frequency

$$\omega = \Omega_{\mathrm{m}} = \frac{\Omega_2(1+b)}{1+bg}.$$

(2.140)

This mode also has a resonance in k_z at Ω_1 (outside the scale in Figure 2.7), that is, it is an ion cyclotron mode (in the lighter ion).

In the negative g case, there are also three positive frequency solutions of Eq. (2.134) for given α, as shown in Figure 2.7(b). In this case the lowest frequency mode is right–hand circularly polarized, has a resonance in k_z at $\omega = \Omega_2$, and has the dispersion relation at low k_z as given by Eq. (2.135). The next highest frequency mode is left–hand circularly polarized, with a resonance at $\omega = \Omega_1$, and with a dispersion relation at low k_z given by Eq. (2.138). Again, for $\omega \ll \Omega_2$ this mode has the same dispersion relation (2.137) as the right–hand mode, and the two modes together can form the linearly polarized Alfvén wave with phase velocity v_{AT}. This mode also has a resonance at the cyclotron frequency of the positively–charged lighter ion.

The highest frequency mode is an R–mode, which becomes the fast Alfvén wave at high frequencies, with a dispersion relation for low k_z given by

$$f \simeq \frac{|g|(1+b)}{1-b|g|} + \frac{|b/|g| - 1|}{2(1-b|g|)^2}\alpha^2$$

(2.141)

and there exists a k_z cutoff at

$$\omega = \Omega_{\mathrm{m}} = \frac{\Omega_2(1+b)}{1-b|g|}.$$

(2.142)

An alternative way of illustrating the dispersion relation is to plot the square of the refractive index against the frequency. The refractive indices for parallel propagating waves are given for a general frequency range by Eq. (2.66). We have, directly from Eq. (2.134), or from Eq. (2.66) in the large $|P|$ and $v_{\mathrm{A}} \ll c$ limit,

$$n_{\pm}^2 = \frac{c^2}{v_{\mathrm{A}}^2}\left[\frac{g(1+b) \pm f(1+bg)}{(1 \pm f)(g \pm f)}\right]$$

(2.143)

where $+, -$ correspond to right and left–hand circularly polarized modes respectively. For propagation parallel to the magnetic field, $n_0^2 = n_{\pm}^2 v_{\mathrm{A}}^2/c^2$ is plotted against f for a positively charged heavier ion ($g > 0$) in Figure 2.8(a), and for a negatively charged heavier ion ($g < 0$) in Figure 2.8(b). The left–hand polarized (L) and right–hand polarized (R) waves are indicated. A resonance occurs at the cyclotron frequency of the heavier ion in each case; a resonance of the L–wave also occurs at the cyclotron frequency of the lighter ion ($f = 1$), but is outside the scale of Figure 2.8. We note that for positive g there is a point where the refractive indices of the oppositely polarized waves are equal, as is also indicated in Figure 2.7(a). If the waves propagate at a nonzero angle to the magnetic field, the dispersion curves are found to reconnect near the crossover point, with a left–hand polarized mode changing continuously to a right–hand polarized mode (Melrose 1986).

The case where the waves propagate obliquely to the equilibrium magnetic field in a multi–ion plasma is discussed further, in connection with nonuniform plasmas, in Chapter 3 (see also the treatment of Melrose (1986)). The additional features of parallel and oblique waves in plasmas with negatively charged dust grains are discussed in Chapters 7 and 8.

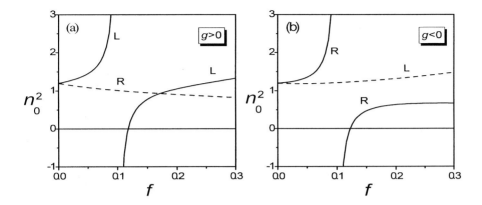

Figure 2.8: The square of the scaled refractive index, $n_0^2 = n^2 v_A^2/c^2$, is plotted against the normalized frequency $f = \omega/\Omega_1$ in a two-ion-species plasma, for the left–hand polarized (L) and right–hand polarized (R) parallel propagating modes. The density ratio b is 0.2. (a) The secondary species is positively charged, and $g = 0.1$. (b) The secondary species is negatively charged, and $g = -0.1$.

2.7 Kinetic Theory of Waves

We now consider the theory of waves in a collisionless plasma, allowing for motions of the charged particles about their mean (fluid) velocities. The fluid model used in the previous sections assumes thermal velocity distributions of particles, and reduces to equations describing the behaviour of fluid elements with well–defined temperatures and pressures. On the other hand, the dielectric tensor approach, using the Vlasov equation to describe the evolution of the particle velocity distribution functions, enables us to consider both thermal and nonthermal velocity distributions of the ions and electrons, as well as describing wave-particle resonance phenomena. The simplest case to consider first is that of parallel propagation of the waves.

2.7.1 Parallel Propagation

The dielectric tensor for waves propagating parallel to the equilibrium magnetic field, with wavenumber $k_z = k$, can be derived from the dielectric tensor components given in the Appendix, using $k_\perp = 0$:

$$K_{ij}(\omega, k) = \begin{bmatrix} K_{11} & K_{12} & 0 \\ -K_{12} & K_{11} & 0 \\ 0 & 0 & K_{33} \end{bmatrix}. \tag{2.144}$$

The resulting collisionless dispersion equation for transverse electromagnetic waves follows from setting the determinant Λ in Eq. (1.40) to zero, retaining only the components in the upper left 2×2 matrix in Eq. (2.144). It can be written in a form analogous to the cold–

plasma forms Eqs. (2.93) and (2.94),

$$n_{\pm}^2 = R_{\pm} \tag{2.145}$$

for the two oppositely circularly polarized modes with refractive indices n_+ (right–hand) and n_- (left–hand). Longitudinal electrostatic modes are described by the equation $K_{33} = 0$.

The functions $R_{\pm}$ for an arbitrary (but nonrelativistic) zero–order momentum distribution function $f_\alpha(p)$ for particles of the species α, satisfying the normalization condition (Eq. A.1), have the following relatively simple forms:

$$R_{\pm} = 1 + \sum_\alpha \frac{\omega_{p\alpha}^2}{2\omega^2 n_\alpha} \int \mathrm{d}^3 p \, p_\perp \left(k v_\perp \frac{\partial f_\alpha}{\partial p_z} + (\omega - k v_z) \frac{\partial f_\alpha}{\partial v_\perp} \right) / (\omega \pm \Omega_\alpha - k v_z) \tag{2.146}$$

where v_z is the particle velocity in the magnetic field direction. If the distribution functions are isotropic, with $f_\alpha(p) = f_\alpha(p)$, Eq. (2.146) becomes

$$R_{\pm} = 1 - \sum_\alpha \frac{\omega_{p\alpha}^2}{\omega n_\alpha} \int \mathrm{d}^3 p \, p \frac{f_\alpha(p)}{\omega \pm \Omega_\alpha - k v_z}. \tag{2.147}$$

For cold particles with zero mean velocity, with delta function momentum distribution functions, Eq. (2.147) reduces to Eq. (2.57), with the resulting cold–plasma Alfvén and fast magnetoacoustic modes described by Eqs. (2.93) and (2.94).

Damping of the waves due to wave-particle resonances follows from the fact that the resonant denominator gives rise to a pole in the integrand in Eq. (2.146) or Eq. (2.147). Let us assume that the momentum distribution function is the product of a parallel part and a perpendicular part:

$$f_\alpha(p) = f_\alpha(p_z) f_\alpha(p_\perp) \quad \text{with} \quad \int \mathrm{d}^2 p_\perp f_\alpha(p_\perp) = 1. \tag{2.148}$$

In evaluating the integrals in Eq. (2.146) or Eq. (2.147), if the imaginary part of the frequency γ satisfies $|\gamma| \ll |\omega_r|$, where ω_r is the real part of the frequency, we may approximate the contour of integration in the complex momentum plane, prescribed by Landau (e.g. Melrose (1986)), by a straight line along the p_z axis that excludes the pole at $v_z = \omega/k \pm \Omega_\alpha$, and a semicircular path around the pole. The residue at the pole gives an imaginary contribution to the integral. Writing $\omega^2 \simeq \omega_r^2 + 2i\omega_r\gamma$ and equating the imaginary terms in Eq. (2.145), the approximate result for γ is

$$\gamma = \frac{\pi}{2\omega_r} \sum \frac{\omega_{p\alpha}^2}{n_\alpha} \left[\frac{\langle v_\perp^2 \rangle}{2} \frac{\partial f_{\alpha z}}{\partial p_z} \pm \frac{m_\alpha \Omega_\alpha}{k} f_{\alpha z} \right]_{v_z = (\omega_r \pm \Omega_\alpha)/k} \tag{2.149}$$

where the angular brackets denote an average over the distribution function $f_\alpha(p_\perp)$, and the $+(-)$ sign corresponds to right (left)–hand polarization of the wave.

For a plasma made up of charged species with anisotropic temperatures $T_\parallel$ and $T_\perp$, respectively parallel and perpendicular to the magnetic field, that is, with bi-Maxwellian distribution functions of the form (Eq. A.8), parallel propagating waves at frequencies well below the

electron cyclotron frequency are described by the following dispersion equations, derived
from Eqs. (2.145) and (2.146) (Melrose 1986):

$$n_\pm^2 = 1 - \sum_i \frac{\omega_{pi}^2}{\omega^2}\left[\frac{\omega}{\Omega_i} - z_{\pm 1}^i Z(z_{\pm 1}^i)\left\{\mp\frac{\omega}{\omega \pm \Omega_i} + A_i\right\} - A_i\right] \tag{2.150}$$

where the sum is over the ions only (i.e. excluding the electrons).

Here A_i measures the imbalance in the parallel and perpendicular temperatures,

$$A_i = \frac{T_{i\perp}}{T_{i\|}} - 1 \tag{2.151}$$

and

$$z_{\pm 1}^i = \frac{\omega \pm \Omega_i}{\sqrt{2}V_{i\|}k} \tag{2.152}$$

is the argument of the plasma dispersion function Z defined in Eq. (A.13), where $V_{i\|}$ is the
thermal speed (Eq. 1.7) of each species of ion parallel to the magnetic field. If all the ions
are positively charged, the fast wave (right–hand polarized) is described by the positive sign,
and the slow Alfvén wave, or ion cyclotron wave (left–hand polarized), is described by the
negative sign in Eq. (2.150). As the temperatures become small, we have $|z_{\pm 1}^i| \gg 1$, and
using the asymptotic form Eq. (A.15) for Z, we obtain (for a single ion species) the cold
plasma results (2.93)-(2.95).

The imaginary contribution to the dispersion equation (2.150) arising from the resonant
denominator in the integrand of Eq. (2.147) is contained in the complex function Z. The
resulting approximate collisionless damping rate Eq. (2.149), for Maxwellian distributions
with isotropic temperatures for all species α, can be derived directly from Eq. (2.149) as

$$\gamma = -\frac{\pi}{2k}\sum_\alpha \frac{\omega_{p\alpha}^2}{\sqrt{2\pi}V_\alpha}\exp\left[\frac{-(\omega_r \pm \Omega_\alpha)^2}{2k^2 V_\alpha^2}\right]. \tag{2.153}$$

In the case of $\omega \ll \Omega_e$ and for a single ion species, the damping rate of the ion cyclotron wave
becomes (Stix 1992)

$$\gamma = -\omega_r\sqrt{\frac{\pi}{2}}\frac{\omega_r}{(2\Omega_i - \omega_r)}\frac{\omega_r^2\Omega_i^2}{k^4 v_A^4}\frac{\Omega_i}{|k|V_i}\exp\left[\frac{-(\omega_r - \Omega_i)^2}{2k^2 V_i^2}\right]. \tag{2.154}$$

This *cyclotron resonance* damping occurs because of those particles that experience a Doppler–
shifted wave frequency at $\pm\Omega_\alpha$. For oblique propagation, the cyclotron damping becomes
large when $\omega \simeq |s\Omega_\alpha|$ for any harmonic number s. For the parallel propagating left–hand
polarized ion cyclotron mode, cyclotron damping occurs when ω approaches Ω_i. Nonthermal
momentum distribution functions, for example those describing mean drifts of a population
of particles relative to others, can give rise to positive values of γ, that is, to instabilities of
the waves (see Chapter 5). Further discussion of parallel propagation of the waves, as well as
oblique propagation, is given in the following sections, for different plasma beta regimes.

2.7.2 Low Plasma Beta

Let us now consider oblique propagation of the waves, at the angle θ to the magnetic field, in a warm plasma, but with a low plasma beta. We assume a plasma with one ion species, with Maxwellian distribution functions for the ions and electrons, isotropic temperatures (but different ion and electron temperatures) and no mean drifts. The ion and electron partial pressures can be written in terms of the respective temperatures by means of the perfect gas law, and the plasma, ion and electron betas (β, β_i and β_e) are then defined by Eqs. (1.20) and (1.23).

In this section we assume that the magnetic energy is dominant over the particle energy in the plasma, so that $\beta = \beta_i + \beta_e \ll 1$. The dielectric tensor can be written (from Eq. (A.10) with $A_\alpha = 0$ and $U_\alpha = 0$),

$$K_{ij}(\omega, \boldsymbol{k}) = \delta_{ij} + \sum \frac{\omega_p^2}{\omega^2}\left[2z_0^2 b_i b_j + z_0 \sum_{s=-\infty}^{\infty} Z(z_s) N_{ij}(\omega, \boldsymbol{k}, s)\right] \tag{2.155}$$

where the first sum runs over particle species α, Z is the plasma dispersion function (Eq. A.13) and the components of the tensor N_{ij} for each α are defined in Eq. (A.16). The argument of the plasma dispersion function is

$$z_s^\alpha = \frac{\omega - s\Omega_\alpha}{\sqrt{2}k_z V_\alpha}. \tag{2.156}$$

In the case of oblique propagation it is important to compare the size of the wavelength perpendicular to the magnetic field with the Larmor radius of the charged particles, where the Larmor radius $\rho_\alpha = V_\alpha/\Omega_\alpha$ is the radius of gyration about the magnetic field of a particle with the thermal speed V_α defined in Eq. (1.7). If the Larmor radius is small relative to the perpendicular wavelength, that is if

$$\lambda_\alpha = k_\perp^2 \rho_\alpha^2 \ll 1 \tag{2.157}$$

the calculation of the dielectric tensor is simplified because the expansions of $e^{-\lambda}I_s(\lambda)$ for small λ as given in the Appendix can be used. Because of the small electron mass and the resulting small electron Larmor radius, we shall assume in this book that it is always the case that $\lambda_e \ll 1$. We shall also assume in this section that the perpendicular wavelength is larger than the ion Larmor radius, such that $\lambda_i \ll 1$, and only the dominant terms in the dielectric tensor components in powers of λ_i are retained. Higher order terms in λ_i are included when we discuss waves in high–β plasmas and kinetic Alfvén waves in later sections.

The terms in the dielectric tensor depending on the parallel wavenumber k_z can also be simplified in some circumstances. In the frequency regime $\omega \lesssim \Omega_i \ll \Omega_e$ considered here, the phase velocities of the Alfvén and magnetoacoustic waves are of the order of the Alfvén speed v_A. Noting that we can write

$$\beta_i = \frac{p_i}{(B^2/2\mu_0)} = 2\frac{V_i^2}{v_A^2} \tag{2.158}$$

where V_i is the ion thermal speed defined in Eq. (1.7), and that by assumption $\beta_i \ll 1$, we have

$$|\omega/k_z| \simeq v_A = \sqrt{2/\beta_i}\, V_i \gg V_i. \tag{2.159}$$

This implies, using Eq. (2.156) and provided that ω is not close to $|s\Omega_\alpha|$, that $|z_s^i| \gg 1$ for all s. (Thus we are assuming in this section that ion cyclotron damping is negligible.) The first term in the asymptotic representation of Z given in Eq. (A.15) can therefore be used for the ions.

Considering now the electrons, we note that

$$z_0^e \simeq \frac{v_A}{V_e} = \left(\frac{m_e}{m_i}\right)^{1/2} \left(\frac{T_i}{T_e}\right)^{1/2} \sqrt{\frac{2}{\beta_i}} \tag{2.160}$$

that is, the electron thermal speed might be large enough that $|z_0^e|$ can be of order 1 or less, even if $\beta_i \ll 1$. However, $|z_s^e| \gg 1$ for all $|s| > 0$. Thus we shall use the asymptotic form of $Z(z_s^e)$ for $|s| > 0$ but retain the exact form of $Z(z_0^e)$.

If we use all the above approximations, we find that the waves in the low–β plasma have the basic properties described by the cold collisionless plasma theory of Section 2.4, but there are additional terms in the dielectric tensor components that lead to collisionless electron Landau damping. The components of the dielectric tensor may be derived from Eq. (2.155), under the above approximations, to be (Akhiezer *et al.* 1975, Melrose 1986):

$$K_{11} = S = -\frac{c^2}{v_A^2} \frac{\Omega_i^2}{\omega^2 - \Omega_i^2} \tag{2.161a}$$

$$K_{22} = S + K_{22}' = S + \frac{m_e}{m_i} \tan^2\theta \, \frac{Z(z_0^e)}{z_0^e} \tag{2.161b}$$

$$K_{33} = -\frac{c^2}{v_A^2} \frac{\Omega_i^2}{\omega^2} \left(1 - \frac{\omega^2}{v_s^2 k^2 \cos^2\theta}[1 + z_0^e Z(z_0^e)]\right) \tag{2.161c}$$

$$K_{12} = -K_{21} = -iD = -i\frac{c^2}{v_A^2} \frac{\omega\Omega_i}{\omega^2 - \Omega_i^2} \tag{2.161d}$$

$$K_{23} = -K_{32} = -i\frac{c^2}{v_A^2} \frac{\Omega_i}{\omega} \tan\theta \, [1 + z_0^e Z(z_0^e)] \tag{2.161e}$$

where the quantities S and D are as defined in Eqs. (2.91) and (2.92) for the cold plasma. Here

$$v_s = \sqrt{k_B T_e/m_i} \tag{2.162}$$

is the ion-sound or ion-acoustic velocity, which is the phase velocity of the electrostatic ion-acoustic mode. The ion-acoustic mode is the analogue of the ordinary sound wave, and arises in an unmagnetized collisionless plasma. It is lightly damped if $T_e \gg T_i$.

Let us consider the low–frequency limit $\omega \ll \Omega_i$. If we also have a high electron thermal speed, $V_e > v_A$, so that $z_0^e \ll 1$ (but still $\beta \ll 1$), we can use the series form for $Z(z_0^e)$ (Eq. A.14). To dominant order in ω/Ω_i and z_0^e, and neglecting terms of order m_e/m_i (including the imaginary terms leading to damping), we then have

$$K_{11} = K_{22} = \frac{c^2}{v_A^2} \tag{2.163a}$$

$$K_{33} = -\frac{c^2}{v_A^2} \frac{\Omega_i^2}{\omega^2} \left(1 - \frac{\omega^2}{v_s^2 k^2 \cos^2 \theta}\right) \tag{2.163b}$$

$$K_{12} = -K_{21} = i\frac{c^2}{v_A^2} \frac{\omega}{\Omega_i} \tag{2.163c}$$

$$K_{23} = -K_{32} = -i\frac{c^2}{v_A^2} \frac{\Omega_i}{\omega} \tan \theta. \tag{2.163d}$$

To exploit the dependence of the dielectric tensor components given by Eq. (2.161) or Eq. (2.163) on the factor ω/Ω_i, the dispersion equation obtained from Eq. (1.40) may be written explicitly as a sum of terms arranged in increasing order in ω^2/Ω_i^2 (Melrose 1986):

$$[(n^2 \cos^2 \theta - K_{11})\{(n^2 - K_{22})K_{33} - K_{23}^2\}]$$

$$+[n^4 K_{11} \sin^2 \theta - n^2 K_{11} K_{22} \sin^2 \theta + 2n^2 K_{12} K_{23} \sin \theta \cos \theta + K_{12}^2 K_{33}] \tag{2.164}$$

$$-[n^2 K_{12}^2 \sin^2 \theta] = 0.$$

Using the components given by Eq. (2.163), the first term in square brackets in Eq. (2.164) is of order $(\Omega_i/\omega)^2$. Its first factor gives the Alfvén (A) mode,

$$n_A^2 = \frac{K_{11}}{\cos^2 \theta} \quad \text{and} \quad \omega_A = k v_A |\cos \theta|. \tag{2.165}$$

The second factor of the first term gives the collisionless analogues of the fast (F) and slow (S) magnetoacoustic modes, with frequencies of the same form as Eqs. (2.20) and (2.21), but with c_s replaced by the ion-sound velocity v_s.

Collisionless damping of the waves at frequency $\omega \ll \Omega_i$ is due to Landau damping by thermal electrons travelling along the magnetic field. The electrons interact with the E_z component of the wave electric field, so for damping of the Alfvén and magnetoacoustic modes there must be coupling between those modes and the longitudinal mode, which has an E_z component. The coupling depends on the dielectric tensor component K_{33} in Eq. (2.164), producing an E_z component in the Alfvén and magnetoacoustic modes.

If we retain the imaginary term in $Z(z_0^e)$ in K_{33} and K_{23} in Eq. (2.161), the damping rate of the Alfvén mode to lowest order in ω/Ω_i is given by

$$\gamma_A = \omega_A \left(\frac{\pi}{2} \frac{m_e}{m_i}\right)^{1/2} \frac{v_s}{v_A} \frac{\omega^2}{\Omega_i^2} \left(\cot^2 \theta + \frac{\tan^2 \theta}{|1 + z_0^e Z(z_0^e)|^2}\right) e^{-(z_0^e)^2}. \tag{2.166}$$

The damping rate for the fast magnetoacoustic mode is

$$\gamma_F = \omega_F \left(\frac{\pi}{2} \frac{m_e}{m_i}\right)^{1/2} \frac{v_s}{v_A} \frac{\sin^2 \theta}{|\cos \theta|} e^{-(z_0^e)^2} \tag{2.167}$$

and for the slow magnetoacoustic mode:

$$\gamma_S = \omega_S \left(\frac{\pi}{2} \frac{m_e}{m_i} \right)^{1/2} e^{-(z_0^e)^2}. \tag{2.168}$$

Thus at intermediate angles of propagation, the damping of the Alfvén wave is much lower than that of the fast wave, which is lower than that of the slow wave. However if the plasma is nonthermal, such that $T_e \gg T_i$, the slow magnetoacoustic wave is only weakly damped, as for the ion-sound wave in an unmagnetized plasma. For almost parallel propagation, $\theta^2 \lesssim \omega/\Omega_i$, the electron Landau damping is almost the same for the Alfvén and fast waves (Akhiezer *et al.* 1975):

$$\gamma = \frac{1}{4} \left(\frac{\pi}{2} \frac{m_e}{m_i} \right)^{1/2} \frac{v_s}{v_A} \theta^2 \left(1 \pm \frac{\theta^2}{(\theta^4 + 4\omega^2/\Omega_i^2)^{1/2}} \right) e^{-(z_0^e)^2}. \tag{2.169}$$

Let us now consider frequencies up to Ω_i. If we also assume the electron temperature to be low, such that $V_e \ll v_A$, we have $z_0^e \gg 1$ and the asymptotic form (Eq. A.15) for $Z(z_0^e)$ can be used. We then find

$$K_{33} = -\frac{c^2}{v_A^2} \frac{\Omega_i^2}{\omega^2} \left(1 + \frac{m_i}{m_e} \right) \tag{2.170}$$

so that the component K_{33} is very large in magnitude compared with the other tensor components, and dividing Eq. (2.164) through by K_{33} and neglecting terms of order $1/K_{33}$ gives

$$(n^2 \cos^2\theta - K_{11})(n^2 - K_{22}) + K_{12}^2 = 0. \tag{2.171}$$

Equation (2.171) is equivalent, if thermal effects are neglected, to the dispersion equation (2.42) obtained from the Hall model for the cold plasma, with an ion cyclotron wave solution and a fast wave solution. However, the thermal terms present in Eq. (2.171), using the components given by Eq. (2.161), lead to modifications of the dispersion relations, the most important of which occurs at the ion cyclotron frequency.

As the ion cyclotron (slow Alfvén) mode frequency approaches the ion cyclotron frequency, the electron Landau damping rate increases according to

$$\gamma_A \simeq \Omega_i \frac{\sqrt{\pi}}{8} \frac{m_e}{m_i} \tan^2\theta \frac{\omega}{\Omega_i - \omega} \frac{1}{z_0^e |1 + z_0^e Z(z_0^e)|^2} e^{-(z_0^e)^2}. \tag{2.172}$$

However, close to the ion cyclotron frequency when $z_1^i \simeq 0$, ion cyclotron damping dominates for this mode, as was discussed for the parallel propagation case. If we consider a wave of real frequency and fixed angle of propagation θ, suffering spatial damping as the frequency approaches the ion cyclotron frequency (when $z_1 \simeq 0$), the real and imaginary parts of the wavenumber (or refractive index) become of the same order (Melrose 1986), with

$$k = \frac{\Omega_i}{(v_A^2 V_i)^{1/3}} \frac{\sqrt{3} + i}{2} \left[\sqrt{\frac{\pi}{8}} \frac{1 + \cos^2\theta}{\cos^3\theta} \right]^{1/3}. \tag{2.173}$$

2.7.3 High Plasma Beta

We now consider the kinetic theory of a plasma with appreciable ion kinetic energy density compared with magnetic energy density, that is with β_i not small. Waves in a high β_i plasma have been analysed independently in the fusion context (Spies & Li 1990) and in the astrophysical context (Achterberg & Blandford 1986), such as the hot component of the interstellar medium. The waves in the galactic medium may serve to heat the gas, to transport energy over large distances, or to scatter energetic particles, as we shall see in Chapter 8. The general β case has also been considered in the space physics context, such as the high–β solar wind, by Lysak & Lotko (1996).

In our treatment of the waves in this case, we assume as usual that the electron Larmor radius is very small compared with the perpendicular wavelength, that is $\lambda_e = 0$, and neglect terms in the dielectric tensor components of first order in λ_e. If however the electrons are assumed to be hot, and the ions are cold, such that $\beta \simeq \beta_e$, and β_e is not small, the resulting dispersion equation is similar to that derived from fluid theory for a high–β plasma (Eq. 2.52), but with modifications due to electron Landau damping and ion cyclotron damping terms (Akhiezer *et al.* 1975). In that case c_s in Eq. (2.52) is replaced by the ion-sound speed v_s. We devote this section, however, to the case of a high ion beta.

The ion Larmor radius is taken to be small, such that $\lambda_i \ll 1$, but we retain the first–order terms in λ_i in the dielectric tensor. The assumption $\lambda_i \ll 1$ puts a condition on the angle of propagation:

$$\tan^2 \theta = \frac{k_\perp^2}{k_z^2} \ll \frac{\Omega_i^2}{\omega^2} \frac{1}{\beta_i} \tag{2.174}$$

assuming that $\omega \simeq v_A k_z$. Even for $\beta_i \simeq 1$, for low frequencies the condition Eq. (2.174) still allows a large range of angles of propagation.

We use the dielectric tensor in the form of Eq. (2.155) for Maxwellian velocity distribution functions, and assume isotropic temperatures (anisotropic temperatures have been included by Achterberg & Blandford (1986). If we assume low frequency, and long wavelength parallel to the magnetic field, such that $\omega \ll \Omega_i$, then

$$|z_s^i| \simeq \frac{|s\Omega_i|}{V_i k_z} \simeq \frac{|s\Omega_i|}{\omega} \frac{v_A}{V_i} \simeq \frac{|s\Omega_i|}{\omega} \beta_i^{-1/2} \gg 1 \tag{2.175}$$

for $s \neq 0$, provided $\beta_i \ll \Omega_i^2/\omega^2$. As usual, we have $|z_s^e| \gg 1$ for $s \neq 0$. We can then use the asymptotic expansion for the plasma dispersion function $Z(z_s^\alpha)$ for the harmonics $s \neq 0$ for both ions and electrons. However, as yet we make no assumption about either z_0^i or z_0^e.

It is then useful to separate out the $s = 0$ terms in the expansion to third order in $1/|z_s^\alpha|$ of the dielectric tensor given by Eq. (A.10), which yields (suppressing the species index α):

$$K_{ij}(\omega, \boldsymbol{k}) =$$

$$\delta_{ij} + \sum \frac{\omega_p^2}{\omega^2} \left[2z_0^2 b_i b_j + z_0 Z(z_0) N_{ij}(0) - z_0 \sum_{s \neq 0} \left(\frac{1}{z_s} + \frac{1}{2z_s^3} \right) N_{ij}(s) \right]. \tag{2.176}$$

Using the following relation involving the modified Bessel functions :

$$e^{-\lambda} \sum_{s=1}^{\infty} I_s(\lambda) = \frac{1}{2} \left(1 - e^{-\lambda} I_0(\lambda)\right) \tag{2.177}$$

assuming a single ion species, and retaining terms of order ω^2/Ω_i^2, we find that

$$K_{11} = 1 + \frac{c^2}{v_A^2} \frac{1}{\lambda_i} \left[1 - e^{-\lambda_i} I_0(\lambda_i) + \sum_{s=1}^{\infty} \frac{e^{-\lambda_i} I_s(\lambda_i)}{s^2} \left(2\frac{\omega^2}{\Omega_i^2} + 6\frac{V_i^2 k_z^2}{\Omega_i^2}\right)\right]. \tag{2.178}$$

Then, using Eqs. (A.23) and (A.24), we have to first order in λ_i,

$$K_{11} = 1 + \frac{c^2}{v_A^2} \left[1 + \frac{\omega^2}{\Omega_i^2} - \frac{3}{4}\lambda_i + 3\frac{V_i^2 k_z^2}{\Omega_i^2}\right]. \tag{2.179}$$

For the other components we find, to the same order,

$$K_{22} = 1 + \frac{c^2}{v_A^2} \left[1 + \frac{\omega^2}{\Omega_i^2} - \frac{11}{4}\lambda_i + 3\frac{V_i^2 k_z^2}{\Omega_i^2} + 2\frac{\Omega_i^2}{\omega^2} z_0^i Z(z_0^i)\lambda_i\right] \tag{2.180}$$

$$K_{12} = i\frac{c^2}{v_A^2} \frac{\omega}{\Omega_i} \left[1 + \frac{V_i^2 k_z^2}{\omega^2} - \frac{3}{2}\frac{V_i^2 k_\perp^2}{\omega^2}\right] \tag{2.181}$$

$$K_{13} = -\frac{c^2}{v_A^2} 2\cot\theta\, \lambda_i \tag{2.182}$$

$$K_{23} = \frac{c^2}{v_A^2} \tan\theta \left[i\frac{\Omega_i}{\omega} z_0^i Z(z_0^i) + i\frac{V_i^2 k_z^2}{\omega\Omega_i} \left(1 - \frac{3}{2}\tan^2\theta\right)\right]. \tag{2.183}$$

The final component may be written in terms of the ion and electron Debye lengths λ_{Di} and λ_{De} as

$$K_{33} = 1 + \frac{1}{k_z^2 \lambda_{Di}^2} \left[1 + z_0^i Z(z_0^i)\right] + \frac{1}{k_z^2 \lambda_{De}^2} \left[1 + z_0^e Z(z_0^e)\right] + \frac{c^2}{v_A^2}\lambda_i. \tag{2.184}$$

If we now consider the limit of high ion kinetic energy density compared with magnetic energy density, that is $\beta_i \gg 1$, we have $z_0^i \ll 1$, and we can use the small argument expansion of Z from Eq. (A.14):

$$Z(z) \simeq i\pi^{1/2} - 2z_0. \tag{2.185}$$

Then we have

$$K_{22} = 1 + \frac{c^2}{v_A^2} \left[1 + \frac{\omega^2}{\Omega_i^2} - \frac{11}{4}\lambda_i + 3\frac{V_i^2 k_z^2}{\Omega_i^2} - 2\tan^2\theta - i2\pi^{1/2}\frac{\Omega_i^2}{\omega^2} z_0^i \lambda_i\right] \tag{2.186}$$

and

$$K_{23} = \frac{c^2}{v_A^2} \tan\theta \left[-\pi^{1/2}\frac{\Omega_i}{\sqrt{2}V_i k_z} - i\frac{\omega\Omega_i}{V_i^2 k_z^2} + i\frac{V_i^2 k_z^2}{\omega\Omega_i}\left(1 - \frac{3}{2}\tan^2\theta\right)\right]. \tag{2.187}$$

The imaginary term which appears in the ion dispersion function Eq. (2.185) produces an imaginary term in K_{22} and a real term in K_{23}, leading to damping of the waves. This damping is due to the interaction of the ions, travelling along the magnetic field, with the wave. However, this is not an ion cyclotron ($s \neq 0$) interaction (since by assumption $\omega \ll \Omega_i$), but is rather an $s = 0$ interaction analogous to electron Landau damping. An important difference is that the ions interact with the B_z component of the wave magnetic field in the direction of the equilibrium magnetic field, rather than with E_z. The explanation of the physical basis of this damping, which is called *transit time magnetic damping*, uses the fact that the equivalent magnetic moment of the ion, $\mu = m_i v_\perp^2 / 2B_0$, is an adiabatic invariant for a particle in a wave of frequency $\omega \ll \Omega_i$. The magnetic moment is said to interact with the parallel gradient of the magnetic field, with the particle accelerating in the parallel direction (Stix 1992).

For $\beta \gg 1$ the small argument form of Z can be used for both ions and electrons in Eq. (2.184), and the ion Landau damping term (due to interaction of the thermal ions with E_z) dominates the electron Landau damping term in the imaginary part of K_{33}:

$$K_{33} = 1 + \frac{c^2}{v_A^2} \left[\frac{\Omega_i^2}{V_i^2 k_z^2} \left(1 + \frac{T_i}{T_e} + i\pi^{1/2} z_0^i \right) + \lambda_i \right]. \tag{2.188}$$

The component K_{33} is generally much larger than the other components, so that the parallel electric field is suppressed, and the dispersion equation to lowest order in $1/\beta_i$ has the reduced form of Eq. (2.171). Using Eqs. (2.179), (2.181) and (2.186) in Eq. (2.171), and assuming propagation at small angles to the magnetic field, we obtain the dispersion equation (Foote & Kulsrud 1979):

$$\left(\omega^2 - v_A^2 k_z^2 \right) \left(\omega^2 + 2i\pi^{1/2} \beta_i^{1/2} \omega v_A k_z \tan^2 \theta - v_A^2 k_z^2 \right) = \left(\frac{\omega^2}{\Omega_i^2} \right) V_i^4 k_z^4. \tag{2.189}$$

The imaginary term in Eq. (2.189) corresponds to the transit time magnetic damping.

Parallel Propagation

Let us first consider the case of parallel propagation, $\theta = 0$. The dispersion equation (2.189) can then be written as

$$\omega^2 - v_A^2 k_z^2 \mp \frac{V_i^2 k_z^2}{\Omega_i} \omega = 0. \tag{2.190}$$

This dispersion equation can also be obtained using a modification of the MHD model to include finite ion Larmor radius effects (Hamabata 1993), by modifying the ion pressure tensor with effective "viscosity" terms where, in the collisionless plasma, the ion Larmor radius replaces the usual collisional mean free path (Thompson 1961).

The frequencies of the two, oppositely circularly polarized, modes are given from Eq. (2.190) by

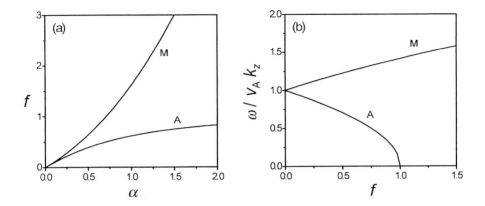

Figure 2.9: (a) The normalized frequency $f = \omega/v_A k_0$ plotted against the normalized wavenumber $\alpha = k_z/k_0$ in a high β_i plasma, for the left–hand polarized Alfvén (A) and right–hand polarized magnetoacoustic (M) parallel propagating modes. (b) The normalized phase velocity plotted against f for the two modes.

$$\omega_\pm = \pm \frac{V_i^2 k_z^2}{2\Omega_i} + \left[\left(\frac{V_i^2 k_z^2}{2\Omega_i} \right)^2 + v_A^2 k_z^2 \right]^{1/2}$$

$$= v_A k_z \left[\pm \frac{k_z}{2k_0} + \left(\frac{k_z^2}{4k_0^2} + 1 \right)^{1/2} \right] \tag{2.191}$$

where we define a characteristic wavenumber (Foote & Kulsrud 1979),

$$k_0 = \frac{\Omega_i v_A}{V_i^2}. \tag{2.192}$$

The dispersion relations are shown in Figure 2.9(a), as a plot of the normalized frequency $f = \omega/v_A k_0$ against the normalized wavenumber $\alpha = k_z/k_0$.

The upper sign in Eq. (2.191) corresponds to the right–hand polarized magnetoacoustic (M) or fast wave, and the lower sign corresponds to the left–hand polarized slow Alfvén (A) wave. A resonance in k_z occurs for the slow Alfvén wave at the frequency

$$\omega_- = \omega_0 = \frac{2\Omega_i}{\beta_i} = v_A k_0. \tag{2.193}$$

However, remembering that terms of order ω/Ω_i are neglected in Eq. (2.191), there is no ion cyclotron resonance for the slow Alfvén wave. Also, there is no transit time magnetic damping, because for parallel propagation there is no wave magnetic field component parallel to the equilibrium magnetic field.

The phase velocity of the two parallel propagating modes can be obtained from Eq. (2.190) as

$$v_{\text{ph}} = \frac{\omega_{\pm}}{k_z} = v_A \left(1 \pm \frac{\beta_i}{2}\frac{\omega_{\pm}}{\Omega_i}\right)^{1/2}. \tag{2.194}$$

We then see that for the slow Alfvén mode, $v_{\text{ph}} \to 0$ at the resonance frequency ω_-. Thus from Eqs. (2.191) and (2.194) we find that in a high–β plasma, the two oppositely circularly polarized parallel propagating waves have significantly different dispersion relations, with strong dispersion for $k_z \gtrsim 2k_0$, compared with the low–β Alfvén wave. The right–hand wave propagates with a phase velocity greater than v_A, while the left–hand wave, for $\omega < \omega_0$, has phase velocity less than v_A.

The physical basis for this mode splitting into two dispersive waves, even with frequency well below the ion cyclotron frequency, is that since the ion thermal speed is much higher than v_A for $\beta_i > 1$, the ions do not have time to develop their full Hall drift ($E \times B$) motion in one period of the wave, and so there is a lack of cancellation between the ion and electron Hall currents. This is reflected in the fact that the off–diagonal elements in the dielectric tensor are nonzero, as is the case for a low–β plasma at frequencies approaching the ion cyclotron frequency, or for a plasma with heavy ions or dust grains carrying a proportion of the charge. The total Hall current is comparable with the polarization current parallel to E when

$$\frac{k_z^2 V_i^2}{\Omega_i^2} \simeq \frac{\omega}{\Omega_i} \tag{2.195}$$

that is when $\omega \simeq \omega_0$.

Even though there is no transit time magnetic damping of the two parallel propagating modes, ion cyclotron damping must in fact be taken into account, even at frequencies much lower than the ion cyclotron frequency, as shown by Achterberg (1981). To show this, we use the following kinetic dispersion equation for parallel propagating waves in a single ion species, isotropic temperature plasma, derived from the general dispersion equation (2.150):

$$\frac{k_z^2 c^2}{\omega^2} = 1 + \frac{\omega_{\text{pi}}^2}{\omega^2}\left[\frac{\omega}{\omega \pm \Omega_i}z_{\pm1}^i Z(z_{\pm1}^i) \pm \frac{\omega}{\Omega_i}\right]. \tag{2.196}$$

Assuming the frequency is well below the ion cyclotron frequency, such that

$$|z_{\pm}^i| \simeq \Omega_i/\omega\beta_i^{1/2} \gg 1 \tag{2.197}$$

Eq. (2.196) yields a modification of the dispersion equation (2.190) with a cyclotron damping term:

$$\omega^2 - v_A^2 k_z^2 \mp \frac{V_i^2 k_z^2}{\Omega_i}\omega + i\left(\frac{\pi}{2}\right)^{1/2}\frac{\Omega_i^2\omega}{V_i k_z}\exp\left(-\frac{\Omega_i^2}{2V_i^2 k_z^2}\right) = 0. \tag{2.198}$$

In terms of the dimensionless frequency f and wavenumber α, Eq. (2.198) becomes

$$f^2 - \alpha^2(1 \pm f) + i\left(\frac{\pi}{2}\right)^{1/2}\left(\frac{\beta_i}{2}\right)^{3/2}\frac{f}{\alpha}\exp\left(-\frac{\beta_i}{4\alpha^2}\right) = 0. \tag{2.199}$$

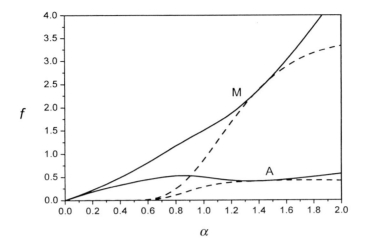

Figure 2.10: The real (solid curves) and imaginary (dashed curves) parts of the normalized frequency $f = \omega/v_A k_0$ plotted against the (real) normalized wavenumber $\alpha = k_z/k_0$ in a high–β_i plasma, for the left–hand polarized Alfvén (A) and right–hand polarized magneto-acoustic (M) parallel propagating modes. Here $\beta_i = 10$.

The frequencies and damping rates of the two modes derived from Eq. (2.199) are shown in Figure 2.10, for the case $\beta_i = 10$. As β_i increases, the dispersion curves approach the case shown in Figure 2.9.

The imaginary cyclotron damping term in Eq. (2.198) becomes important when $\Omega_i^2/2V_i^2 k_z^2$ becomes of order one, that is, when the parallel wavelength is of the order of the ion Larmor radius:

$$k_z \simeq \frac{\Omega_i}{\sqrt{2}V_i} = \left(\frac{\beta_i}{2}\right)^{1/2} k_0 \tag{2.200}$$

and $\omega \simeq \Omega_i/\beta_i$ for the Alfvén mode, that is, at the resonance frequency where the phase velocity goes to zero. Thus cyclotron damping can occur at frequencies well below the ion cyclotron frequency in a high–β plasma.

The frequencies and damping rates of the parallel propagating A and M modes have simple forms in the limits of low and high parallel wavenumber (Achterberg 1981). For $k_z \ll 2k_0$, the frequencies are

$$\omega_{A,M} = v_A k_z \tag{2.201}$$

and the damping rates are

$$\gamma_{A,M} = -\left(\frac{\pi}{8}\right)^{1/2} \frac{\Omega_i^2}{|k_z|V_i} \exp\left(-\frac{\Omega_i^2}{2V_i^2 k_z^2}\right). \tag{2.202}$$

For $k_z \gg 2k_0$, the frequencies are

$$\omega_A = \omega_0, \quad \omega_M = \frac{V_i^2 k_z^2}{\Omega_i} \tag{2.203}$$

and the damping rates are

$$\gamma_A = -\left(\frac{\pi}{2}\right)^{1/2} \left(\frac{\Omega_i}{|k_z|V_i}\right)^3 \beta_i^{-1} \Omega_i \exp\left(-\frac{\Omega_i^2}{2V_i^2 k_z^2}\right) \tag{2.204}$$

and

$$\gamma_M = -\left(\frac{\pi}{2}\right)^{1/2} \frac{\Omega_i^2}{|k_z|V_i} \exp\left(-\frac{\Omega_i^2}{2V_i^2 k_z^2}\right). \tag{2.205}$$

Thus the relative damping rate $|\gamma|/\omega$ is the same for the two modes. For $k_z \gg k_0$, $|\gamma|/\omega$ is proportional to

$$k_z^{-3} \exp[-\Omega_i^2/(2V_i^2 k_z^2)]$$

and reaches a maximum at $k_z = (\beta_i/3)^{1/2} k_0$. Parallel propagation and damping in high–β plasmas at frequencies extending up to the electron cyclotron frequency has been considered by Davila & Scott (1984).

Oblique Propagation

Returning to the case of oblique propagation, it is found that the waves propagating at a nonzero angle to the magnetic field are essentially plane–polarized, and are subject to strong transit time magnetic damping (Achterberg & Blandford 1986) (we neglect the ion cyclotron damping here). Eq. (2.189) can be written

$$D_A D_M = D_C \tag{2.206}$$

where D_A and D_M are the first and second brackets of the left–hand side of Eq. (2.189). D_C is a coupling term that couples together the "pure" low–frequency Alfvén mode (described by $D_A = 0$) and magnetoacoustic mode (described by $D_M = 0$). The magnetoacoustic mode is damped for high β and oblique propagation by transit time damping, so the Alfvén mode is also damped because of the coupling term. The damping becomes strong when the imaginary term in Eq. (2.189) is comparable in magnitude with the real terms, that is when

$$\tan\theta > (v_A/V_i)^{1/2} = \beta_i^{-1/4}. \tag{2.207}$$

For example, when $\beta_i \sim 10^4$, the maximum angle of lightly damped propagation is $\simeq 4°$.

Equation (2.189) was rewritten by Achterberg & Blandford (1986) in the form

$$(\nu^2 + 2i\sigma l\nu - 1)(\nu^2 - 1) = l^2\nu^2 \tag{2.208}$$

where the normalized phase velocity is

$$\nu = \omega/v_A k_z \tag{2.209}$$

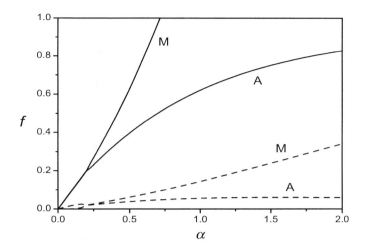

Figure 2.11: The real (solid curves) and imaginary (dashed curves) parts of the normalized frequency $f = \omega/v_A k_0$ plotted against the (real) normalized wavenumber $\alpha = k_z/k_0$ in a high-β_i plasma, for the Alfvén (A) and magnetoacoustic (M) obliquely propagating modes. The parameter s measuring the transit time magnetic damping, and depending on β_i and the angle of propagation θ, is 0.2.

the normalized wavenumber is

$$l = \frac{v_A k_z}{\Omega_i} \frac{\beta_i}{2} \tag{2.210}$$

and

$$\sigma = \pi^{1/2} \beta_i^{1/2} \theta^2 / |l| \tag{2.211}$$

where we have used the result $\tan\theta = |k_x/k_z| \simeq \theta$ for small angles of propagation.

Now Eq. (2.208) may be factored to give (Achterberg & Blandford 1986)

$$\nu^2 + (i\sigma \pm (1 - \sigma^2)^{1/2}) l\nu - 1 = 0. \tag{2.212}$$

Thus there are four independent modes for oblique propagation with frequencies given by the solutions of Eq. (2.212), including reactive, that is nonpropagating, modes. Eq. (2.189) can also be written in terms of the normalized frequency and wavenumber defined above:

$$(f^2 - \alpha^2)(f^2 + 2isf\alpha - \alpha^2) = f^2\alpha^2 \tag{2.213}$$

where s is a parameter measuring the strength of the transit time magnetic damping:

$$s = \pi^{1/2} \beta_i^{1/2} \theta^2. \tag{2.214}$$

The dispersion relations and damping rates of the two positive frequency solutions of Eq. (2.213) are shown in Figure 2.11.

The obliquely propagating waves in a high–β plasma, damped by transit time magnetic damping, have been discussed in detail by Achterberg & Blandford (1986) in the context of a discussion of the transmission of waves through a strong shock wave front in a high–β astrophysical plasma, with its implications for cosmic ray acceleration. The least damped mode for small angles of propagation is found to be the modified Alfvén wave, as shown in Figure 2.11. The detailed behaviour of the polarization of the waves at high β was also studied by Gary (1986), and a comparison of the properties of the waves using the Hall-MHD and kinetic theories was made by Krauss-Varban, Omidi & Quest (1994).

2.8   Kinetic Alfvén Wave and Inertial Alfvén Wave

We shall see in the next chapter that at the Alfvén resonance point in a nonuniform plasma, strong mode conversion of the fast magnetoacoustic wave propagating obliquely to the equilibrium magnetic field can occur. The wave mode converts to a short perpendicular wavelength mode. This process corresponds to a coupling of the magnetoacoustic mode to the purely shear Alfvén wave that is modified by nonideal effects. In this section we consider the modification of the dispersion relation of the shear Alfvén wave in a uniform plasma, due to a large perpendicular wavenumber, leading to a short–wavelength mode called the *quasi–electrostatic wave* (QEW). The important role of the Alfvén resonance in the heating of plasmas and the damping of surface waves, and the circumstances of propagation of the QEW after mode conversion at the Alfvén resonance, are considered more fully in Chapters 3 and 4.

The QEW modes have a distinctly different dispersion relation to the Alfvén wave dispersion relation only when the wavelength perpendicular to the magnetic field is very short; in other words, when the wavevector is almost perpendicular to the magnetic field, and the phase front is almost parallel to the field. However, we shall see that the group velocity of these waves is almost parallel to the field, just as for the ideal MHD shear Alfvén wave. The QEW is further classified as the *kinetic Alfvén wave* (KAW) for warm plasmas, and the *inertial Alfvén wave* (IAW) for cold plasmas. We shall consider the low plasma β case here, which is of relevance to tenuous plasmas in strong magnetic fields, such as in Tokamak fusion devices and some space plasmas.

An early reference to the effects of a large perpendicular wavenumber on the Alfvén wave in the warm plasma case is a paper by Stefant (1970), where it was pointed out that when the perpendicular wavelength becomes comparable to the ion Larmor radius ($\lambda_i \sim 1$), ions can no longer follow the magnetic lines of force, whereas electrons are still attached to the lines of force due to their small Larmor radius. Thus charge separation is produced and there is coupling of the Alfvén wave to the electrostatic longitudinal mode, producing the KAW mode. This coupling can lead to strong Landau ($s = 0$) damping of the KAW. Charge separation can also occur if electron inertia is significant on the short spatial length–scale perpendicular to the magnetic field, leading to the IAW mode in a cold plasma.

Both the KAW and IAW modes have a significant electric field parallel to the background magnetic field, with the consequence that they can interact efficiently with electrons and ions. As a result, the KAW and IAW modes can play roles in the formation of auroral beams and in the generation of fast ions in the auroral ionosphere, and in the transport of energy in magnetic confinement devices (Morales & Maggs 1997). The KAW and IAW modes can each play roles in such plasmas, but at different locations, because of the strong spatial variation of the plasma parameters such as temperature. To treat the properties of KAW and IAW modes we can employ fluid theory, the "two–potential" theory, and kinetic theory, each of which we now consider. Finally, we treat the problem of the excitation of KAW and IAW modes from a spatially localized wave source in a uniform plasma.

2.8.1   Fluid Theory

We take an infinitely conducting plasma with nonzero pressure, and include the Hall term and electron inertia in the two–fluid model of Section 1.4. The Ohm's law we use is Eq. (1.31), retaining the Hall term and the electron inertia and pressure gradient terms, but with the resistivity η assumed zero. Upon taking the curl of Eq. (1.31), the electron pressure gradient term is eliminated.

We then obtain the following differential equations in the characteristic variables:

$$\frac{\partial B_{1z}}{\partial t} - \frac{c^2}{\omega_{pe}^2}\nabla^2\frac{\partial B_{1z}}{\partial t} + B_0\left(\nabla\cdot\mathbf{v}_1 - \frac{\partial v_{1z}}{\partial z}\right) + \frac{v_A^2}{\Omega_i}\frac{\partial J_{1z}}{\partial z} = 0 \qquad (2.215)$$

and

$$\mu_0\frac{\partial J_{1z}}{\partial t} - \mu_0\frac{c^2}{\omega_{pe}^2}\nabla^2\frac{\partial J_{1z}}{\partial t} - B_0\frac{\partial \zeta_{1z}}{\partial z} + \frac{v_A^2}{\Omega_i}\frac{\partial}{\partial z}\nabla^2 B_{1z} = 0 \qquad (2.216)$$

where the terms involving the ratio c^2/ω_{pe}^2 derive from the electron inertia term in Eq. (1.31).

These equations, together with Eqs. (2.10), (2.12), (2.14) and (2.15), lead to the following dispersion equation:

$$\left[\omega^2\left(1 + \frac{c^2 k^2}{\omega_{pe}^2}\right) - v_A^2 k_z^2\right]\left[\omega^2\left(1 + \frac{c^2 k^2}{\omega_{pe}^2}\right)(\omega^2 - c_s^2 k^2) - v_A^2 k^2(\omega^2 - c_s^2 k_z^2)\right]$$

$$= \left(\frac{\omega}{\Omega_i}\right)^2 v_A^4 k^2 k_z^2(\omega^2 - c_s^2 k^2). \qquad (2.217)$$

We are interested in short perpendicular wavelengths, and assume that $k_y = 0$. Then in the limit $k_x^2 \gg k_z^2$, and assuming $\omega^2 \lesssim v_A^2 k_z^2$, Eq. (2.217) can be written

$$\left[1 + \delta_e^2 k_x^2 - \frac{v_A^2 k_z^2}{\omega^2}\right]\left[1 + \frac{c_s^2}{v_A^2}\left(1 + \delta_e^2 k_x^2 - \frac{v_A^2 k_z^2}{\omega^2}\right)\right] = \frac{v_A^2 k_z^2}{\omega^2}\left(\lambda_s - \frac{\omega^2}{\Omega_i^2}\right). \qquad (2.218)$$

Here

$$\lambda_s = k_x^2 \rho_s^2 \qquad (2.219)$$

where $\rho_s = c_s/\Omega_i$ is the Larmor radius using the sound speed c_s, and

$$\delta_e = c/\omega_{pe} \qquad (2.220)$$

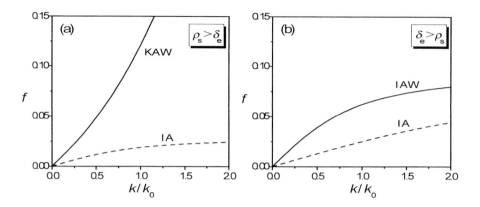

Figure 2.12: The normalized frequency $f = \omega/v_A k_0$ of the modified Alfvén wave and the ion-acoustic wave (IA) plotted against the normalized wavenumber k/k_0, with k_0 a reference wavenumber, for an angle of propagation of $85°$, and $c_s^2/v_A^2 = 0.1$. (a) $r = k_0\rho_s = 1$ and $d = k_0\delta_e = 0.1$, and the modified Alfvén wave is the kinetic Alfvén wave (KAW), (b) $r = 0.1$ and $d = 1$, and the modified Alfvén wave is the inertial Alfvén wave (IAW).

is the electron collisionless skin depth, sometimes called the electron inertial length. The two positive frequency solutions of Eq. (2.218) are shown in Figure 2.12, plotted against k, the magnitude of the wavenumber, for a propagation angle to the magnetic field of $\theta = 85°$. The two cases of $\rho_s > \delta_e$ and $\delta_e > \rho_s$ are illustrated. Here a normalized frequency, $f = \omega/v_A k_0$, has been used, where k_0 is a reference wavenumber. The importance of the Larmor radius effect and the electron inertia effect is measured by the parameters $r = k_0\rho_s$ and $d = k_0\delta_e$ respectively.

The dispersion equation (2.218) is written in a form that shows the coupling of the Alfvén wave (first factor on the left side of the equation) with a sound wave, or slow magnetoacoustic wave (second factor on the left), both modified by an electron inertia term $\delta_e^2 k_x^2$. We henceforth assume that $\omega/\Omega_i \ll \lambda_s$. Then the term coupling the Alfvén wave and the sound wave (the right side of the equation) is proportional to the dimensionless parameter λ_s, which measures the size of the Larmor radius relative to the perpendicular wavelength. In a collisionless plasma with $T_e \gg T_i$, the second factor on the left–hand side of Eq. (2.218) describes the lightly damped ion-acoustic wave with $c_s = v_s$, the ion-acoustic velocity.

For a low–β plasma ($c_s \ll v_A$) and $\omega \simeq v_A|k_z|$, the second factor on the left–hand side of Eq. (2.218) is approximately unity, and the resulting dispersion equation describes the modified Alfvén wave, or "quasi–electrostatic wave" (QEW):

$$\omega^2 = v_A^2 k_z^2 \frac{1 + \lambda_s}{1 + \delta_e^2 k_x^2}. \tag{2.221}$$

Noting the identity

$$\delta_e^2 = \frac{c^2}{\omega_{pe}^2} = \frac{m_e}{m_i}\frac{v_A^2}{\Omega_i^2} \tag{2.222}$$

we find that the ratio of the term λ_s (involving the Larmor radius ρ_s) in Eq. (2.221) to the electron inertia term is given by

$$\frac{\rho_s^2 \omega_{pe}^2}{c^2} = \frac{\gamma}{2} \beta \frac{m_i}{m_e} \qquad (2.223)$$

so that the Larmor radius term is dominant over the electron inertia term when the following condition holds:

$$\beta \gtrsim m_e / m_i . \qquad (2.224)$$

If the electrons are hot, such that $T_e \gg T_i$, condition Eq. (2.224) is equivalent to the condition

$$V_e \gtrsim v_A . \qquad (2.225)$$

In that case we have the dispersion equation

$$\omega^2 = v_A^2 k_z^2 (1 + \lambda_s). \qquad (2.226)$$

Thus the low–frequency Alfvén wave dispersion relation has been modified because the perpendicular wavelength is short enough to be comparable with the thermal ion Larmor radius. In this case the wave is called the kinetic Alfvén wave (KAW), and its dispersion relation is illustrated in Figure 2.12(a).

In the opposite case,

$$\beta \lesssim m_e / m_i \qquad (2.227)$$

which is equivalent to the condition

$$V_e \lesssim v_A \qquad (2.228)$$

if $T_e \gg T_i$, the Alfvén wave dispersion relation is modified if the perpendicular wavelength is comparable with the electron inertial length δ_e, and the wave is called the inertial Alfvén wave (IAW), satisfying the dispersion equation

$$\omega^2 = \frac{v_A^2 k_z^2}{1 + \delta_e^2 k_x^2} \qquad (2.229)$$

with its dispersion relation illustrated in Figure 2.12(b).

An important point to note from the dispersion relations (2.226) and (2.229) is that, whereas the ideal MHD shear Alfvén wave dispersion relation is independent of the perpendicular wavenumber, the Larmor radius and electron inertia effects introduce a dependence on that wavenumber component. One consequence, explored further in the next chapter, is that wave energy can propagate obliquely to the magnetic field, rather than strictly parallel to the field. In other words, the KAW and IAW have nonzero group velocity components transverse to the magnetic field. For either type of wave, the group velocity component along the field is equal to the phase velocity along the field:

$$v_{gz} = \frac{d\omega}{dk_z} = \frac{\omega}{k_z} . \qquad (2.230)$$

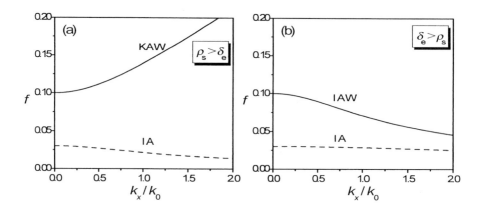

Figure 2.13: The normalized frequency $f = \omega/v_A k_0$ of the modified Alfvén wave and the ion-acoustic wave (IA) plotted against the normalized perpendicular wavenumber k_x/k_0, for the parallel wavenumber $k_z/k_0 = 0.1$, and $c_s^2/v_A^2 = 0.1$. (a) $r = 1$ and $d = 0.1$, and the modified Alfvén wave is the kinetic Alfvén wave (KAW), (b) $r = 0.1$ and $d = 1$, and the modified Alfvén wave is the inertial Alfvén wave (IAW).

The transverse group velocity components derived from Eqs. (2.226) and (2.229) are:

$$v_{gx} = \frac{\omega}{k_x} \frac{\lambda_s}{1 + \lambda_s} \qquad \text{(KAW)} \tag{2.231}$$

$$v_{gx} = -\frac{\omega}{k_x} \frac{\delta_e^2 k_x^2}{1 + \delta_e^2 k_x^2} \qquad \text{(IAW)}. \tag{2.232}$$

Thus if $|k_x| \gg |k_z|$, either wave transports energy very slowly in the transverse direction. The (normalized) frequencies of each of the waves are plotted against k_x/k_0, for fixed k_z, in Figure 2.13.

We note from Eq. (2.232) and from the negative slope of the IAW dispersion curve in Figure 2.13(b) that the IAW is a *backward* wave, with group velocity component perpendicular to the magnetic field of opposite sign to the wavevector component. The KAW on the other hand is a *forward* propagating wave, with the same sign of both components. The KAW and IAW are compressive waves, that is, there is a perturbation of the plasma density, in contrast to the ideal MHD Alfvén wave. The waves are elliptically polarized, with right–hand sense rotation of the transverse electric field, in contrast to the linearly polarized ideal MHD Alfvén wave (Gary 1986, Hollweg 1999).

Let us now neglect the electron inertia term in Eq. (2.218) and assume the frequency is low, such that the phase velocity is approximately equal to c_s, which is much less than v_A. Retaining the dominant terms in Eq. (2.218), we then obtain the dispersion relation of the sound wave, modified by the ion Larmor radius effect:

$$\omega^2 = \frac{c_s^2 k_z^2}{1 + \lambda_s}. \tag{2.233}$$

If $T_e \gg T_i$, we have the ion-acoustic wave with v_s replacing c_s in Eq. (2.233). The dispersion relations of the ion-acoustic wave for the case $c_s/v_A^2 = 0.1$ are shown in Figures 2.12 and 2.13. Note that it follows from Eq. (2.233) that the sound wave is also a backward wave with respect to propagation across the magnetic field, as is also seen from the negative slope of the ion-acoustic dispersion curves in Figure 2.13.

If we now remove the assumption of a low–β plasma, the second factor on the left–hand side of Eq. (2.218) gives rise to the slow magnetoacoustic wave, which modifies the Alfvén wave via the coupling term on the right–hand side of Eq. (2.218), as discussed by Hollweg (1999). In particular, there is now a wave magnetic field component B_{1z} parallel to the equilibrium field, so that transit time magnetic damping plays a role, just as for the general case of oblique propagation in a high–β plasma discussed in the previous section.

2.8.2   Parallel Electron Temperature Effects

Another variant of a fluid model of the KAW and IAW modes starts from the collisionless fluid equations for ions and electrons (1.5), but allows for pressure variations only in the electron fluid, parallel to the magnetic field (Stix 1992). This model is best cast in terms of the dielectric tensor approach, and provides an extension of the cold plasma theory to include electron temperature effects. The ions are assumed cold, so the pressure gradient term in the ion fluid equation of motion is neglected. The electron pressure gradient terms, as well as the inertial terms, in the components of the electron equation of motion perpendicular to the magnetic field are also neglected in comparison with the electric and magnetic forces. However, the electron pressure gradient and the electron inertia term are retained in the component of the equation of motion parallel to the magnetic field (equivalent to retaining those terms, as well as the Hall term, in the generalized Ohm's law Eq. (1.31)).

The result is a dielectric tensor of the cold–plasma form Eq. (2.55), with S and D given by Eqs. (2.91) and (2.92) (for $\omega \ll \Omega_e$), but with

$$P = 1 - \frac{\omega_{pi}^2}{\omega^2} - \frac{\omega_{pe}^2}{(\omega^2 - V_e^2 k_z^2)}. \tag{2.234}$$

In the nonrelativistic case, and neglecting terms of order m_e/m_i, we can neglect the first two terms on the right–hand side of Eq. (2.234).

It is useful to employ the scaled dielectric tensor components u_1, u_2 and u_3 defined in Eqs. (2.73) and (2.74). In this case these components are:

$$u_1 = \frac{\omega^2}{v_A^2(1-f^2)}, \qquad u_2 = f u_1 \tag{2.235}$$

$$u_3 = -\frac{\omega^2 \omega_{pe}^2}{c^2(\omega^2 - V_e^2 k_z^2)}. \tag{2.236}$$

The dispersion equation (2.76), with tensor components given by Eqs. (2.235) and (2.236), is valid for frequencies approaching the ion cyclotron frequency. However, in the low–frequency limit, we have u_2 (and D) $\to 0$, and the dispersion equation becomes

$$\left(u_1 - k_z^2 - \frac{u_1}{u_3} k_x^2 \right) (u_1 - k^2) = 0 \tag{2.237}$$

with $u_1 = \omega^2/v_{\mathrm{A}}^2$. In terms of the refractive index and the angle of propagation, the dispersion equation is

$$\left(n^2 \cos^2\theta - S + \frac{S}{P}n^2 \sin^2\theta\right)(n^2 - S) = 0. \tag{2.238}$$

Setting the first bracket on the left–hand sides of Eq. (2.237) or Eq. (2.238) to zero yields the modified Alfvén wave dispersion relation (2.221) with the ion Larmor radius ρ_{s} based on the ion-acoustic speed v_{s}, which leads to the KAW and IAW modes discussed above. The second bracket in the equations yields the fast wave dispersion relation.

This model, with ion cyclotron terms retained in the functions u_1 and u_2, is also employed in Chapter 3 to describe mode conversion into the KAW or IAW modes at the generalized Alfvén resonance.

2.8.3 Two–Potential Theory

Another approach to deriving the linear KAW and IAW dispersion relations is to use the "two–potential" theory. One version of this theory, which we describe here, uses simplified fluid equations (Hasegawa & Chen 1976, Cramer & Donnelly 1981). If the plasma β is assumed to be very small, and the frequency is low ($\omega \ll \Omega_{\mathrm{i}}$), the Alfvén wave and the ion-acoustic wave will produce no significant compression of the magnetic field, $B_{1z} = 0$. (Because we assume $\omega \simeq v_{\mathrm{A}}k_z$, the fast compressional wave, for $|k_x| \gg |k_z|$, will be cut off.) The wave electric field can be expressed in terms of two potential functions, ϕ and ψ, with

$$\boldsymbol{E}_{1\perp} = -\nabla_\perp\phi \tag{2.239}$$

$$E_{1z} = -\frac{\partial\psi}{\partial z}. \tag{2.240}$$

The electric fields then become, if $k_y = 0$,

$$E_{1x} = -\mathrm{i}k_x\phi, \quad E_{1y} = 0, \quad E_{1z} = -\mathrm{i}k_z\psi. \tag{2.241}$$

As the frequency is much lower than the electron cyclotron frequency, the electron inertia can be set to zero in the y–component of the linearized electron equation of motion (Eq. 1.5 with $\alpha = \mathrm{e}$ and no collisions). We then obtain

$$v_{1ex} = 0. \tag{2.242}$$

The z–component of the electron equation of motion gives

$$p_{1e} - \mathrm{i}\frac{n_0 m_e}{k_z}\frac{\partial v_{1ez}}{\partial t} = n_0 e\psi. \tag{2.243}$$

If the electron thermal speed is taken to be greater than the wave phase speed in the direction of the magnetic field (equivalent to $V_e > v_{\mathrm{A}}$), the electron temperature parallel to the field will be constant, and

$$p_{1e} = n_{1e}k_{\mathrm{B}}T_e = n_{1e}m_i v_{\mathrm{s}}^2 \tag{2.244}$$

where n_{1e} is the perturbation of the electron number density. The electron inertia term in Eq. (2.243) can also be neglected, and so

$$n_{1e} = n_0 e\psi/m_i v_s^2. \tag{2.245}$$

The frequency is assumed well below the electron plasma frequency, and the Debye length is assumed negligibly small compared with the wavelength, so the electrons can move freely along the magnetic field in one wave period. Thus the quasi–neutrality condition on the perturbations of the electron and ion number densities n_{1e} and n_{1i},

$$n_{1e} = n_{1i} \tag{2.246}$$

can be used. It can be shown a posteriori that $(n_{1i} - n_{1e})/n_{1e}$ is of the order of v_A^2/c^2.

The ion continuity equation (Eq. 1.6 with $\alpha =$ i) becomes

$$e\frac{\partial \psi}{\partial t} + m_i v_s^2 (ik_x v_{1ix} + ik_z v_{1iz}) = 0. \tag{2.247}$$

The corresponding electron continuity equation gives

$$v_{1ez} = (k_x/k_z)v_{1ix} + v_{1iz} \tag{2.248}$$

so the z–component of the current density is

$$J_{1z} = -n_0 e(k_x/k_z)v_{1ix}. \tag{2.249}$$

From Ampère's equation, neglecting the displacement current, we have

$$\mu_0 \frac{\partial J_{1z}}{\partial t} = -ik_x^2 k_z(\phi - \psi) \tag{2.250}$$

so elimination of J_{1z} leads to

$$\frac{\partial v_{1ix}}{\partial t} - \frac{ik_x k_z^2}{\mu_0 n_0 e}(\phi - \psi) = 0. \tag{2.251}$$

The x–component of the ion equation of motion gives

$$m_i \frac{\partial v_{1ix}}{\partial t} - e(-ik_x\phi + v_{1iy}B_0) = 0 \tag{2.252}$$

and elimination of ϕ gives

$$m_i \frac{\partial v_{1ix}}{\partial t} - \frac{ev_A^2 k_z^2}{\Omega_i^2}(-ik_x\psi + v_{1iy}B_0) = 0. \tag{2.253}$$

Terms of order $v_A^2 k_z^2/\Omega_i^2$, that is of order ω^2/Ω_i^2, have been neglected in the derivation of Eq. (2.253).

We thus have the following set of four coupled differential equations, in the four variables ψ, v_{1ix}, v_{1iy} and v_{1iz}:

$$\frac{\partial \psi}{\partial t} + \frac{m_i v_s^2}{e}(ik_x v_{1ix} + ik_z v_{1iz}) = 0 \tag{2.254}$$

$$\frac{\partial v_{1ix}}{\partial t} + \frac{v_A^2 k_z^2}{\Omega_i^2}\left(\frac{ik_x e\psi}{m_i} - \Omega_i v_{1iy}\right) = 0 \tag{2.255}$$

$$\frac{\partial v_{1iy}}{\partial t} + \Omega_i v_{1ix} = 0 \tag{2.256}$$

$$\frac{\partial v_{1iz}}{\partial t} + \frac{ik_z e\psi}{m_i} = 0. \tag{2.257}$$

The dispersion equation obtained from this set of equations is

$$\left(1 - \frac{v_A^2 k_z^2}{\omega^2}\right)\left(1 - \frac{v_s^2 k_z^2}{\omega^2}\right) = \frac{v_A^2 k_z^2}{\omega^2}\lambda_s. \tag{2.258}$$

This equation also follows from the dispersion equation derived from the two–fluid theory (Eq. 2.218), in the low–β limit, by neglecting the electron inertia term. Equation (2.258) shows the coupling of the Alfvén wave with the ion-acoustic wave via the coupling term on the right–hand side proportional to λ_s. Setting the second (ion-acoustic) factor on the left–hand side of Eq. (2.258) to unity yields the KAW dispersion relation (2.226), while the modified ion-acoustic wave dispersion relation (2.233) is obtained if $\omega \simeq v_s|k_z|$.

We now consider the low–temperature case, where the electron inertia term is retained in Eq. (2.243) but the electron pressure perturbation is set to zero. This yields, again assuming quasi–neutrality, and using Eq. (2.248), the equation

$$\frac{\partial v_{1ix}}{\partial t} + \frac{k_z}{k_x}\frac{\partial v_{1iz}}{\partial t} - \frac{ik_z^2 e\psi}{k_x m_e} = 0. \tag{2.259}$$

Eqs. (2.255)-(2.257) and (2.259) then yield the IAW dispersion relation (2.229).

An alternative form of the two–potential theory, employed by Stefant (1970) and Hasegawa & Chen (1976), is to use the "drift-kinetic" form of the Vlasov equation to express the parallel ion and electron currents J_{1iz} and J_{1ez} in terms of ϕ and ψ. This approach uses the fluid equations for motion transverse to the magnetic field lines but kinetic theory for the parallel motion, and yields corrections to the KAW and IAW dispersion relations due to the lowest order results of the kinetic theory, which we describe next.

2.8.4 Kinetic Theory

Kinetic theory provides an alternative method of calculating the thermal effects leading to the KAW mode, but it also provides a description of the damping of the mode because of wave-particle interactions. We can use suitable approximations of the dielectric tensor components for the Maxwellian plasma, previously calculated to first order in the ion Larmor parameter λ_i (Eqs. 2.179-2.184). For low frequency, the dielectric tensor component K_{12} can be neglected, and for small λ_i, K_{13} can also be neglected. The kinetic dispersion equation derived from Eq. (2.164) can then be written

$$\left(n^2 \cos^2 \theta - K_{11} + \frac{K_{11}}{K_{33}} n^2 \sin^2 \theta\right) (n^2 - K_{22}) = (n^2 \cos^2 \theta - K_{11}) \frac{K_{23}^2}{K_{33}}. \quad (2.260)$$

For $\beta \ll 1$, but with hot electrons ($V_e \gg v_A$), we have $|z_0^i| \gg 1$ and $|z_0^e| \ll 1$. This implies from Eq. (2.184) that

$$K_{33} = \frac{1}{k_z^2 \lambda_{De}^2} \quad (2.261)$$

if the imaginary terms are neglected.

The term K_{23}^2/K_{33} is of order β_i relative to the other terms in Eq. (2.260), and so may be neglected, leaving the right–hand side of Eq. (2.260) zero. To first order in λ_i and neglecting terms of order ω/Ω_i and β_i, Eq. (2.179) gives

$$K_{11} = \frac{c^2}{v_A^2} \left(1 - \frac{3}{4}\lambda_i\right). \quad (2.262)$$

Setting the first factor on the left–hand side of Eq. (2.260) to zero then gives the dispersion equation for the KAW,

$$\omega^2 = k_z^2 v_A^2 \left[1 + \lambda_i \left(\frac{3}{4} + \frac{T_e}{T_i}\right)\right]. \quad (2.263)$$

Noting that $\lambda_s = \lambda_i T_e/T_i$, we can write Eq. (2.263) as

$$\omega^2 = k_z^2 v_A^2 \left[1 + \lambda_s \left(1 + \frac{4}{3}\frac{T_i}{T_e}\right)\right] \quad (2.264)$$

to make explicit the change in the KAW dispersion equation derived from the fluid theory (Eq. 2.226), because of the kinetic theory.

Retaining the imaginary term in $Z(z_0^e)$ in K_{33} gives the electron Landau damping of the KAW, with the damping rate

$$\gamma = \frac{1}{2} \left(\frac{\pi}{2}\right)^{1/2} \left(\frac{m_e}{m_i}\right)^{1/2} \frac{k_x^2 v_A^2}{\Omega_i^2} |k_z| v_s. \quad (2.265)$$

The ion Landau damping of the KAW is exponentially small for small β, and there is no transit time magnetic damping because $B_{1z} \simeq 0$ for the KAW. The second factor on the left–hand side of Eq. (2.260), set to zero, gives the fast magnetoacoustic wave, whose frequency is $\simeq \Omega_i$ if $\lambda_i \simeq 1$. The electron Landau damping of the KAW has also been verified with a particle–in–cell simulation (Tanaka, Sato & Hasegawa 1987), and may lead to electron acceleration in space plasmas (see Chapter 7).

2.8.5 Localized Alfvén Waves

Consider a spatially localized source of waves of given frequency ω placed in a uniform plasma. If the source couples into the shear Alfvén wave fields, the waves propagate away almost along the magnetic field, with a parallel wavenumber $k_z \simeq \omega/v_A$. If the transverse scale–length of the source is short compared with the parallel wavelength, the resulting wave fields will contain a spectrum of modes with perpendicular wavenumbers $k_\perp$ that are predominantly large compared with k_z. The fields comprise therefore a spectrum of short perpendicular wavelength KAW or IAW modes, depending on the temperature of the plasma.

The quantity that determines how far the path of propagation of energy in each mode diverges from the magnetic field direction is the angle to the field, derived from the ratio of the perpendicular group velocity $v_{g\perp}$ to the parallel group velocity v_{gz}. The wave fields may remain strongly localized in the direction transverse to the magnetic field, in other words localized to a single magnetic field line, if the angle of divergence of the energy propagation path is small. This phenomenon of the localization of Alfvén waves to field lines is thought to have relevance to observations of waves in the Earth's magnetosphere, and has been observed in the laboratory (Morales & Maggs 1997, Gekelman *et al.* 1997, Gekelman 1999) (see Chapters 6 and 7).

The propagation angle for a KAW is derived from Eqs. (2.230) and (2.231), setting $k_\perp = k_x$, and is given by

$$\tan \theta_{\mathrm{K}} = \frac{v_{gx}}{v_{gz}} = \frac{k_z}{k_x} \frac{\lambda_{\mathrm{s}}}{1 + \lambda_{\mathrm{s}}} = \frac{\omega \rho_{\mathrm{s}}}{v_{\mathrm{A}}} \frac{k_x \rho_{\mathrm{s}}}{(1 + k_x^2 \rho_{\mathrm{s}}^2)^{3/2}} \tag{2.266}$$

while for the IAW it is given, from Eqs. (2.230) and (2.232), by

$$\tan \theta_{\mathrm{I}} = -\frac{k_z}{k_x} \frac{\delta_{\mathrm{e}}^2 k_x^2}{1 + \delta_{\mathrm{e}}^2 k_x^2} = -\frac{\omega \delta_{\mathrm{e}}}{v_{\mathrm{A}}} \frac{\delta_{\mathrm{e}} k_x}{(1 + \delta_{\mathrm{e}}^2 k_x^2)^{1/2}}. \tag{2.267}$$

These angles are shown in Figure 2.14 as a function of k_x (normalized to the relevant scale–length), for waves with $\omega \rho_{\mathrm{s}}/v_{\mathrm{A}} = 0.1$ (KAW) and $\omega \delta_{\mathrm{e}}/v_{\mathrm{A}} = 0.01$ (IAW).

There exists a maximum angle of propagation of a KAW of given ω and k_z, as k_x increases, obtained from Eq. (2.266) as

$$\tan \theta_{\mathrm{M}} = \frac{2}{3\sqrt{3}} \left(\frac{\omega \rho_{\mathrm{s}}}{v_{\mathrm{A}}} \right) = \frac{2}{3\sqrt{3}} \frac{\omega}{\Omega_{\mathrm{i}}} \sqrt{\frac{\gamma \beta}{2}} \tag{2.268}$$

corresponding to $k_\perp \rho_{\mathrm{s}} = 1/\sqrt{2}$. This is the maximum angle of spread of the radiation pattern of a KAW emerging from a localized source. For angles of propagation of energy $\theta < \theta_{\mathrm{M}}$, there are two solutions of Eq. (2.266) for $k_\perp$, a long perpendicular wavelength given by

$$k_\perp \rho_{\mathrm{s}} \simeq \frac{v_{\mathrm{A}}}{\omega \rho_{\mathrm{s}}} \tan \theta \tag{2.269}$$

and a short wavelength given by

$$k_\perp \rho_{\mathrm{s}} \simeq \left[\frac{\omega \rho_{\mathrm{s}}}{v_{\mathrm{A}} \tan \theta} \right]^{1/2}. \tag{2.270}$$

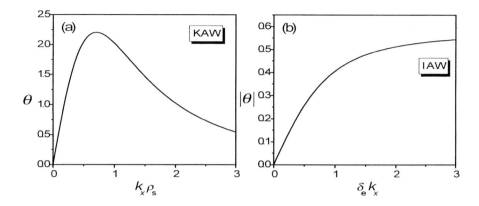

Figure 2.14: (a) The energy propagation angle for a localized source of KAWs, plotted against $k_x \rho_s$. Here $\omega \rho_s / v_A = 0.1$. (b) The angle for a localized source of IAWs, plotted against $\delta_e k_x$. Here $\omega \delta_e / v_A = 0.01$.

The mixture of the two different perpendicular scales at a fixed value of θ gives rise to unusual radiation patterns for the KAW that are described by Morales & Maggs (1997).

The behaviour of the IAW is quite different to that of the KAW (Morales, Loritsch & Maggs 1994). In the limit of large $|k_\perp|$ the angle of propagation of the IAW becomes independent of $k_\perp$, that is, there is a maximum angle given by

$$\tan \theta_C = \frac{\omega \delta_e}{v_A} = \left(\frac{\omega}{\Omega_i} \right) \left(\frac{m_e}{m_i} \right)^{1/2}. \tag{2.271}$$

Most of the energy from a localized source in a cool plasma radiates in a cone about the magnetic field line through the source, close to the angle θ_C. In other words, there is an annular region in the plane cut transverse to the magnetic field, corresponding to the cone, through which most of the Poynting flux flows. The annulus has an approximate width of several skin depths, and the radius of the annulus increases with distance from the source. This behaviour has been verified experimentally (Gekelman *et al.* 1994). The cone has also been called an Alfvén resonance cone (Stasiewicz *et al.* 1997) (see Chapter 3 for a discussion of the Alfvén resonance). Noting that

$$\frac{\tan \theta_M}{\tan \theta_C} = \frac{2}{3\sqrt{3}} \left(\frac{m_i}{m_e} \frac{\gamma \beta}{2} \right)^{1/2} \tag{2.272}$$

and using the condition of Eq. (2.224) for the existence of the KAW, we find that the maximum propagation angle θ_M for the KAW, under the conditions when it propagates, is larger than the cone angle θ_C for the IAW.

3 Waves in Nonuniform Plasmas

3.1 Introduction

As we have seen in Chapter 2, a uniform plasma can support three basic types of magneto-hydrodynamic (MHD) waves, the Alfvén wave and the fast and slow magnetoacoustic waves. Only the fast and Alfvén waves exist in the limit of a cold plasma, the slow wave relying on nonzero particle pressure for propagation.

In real physical environments, the plasma is usually nonuniform to some degree. Laboratory plasmas can have large gradients in density or magnetic field, but often possess symmetries that simplify the analysis of wave propagation. Space and astrophysical plasmas are often similarly nonuniform. The influence of nonuniformities on the waves depends on the size of the characteristic scale–lengths compared to the wavelengths. The question of the propagation of the waves in smoothly nonuniform plasmas is addressed in this chapter.

We find that one important consequence of the nonuniformity of the plasma on Alfvén wave propagation is that there is a coupling between the (low–frequency) shear and compressional waves, in contrast to the uniform plasma, where the modes can propagate independently. This coupling leads to the phenomenon of Alfvén resonance absorption, which is important for both laboratory and astrophysical plasmas. The Alfvén resonance occurs (at low frequency) when the wave phase velocity along the equilibrium magnetic field equals the local Alfvén velocity, as may occur in a region where the equilibrium density or magnetic field are gradually changing in space (Uberoi 1972). Alfvén resonance absorption presents an efficient mechanism for the thermalization of the wave energy, and was first proposed as a means of heating laboratory plasmas by Chen & Hasegawa (1974a), Tataronis (1975) and Tataronis & Grossman (1976). The related resonance absorption of azimuthally symmetric waves in a cylindrical laboratory plasma, which occurs because of Hall effects, was discussed as early as 1966 by Dolgopolov & Stepanov (1966).

Instead of the wave energy building up without limit at the Alfvén resonance point, as in the MHD theory, a consequence of the inclusion of the dissipative and thermal nonideal effects in this process is that the wave will mode convert into other wave modes made possible by the nonideal effects. The wave energy is subsequently absorbed by damping processes, or is transported away from the resonance region. The disadvantage for the theory of resonance absorption of including nonideal effects compared to the MHD approach, is that the starting equations are higher–order fluid differential equations, in the case of finite frequency, resistivity and viscosity effects, or are coupled integro-differential equations in the case of kinetic theory effects. The greater complexity of the equations reflects the greater number of wave modes allowable in the plasma.

In this chapter we first use the uniform plasma dispersion relations obtained in Chapter 2, employing a local approximation, to consider small amplitude wave propagation in plasmas that are smoothly stratified in a direction perpendicular to the equilibrium magnetic field. The ideal MHD and Hall-MHD models are used, for both cold and warm plasmas, and the effects of multiple ion species, and of collisional processes, are also discussed. Then we establish differential equations for the waves in nonuniform plasmas where the local approximation is not valid, using the ideal MHD and cold plasma models. In the final section, the Alfvén resonance process is discussed, for both sharply and gradually varying plasma density profiles.

3.2   Stratified Plasmas

In the uniform plasmas considered in Chapter 2, the only unique direction was provided by the homogeneous equilibrium magnetic field direction. For plane waves propagating at an oblique angle to the magnetic field, it was therefore sufficient to assume nonzero k_x and k_z components of the wavevector, and define the y-axis such that $k_y = 0$. We are often interested in this chapter in waves of fixed frequency propagating in a plasma with a nonuniform density, at an oblique angle to the uniform equilibrium magnetic field (which we assume is along the z-axis). The density is taken to be stratified in a direction perpendicular to the magnetic field (say in the x-direction). In that case, when two directions are uniquely defined, we must for generality also allow for a nonzero k_y. We also consider the case where the equilibrium magnetic field lies in a fixed plane (the y–z plane), but varies in both direction and magnitude in the direction perpendicular to the plane (the x–direction). In both these cases, the wavevector components k_y and k_z can be considered to be constant, but the component k_x is not defined, and in general the full wave differential equation must be solved to find the wave field variation in the x–direction.

However, if the density or magnetic field gradient is very mild, such that the wavelength in the x-direction is small compared with the scale–length of the gradient, it is instructive to employ a local approximation. In this approach we assume that the uniform plasma dispersion relations derived in Chapter 2, generalized to include a nonzero k_y (by replacing k_x^2 by $k_x^2 + k_y^2$), are valid at any point in the plasma, using the local values of the magnetic field, plasma density and temperature. For a wave of fixed ω, k_y and k_z, this leads to a value of k_x^2 that is a function of x. We make this approximation in this section, for waves described by the different fluid models introduced in Chapters 1 and 2, before considering the more valid differential equation approach in Section 3.3.

3.2.1   Ideal MHD

Let us consider the Alfvén and magnetoacoustic wave dispersion equations obtained from the ideal MHD model of Section 2.2. We now let the plasma density be a local function $\rho(x)$ of the coordinate x in a direction transverse to the magnetic field, which we initially assume to be uniform and to lie along the z–axis. (The case where the magnetic field is also a function of x is considered in Section 3.3.) The Alfvén speed v_A is therefore also a function of x. If the temperature also varies with x, the sound speed c_s will be a function of x as well.

We note first that the local shear Alfvén wave dispersion relation (2.19) in the ideal MHD model, for a wave of given ω, k_y and k_z, is given by

$$\omega = v_A(x)|k_z| \tag{3.1}$$

and so is independent of k_x. In a spatially varying density profile, Eq. (3.1) can only be satisfied at discrete points. Thus it is sometimes said that the low–frequency shear Alfvén wave can exist localized on a single magnetic field line where Eq. (3.1) is satisfied. However, we find later that a plasma nonuniformity will couple the shear Alfvén wave to the compressional magnetoacoustic wave branch, with the result that the shear wave dispersion relation in effect does depend on k_x.

We can derive the following equation for the local value of k_x^2 for the fast and slow waves in the ideal MHD model with nonzero plasma β, and including a nonzero k_y, from Eq. (2.18):

$$k_x^2 = -k_y^2 + \frac{(\omega^2 - c_s^2 k_z^2)(\omega^2 - v_A^2 k_z^2)}{(v_A^2 + c_s^2)(\omega^2 - v_T^2 k_z^2)} \tag{3.2}$$

with

$$v_T = \frac{c_s v_A}{(v_A^2 + c_s^2)^{1/2}}. \tag{3.3}$$

A cutoff in k_x is said to occur when the frequency is such that $k_x = 0$. Alternatively, for a wave of fixed frequency, there may be a spatial position in the density, temperature or magnetic field profile where the value of v_A or c_s is such that $k_x = 0$. From Eq. (3.2) it is evident that there are two cutoff frequencies; if $k_y = 0$ they are the frequencies for parallel propagation of the slow and fast magnetoacoustic waves:

$$\omega = c_s|k_z| \quad \text{and} \quad \omega = v_A|k_z|. \tag{3.4}$$

If the frequency is fixed and the plasma β is constant, Eq. (3.2) with $k_x = 0$ becomes a quadratic equation in v_A^2, indicating two spatial positions where a cutoff occurs.

A single resonance in k_x given by Eq. (3.2), where $k_x \to \infty$, occurs when the denominator in Eq. (3.2) goes to zero, that is, at the frequency given by

$$\omega^2 = v_T^2 k_z^2 = \frac{c_s^2 v_A^2 k_z^2}{(v_A^2 + c_s^2)} = \frac{c_s^2 k_z^2}{1 + \gamma\beta/2} \tag{3.5}$$

or, for a fixed wave frequency, at the spatial position where Eq. (3.5) is satisfied. This resonance is called the *cusp, or compressive, resonance* (Tataronis & Grossman 1976, Wentzel 1979a). The cusp resonance is effectively a resonance of the slow magnetoacoustic wave, and is so named because of the cusp–shaped characteristic phase velocity surfaces of the slow mode (see for example Shercliff 1965).

Another resonance in k_x, the *Alfvén resonance*, corresponding to the frequency of the shear Alfvén wave, and given by the condition

$$\omega = v_A|k_z| \tag{3.6}$$

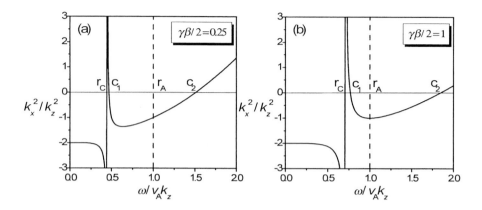

Figure 3.1: The square of the perpendicular wavenumber k_x^2, normalized by k_z^2, plotted against the normalized phase speed $\omega/v_A k_z$ along the magnetic field, using the ideal MHD model. Here $k_y/k_z = 1$. The compressive and Alfvén resonances are denoted by r_C and r_A respectively, and the two cutoffs are denoted by c_1 and c_2. (a) $\gamma\beta/2 = c_s^2/v_A = 0.25$, (b) $\gamma\beta/2 = 1$.

is found to occur when the MHD model is extended. It may be thought that the Alfvén resonance does not occur in the ideal MHD case because the condition of Eq. (3.6) does not lead to a singularity of k_x in Eq. (3.2). In fact, however, we shall see that if the original low–frequency fluid spatial differential equations in the nonuniform plasma are used for a wave of fixed frequency, the Alfvén resonance (Eq. 3.6) is found to occur, in the form of a singularity in the equations, provided $k_y \neq 0$.

This is also expressed by noting that the wavenumber defined in Eq. (3.2) in effect corresponds to the magnetoacoustic mode; at a spatial position where the Alfvén resonance occurs (i.e. where a localized shear Alfvén wave can exist), mode conversion of the magnetoacoustic wave to the shear Alfvén mode occurs, as is shown in Section 3.2.2. It is interesting to note that in an incompressible plasma ($\gamma \to \infty$), or for very large β, $v_T \to v_A$ and the cusp resonance occurs at $\omega = v_A |k_z|$, that is, it is degenerate with the Alfvén resonance. The x–wavenumber in the incompressible case satisfies $k_x^2 + k_y^2 + k_z^2 = 0$ (except at the cusp resonance), and so is imaginary.

Using Eq. (3.2), a plot of k_x^2/k_z^2 against the normalized phase velocity along the magnetic field, $\omega/v_A k_z$, is shown in Figure 3.1, for two values of β, and for $k_y/k_z = 1$. The two cutoffs and the two resonances are indicated. Note that as β decreases, the first cutoff moves close to the compressive resonance. It is useful, for a wave of fixed frequency and fixed k_z and k_y, to regard the horizontal axis as being proportional to the square root of the plasma density. A wave can then be pictured in this local approximation as travelling from right to left in a decreasing density profile, encountering a succession of cutoffs and resonances, and becoming propagating ($k_x^2 > 0$) or evanescent ($k_x^2 < 0$) in different regions. It should be noted that the Alfvén resonance is encountered in a region where the wave is evanescent in the x–direction. This result leads to Alfvén resonance damping of surface waves, as shown in Chapter 4. Of course the detailed behaviour of the wave at the cutoffs and resonances must be

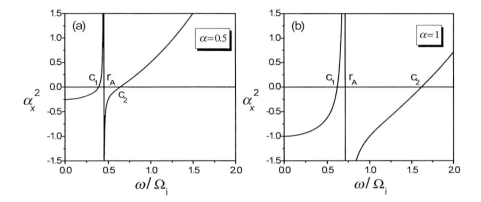

Figure 3.2: The square of the normalized perpendicular wavenumber $\alpha_x^2 = v_A^2 k_x^2/\Omega_i^2$, plotted against the normalized frequency ω/Ω_i, in the cold plasma Hall model. Here $k_y = 0$. The Alfvén resonance is denoted by r_A, and the two cutoffs are denoted by c_1 and c_2. (a) $\alpha = v_A k_z/\Omega_i = 0.25$, (b) $\alpha = 1$.

computed using the wave differential equations.

The existence of the Alfvén resonance has been the subject of some controversy in the literature (Bellan 1994, 1995, 1996a,b; Goedbloed & Lifschitz 1995; Rauf & Tataronis 1995; Ruderman, Goossens & Zhelyazkov 1995). The controversy revolves essentially around whether a resonance can still be said to exist if a nonideal effect removes the singularity of k_x. The singularity survives some nonideal effects. Thus the Alfvén resonance can be regarded as the degenerate low–frequency limit of the generalized Alfvén resonance (including a finite ion cyclotron frequency) that does occur as an infinity of k_x in the magnetoacoustic mode, to be discussed in the following section. On the other hand, other effects such as collisions and electron inertia remove the singularity at the Alfvén resonance. The basic difference, as will be discussed in subsequent sections, is that if the singularity is not removed, wave energy can in principle accumulate without limit at the "Alfvén resonant point" defined by Eq. (3.6) (thus eventually violating the assumption of linear wave fields), but if the singularity is removed, strong damping and/or mode conversion into a modified shear Alfvén wave can occur. In both cases, the magnetoacoustic wave loses energy as it propagates through the Alfvén resonant point. All of these points are addressed in later sections and chapters.

3.2.2   Hall MHD

Cold Plasma

To allow for the wave frequency to approach the ion cyclotron frequency, we consider the Hall-MHD model for a single–ion plasma, with the Hall term included in Ohm's law (Eq. 1.28).

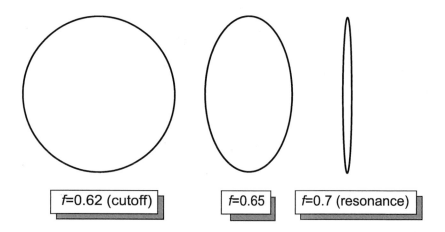

Figure 3.3: The polarization ellipse in the cold plasma Hall model, for normalized frequencies $f = 0.62$, $f = 0.65$ and $f = 0.7$. Here $\alpha = 1$.

We initially assume a zero β plasma. The local value of k_x^2 is given, from Eq. (2.42), by

$$k_x^2 = -k_y^2 - k_z^2 + \frac{\omega^2 (\omega^2 - v_A^2 k_z^2)}{v_A^2 (\omega^2 - v_A^2 k_z^2 (1 - f^2))} \tag{3.7}$$

where

$$f = \omega/\Omega_i. \tag{3.8}$$

There is no cusp resonance in k_x because $\beta = 0$, but there is now a resonance in k_x which occurs at a frequency defined by

$$\omega^2 = \frac{v_A^2 k_z^2}{(1 + v_A^2 k_z^2/\Omega_i^2)} = \frac{v_A^2 k_z^2}{(1 + \alpha^2)} \tag{3.9}$$

with

$$\alpha = \frac{v_A k_z}{\Omega_i}. \tag{3.10}$$

Eq. (3.9) defines the Alfvén resonance, generalized for finite ion cyclotron frequency; it is sometimes called the perpendicular ion cyclotron resonance (Stix 1992).

In the limit of low wavenumber ($\alpha \to 0$) and low frequency ($f \to 0$), the generalized Alfvén resonance reduces to Eq. (3.6). The dispersion equation (3.7) also gives two cutoff frequencies in k_x, which, for $k_y = 0$, correspond to the parallel propagating left–hand circularly polarized ion cyclotron and right–hand circularly polarized fast waves with frequencies given by Eq. (2.45).

Figure 3.2 is a plot of $\alpha_x^2 = v_A^2 k_x^2/\Omega_i^2$ against f, showing the generalized Alfvén resonance lying between the two cutoff frequencies. For a given positive value of k_x^2, a circumstance that arises if a uniform plasma is bounded in the x–direction, giving a standing wave in that direction, there are generally two solutions for the frequency. The higher frequency, above the upper cutoff, corresponds to the fast mode, and the lower frequency, between the lower cutoff and the resonance, corresponds to the ion cyclotron mode, identified in Eq. (2.44). As α tends to zero, which is equivalent to the ion cyclotron frequency increasing for given v_A and k_z, the frequency range of the ion cyclotron mode shrinks to zero just below the resonance frequency, and the cutoff-resonance pair merges, leaving the singular Alfvén resonance shown in Figure 3.1. In that limit the ion cyclotron mode essentially has an indeterminate k_x, has a frequency given by the Alfvén resonance condition Eq. (3.6), and can be identified with the low–frequency shear Alfvén wave.

It is sometimes said that the generalized resonance condition Eq. (3.9) is the dispersion relation of the "shear Alfvén wave"; this is strictly only true in the low–frequency limit, when the linearly polarized shear Alfvén wave exists with no dependence on the perpendicular wavenumber. The polarization ellipse (in the plane perpendicular to the equilibrium magnetic field) of the ion cyclotron mode is shown in Figure 3.3, corresponding to the conditions in Figure 3.2(b) ($\alpha = 1$), indicating the tendency from circular polarization to linear polarization as the resonance is approached from the lower cutoff.

To show the effect of a varying plasma density on a wave of fixed ω, k_y and k_z in the cold plasma Hall model, as in Figure 3.1 for the ideal MHD model, we plot k_x^2/k_z^2 against $\Omega_i/v_A k_z$ in Figure 3.4. Here the two cases of $f = 0.9$ and $f = 1.1$ are shown. A wave travelling from right to left in an increasing density profile encounters the cutoff-resonance-cutoff triplet, provided $\omega < \Omega_i$. If $\omega > \Omega_i$, the generalized Alfvén resonance (and the lower cutoff) is never encountered.

Low Plasma Pressure

If we now introduce thermal effects into the Hall-MHD model, it is found that the perpendicular wavenumber k_x is limited in size for any real frequency, and there is no longer a true resonance. For given real frequency and k_z, it is found that there are two values of k_x^2, rather than the one value given by Eq. (3.7), that is, there are two distinct modes. At the resonance frequency given by Eq. (3.9), the two values of k_x become comparable in size, and mode conversion can occur in a stratified plasma. An instructive way of modelling this process is to use the finite parallel electron temperature modification of the Hall-MHD model described in Section 2.8.2. This model retains finite ion cyclotron frequency effects, but also describes the KAW and IAW modes. Thus we use a dielectric tensor of the form of Eq. (2.75), with the scaled dielectric tensor components u_1, u_2 and u_3 defined in Eqs. (2.235) and (2.236).

The dispersion equation (2.76) resulting from the wave equation with the dielectric tensor (Eq. 2.75), including a nonzero k_y, may be written as

$$u_1(k_x^2 + k_y^2)^2 - \left[(u_3 + u_1)(u_1 - k_z^2) - u_2^2\right](k_x^2 + k_y^2)$$
$$+u_3\left[(u_1 - k_z^2)^2 - u_2^2\right] = 0. \tag{3.11}$$

The normalized components of the dielectric tensor may be written, for a singly charged ion

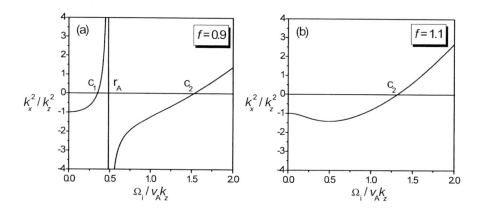

Figure 3.4: The square of the normalized perpendicular wavenumber k_x^2/k_z^2, plotted against $\Omega_i/v_A k_z$, in the cold plasma Hall model. Here $k_y = 0$. The Alfvén resonance is denoted by r_A, and the two cutoffs are denoted by c_1 and c_2. (a) $f = \omega/\Omega_i = 0.9$, (b) $f = 1.1$.

species,

$$\frac{u_1}{k_z^2} = \frac{f^2}{\alpha^2(1-f^2)}, \qquad \frac{u_2}{k_z^2} = \frac{f^3}{\alpha^2(1-f^2)} \tag{3.12}$$

$$\frac{u_3}{k_z^2} = -\frac{m_i}{m_e}\frac{f^2}{\alpha^2}\frac{1}{(f^2 - (V_e^2/v_A^2)\alpha^2)} \tag{3.13}$$

where α is defined in Eq. (3.10). Eq. (3.11) is quadratic in $k_x^2 + k_y^2$, yielding two complex solutions $k_x^2 = k_\pm^2$, where we take $|k_+^2| \geq |k_-^2|$. Using the fact that $|u_3/u_1| \simeq m_i/m_e \gg 1$, the two approximate solutions for k_x^2 are

$$k_-^2 = -k_y^2 + \left\{(u_1 - k_z^2)^2 - u_2^2\right\}/(u_1 - k_z^2) \tag{3.14}$$

$$k_+^2 = -k_y^2 + u_3(u_1 - k_z^2)/u_1. \tag{3.15}$$

These solutions are valid provided the condition

$$u_1 - k_z^2 = 0 \tag{3.16}$$

which is equivalent to the generalized Alfvén resonance condition Eq. (3.9), is not satisfied. If on the other hand the generalized Alfvén resonance condition is approximately satisfied, the wavenumbers are given by

$$k_\pm^2 \simeq -k_y^2 \mp f|u_1 u_3|^{1/2} \tag{3.17}$$

so there is no longer a true Alfvén resonance where $k_x \to \infty$.

The cusp resonance defined by Eq. (3.5) is not present because β is assumed small and the pressure gradient force is neglected in the ion equation of motion. Eq. (3.14) is the same as

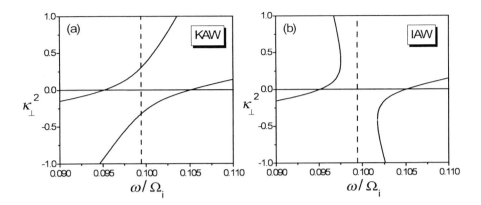

Figure 3.5: The square of the normalized perpendicular wavenumber $\kappa_\perp^2 = (k_x^2 + k_y^2)/k_z^2$, plotted against ω/Ω_i, in the Hall model with electron inertia and electron pressure. The Alfvén resonance is shown by the dashed lines. Here $m_i/m_e = 0.2$ to exaggerate the shape of the curves. (a) the KAW case, with $V_e^2 = 3v_A^2$, (b) the IAW case, with $V_e = 0$.

the wavenumber expression in Eq. (3.7) if the explicit expressions given in Eqs. (2.235) and (2.236) for u_1 and u_2 are used, so that the wave associated with wavenumber k_- is simply the fast zero–β compressional mode with ion cyclotron effects, while the wave with wavenumber k_+ given by Eq. (3.15) is the short–wavelength quasi–electrostatic wave (QEW) discussed in Section 2.8.2 (see Eq. 2.237). Again, depending on the plasma β, the QEW is further described as either a KAW or IAW mode.

It is useful to plot the normalized squares of the two perpendicular wavenumber solutions of Eq. (3.11) as functions of ω, for fixed α. If the electron inertia and electron pressure are neglected, that is, when $|u_3| \to \infty$, there is a single solution k_-^2 given by Eq. (3.14), which has the properties shown in Figure 3.2, with a sharp resonance lying between two cutoffs. In Figure 3.5(a) are plotted the real part of the two solutions for $\kappa_{\perp\pm}^2 = (k_\pm^2 + k_y^2)/k_z^2$ when electron inertia and electron pressure are retained, and the electrons are warm, $V_e > v_A$, while in Figure 3.5(b) is the case where the electron temperature is low, $V_e < v_A$, and the sign of u_3 is reversed. (In these figures the ratio m_i/m_e used in u_3 is exaggeratedly small to show the qualitative features: for realistic values of the ratio, $|k_+^2| \gg |k_-^2|$ except very near the resonance.)

In each of the two cases shown in Figure 3.5, the singularity in $\kappa_\perp^2$ at the Alfvén resonance frequency (shown by the dashed line) is removed. In the case where the electron pressure is dominant (Figure 3.5(a)), two modes exist for every frequency, either purely propagating ($\kappa_{\perp\pm}^2 > 0$) or evanescent ($\kappa_{\perp\pm}^2 < 0$), with the short–wavelength KAW mode propagating at frequencies above the Alfvén resonance frequency. In the case of low electron pressure compared to electron inertia (Figure 3.5(b)), a stop–band appears around the Alfvén resonance frequency in which $\kappa_{\perp\pm}$ are partly evanescent, and the short–wavelength IAW propagates at frequencies below the Alfvén resonance frequency.

If the normalized frequency f is fixed at a value less than unity (so that the Alfvén resonance can be encountered), but the plasma density is varied, the corresponding plots of $\kappa_{\perp\pm}^2$ are similar to those shown in Figure 3.5, with the density increasing in the left-to-right direction. In the warm plasma case with dominant electron pressure (Figure 3.5(a)), the short–wavelength KAW propagates in the high–density region, such as towards the centre of a confined fusion plasma. For the case of dominant electron inertia (Figure 3.5(b)), the short–wavelength IAW propagates in the low–density region of the plasma, such as towards the surface of a confined plasma. The IAW is sometimes called the surface electrostatic wave (SEW) (Stix 1992), because it can propagate in the low–density plasma at the boundary of a fusion plasma, and it can be directly excited by an antenna.

In the Alfvén resonance scheme for heating a confined plasma, a fast compressional wave with dispersion equation (3.14) is launched by an antenna on the low–density edge of the plasma (the left–hand sides of Figures 3.5(a) and (b)) and propagates towards the Alfvén resonance position in the interior of the plasma. There it may undergo mode conversion to the KAW or the IAW. The subsequent collisional or collisionless damping of these short–wavelength modes means that the energy of the compressional wave may be efficiently absorbed. We discuss further the energy absorption and mode conversion processes in later sections.

High Plasma Pressure

We now include a nonzero pressure gradient acting on the ions. A suitable model showing the main effects of the ion pressure is the fluid Hall-MHD model with nonzero plasma β, which includes the pressure gradient term in the ion equation of motion, described in Section 2.3.2. Starting from the dispersion equation (2.52), generalized to include nonzero k_y, a quadratic equation for $k_x^2 + k_y^2$ may be obtained:

$$A(k_x^2 + k_y^2)^2 + B(k_x^2 + k_y^2) + C = 0 \tag{3.18}$$

with

$$A = -f^2 \beta' v_A^6 k_z^2$$

$$B = v_A^2[(1+\beta')\omega^4 - (1 - f^2 + 2\beta')\omega^2 v_A^2 k_z^2 + \beta'(1 - 2f^2)v_A^4 k_z^4] \tag{3.19}$$

$$C = -\omega^6 + (2+\beta')\omega^4 v_A^2 k_z^2 - (1 - f^2 + 2\beta')\omega^2 v_A^4 k_z^4 + \beta'(1 - f^2)v_A^6 k_z^6$$

where $\beta' = c_s^2/v_A^2 = \gamma\beta/2$ and $f = \omega/\Omega_i$.

As in the low–β case of the previous section, there are two solutions for k_x^2 from Eq. (3.18) for given ω and k_z, but no resonances of k_x^2. However, resonance does occur if $c_s \to 0$, that is $\beta \to 0$, in which case we recover the cold Hall model, there is the single solution given in Eq. (3.7) for k_x^2, and the generalized Alfvén resonance (Eq. 3.9) is present. Resonance also occurs if $f \to 0$, because we then recover the low–frequency ideal MHD model, the single solution for k_x^2 is given by Eq. (3.2), and the compressive resonance (Eq. 3.5) occurs.

The presence of both nonzero f and nonzero β, however, removes both resonances by inducing the existence of a second mode that has very large $|k_x^2|$ for small f and small β, that is, a short–wavelength mode as for the case in the previous section. We can again separate

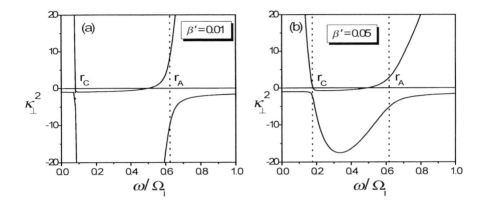

Figure 3.6: The square of the normalized perpendicular wavenumber $\kappa_\perp^2 = (k_x^2 + k_y^2)/k_z^2$, plotted against ω/Ω_i, in the Hall model with ion pressure. The Alfvén resonance (r_A) and compressive resonance (r_C) frequencies are shown by the dotted lines. Here $v_A k_z/\Omega_i = 0.8$. (a) $\beta' = c_s^2/v_A^2 = 0.01$, (b) $\beta' = 0.05$.

the modes, for $f \ll 1$ and $\beta \ll 1$, into the long perpendicular wavelength mode (i.e. the fast magnetoacoustic mode), by setting A to zero in Eq. (3.18):

$$k_-^2 + k_y^2 \simeq -\frac{C}{B} \simeq \frac{\omega^2}{v_A^2} - k_z^2 \tag{3.20}$$

and the short–wavelength mode, by setting C to zero in Eq. (3.18):

$$k_+^2 + k_y^2 \simeq -\frac{B}{A} \simeq \frac{(\omega^2 - v_A^2 k_z^2)(\omega^2 - c_s^2 k_z^2)}{f^2 \beta' v_A^4 k_z^2}. \tag{3.21}$$

Equation (3.21) is the same as the dispersion equation (2.258) (if nonzero k_y is included), and can be derived from Eq. (2.218) if the electron inertia term is neglected.

The behaviour of the two modes in this case is shown in the plots of the two solutions for $(k_x^2 + k_y^2)/k_z^2$ against f in Figure 3.6, as obtained from Eq. (3.18), for a given value of $\alpha = v_A k_z/\Omega_i$. The compressive and Alfvén resonances are shown as almost true resonances in Figure 3.6(a) for $\beta' = 0.01$. As β increases to 0.05 (Figure 3.6), the resonances are smoothed out by the short–wavelength modes. In the vicinity of the Alfvén resonance condition $\omega \simeq v_A|k_z|$, and for low β, the short–wavelength mode given by Eq. (3.21) becomes the KAW mode (Eq. 2.226). The smoothing out of the Alfvén resonance by the short–wavelength KAW is similar to the low–β case shown in Figure 3.5. In the vicinity of the cusp resonance, the short–wavelength mode (Eq. 3.21) becomes the modified sound wave (Eq. 2.233). Note that the sound mode that removes the cusp resonance is a backward propagating mode, as was mentioned in Section 2.8.1, because of the negative gradient of k_+^2 just below the resonance.

The results (Eqs. 3.20 and 3.21) for the two modes are not valid very near the former Alfvén resonance or cusp resonance, where $B \simeq 0$, and the wavenumbers are now limited in size for nonzero f and β. As shown in Figure 3.6, the squares of the perpendicular wavenumbers are equal in magnitude, but are of opposite sign for the two modes:

$$k_x^2 + k_y^2 \simeq \pm \left(\frac{-C}{A}\right)^{1/2}. \tag{3.22}$$

At the Alfvén resonance point we then have

$$k_\pm^2 + k_y^2 \simeq \pm \frac{k_z^2}{\sqrt{\beta'}} \tag{3.23}$$

and near the cusp resonance we have

$$k_\pm^2 + k_y^2 \simeq \pm \frac{\sqrt{\beta'}\,k_z^2}{f}. \tag{3.24}$$

The corresponding plots for a fixed normalized frequency f but varying plasma density are similar to those in Figures 3.6(a) and (b), provided $f < 1$, with the density increasing in the left–to–right direction. A magnetoacoustic wave, launched from the low–density plasma, can mode convert into the short–wavelength modes at both the cusp and Alfvén resonances, as is discussed in later sections. The mode conversion property is reflected in the fact that the waves in a nonuniform plasma are described by fourth–order spatial differential equations, as discussed later.

3.2.3   Multi–Ion Plasmas

We now discuss the propagation of waves obliquely to the magnetic field in cold plasmas with two species of ions. The second ion species may be a positively or negatively charged ionized atom or molecule, or it may be a relatively massive charged dust grain (assuming all the grains have the same charge, usually negative). The frequency range is assumed to encompass the cyclotron frequencies of both the primary and the secondary ion species, but to be much less than the electron cyclotron frequency. Using a local approximation in a plasma of varying density in the x–direction, we investigate the resonances and cutoffs of the perpendicular wavenumber.

The local wavenumber k_x is given by the collisionless cold plasma result (Eq. 2.78), generalized to allow nonzero k_y:

$$k_x^2 = -k_y^2 + \left(\frac{G^2 - H^2}{G}\right) \tag{3.25}$$

where G and H are defined in Eq. (2.79), using u_1 and u_2 from Eqs. (2.132) and (2.133). Cutoffs occur where $k_x^2 = 0$. If $k_y = 0$ also, the cutoffs correspond to the case of propagation of the wave purely parallel to the magnetic field, discussed in Section 2.6.

A resonance in k_x occurs, from Eq. (3.25), when $G = 0$, that is for

$$\alpha^2 = f^2 \left(\frac{1}{1 - f^2} + \frac{bg^2}{g^2 - f^2}\right) \tag{3.26}$$

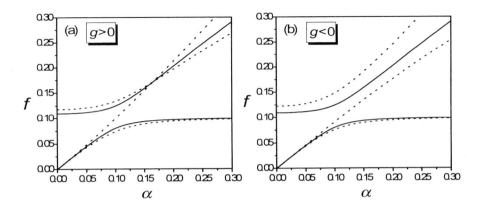

Figure 3.7: The normalized frequency $f = \omega/\Omega_1$ for the resonance conditions (solid lines) and the perpendicular cutoff conditions (dotted lines) for a two ion species plasma, plotted against the normalized wavenumber $\alpha = v_A k_z/\Omega_i$. The density ratio b is 0.2. (a) The secondary species is positively charged, and $g = 0.1$. (b) The secondary species is negatively charged, and $g = -0.1$.

where $f = \omega/\Omega_1$ and $\alpha = v_A k_z/\Omega_1$, with Ω_1 the magnitude of the cyclotron frequency of the primary ion, and the Alfvén speed is defined in terms of the primary ion mass density. The other symbols are as defined in Section 2.6. For a given α, Eq. (3.26) has two solutions for f^2, so there are two resonance frequencies for the cold plasma with two ion species, which are independent of the sign of the charge on the secondary ion. For a cold single ion species plasma ($b = 0$), there is a single resonance frequency from Eq. (3.26), namely the generalized Alfvén resonance given by Eq. (3.9). The resonance frequencies, as well as the three cutoff frequencies obtained by setting the right–hand side of Eq. (3.25) to zero (with $k_y = 0$), are plotted in Figure 3.7 against α for the two cases of a plasma with a positively charged primary ion, and positively and negatively charged secondary ions. The relevant parameters are $g = \pm 0.1$ and $b = 1$. The cutoff frequencies correspond to the parallel propagating wave dispersion relations shown in Figure 2.7. The highest cutoff frequency occurs at $\omega = \Omega_m$ for $k_z = 0$, where Ω_m is defined in Eq. (2.140). It is useful to consider two ranges of α for a discussion of the character of the resonances.

In the limit of long wavelength along the magnetic field, such that $\alpha < |g| < 1$, that is $v_A k_z \ll \Omega_2$ (the left part of Figures 3.7(a) and (b)), the lower of the two resonance frequencies is given by

$$f_A = \frac{1}{(1+b)^{1/2}}\alpha \tag{3.27}$$

or,

$$\omega_A = \frac{v_A k_z}{(1+b)^{1/2}} = v_{AT} k_z \tag{3.28}$$

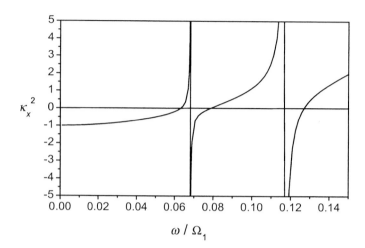

Figure 3.8: The normalized square of the perpendicular wavenumber $\kappa_x^2 = k_x^2/k_z^2$ for a two ion species plasma, plotted against normalized frequency $f = \omega/\Omega_1$, for $\alpha = 0.08$, $b = 0.2$ and $g = 0.1$. A lower (Alfvén) and an upper (ion-ion hybrid) resonance are present, as well as three cutoffs.

which is the Alfvén resonance frequency for the plasma with mass density contributed by both the ion species, and Alfvén speed v_{AT} based on the total mass density. The second (higher) resonance frequency is called the *ion-ion hybrid resonance* frequency, and in the limit $\alpha = 0$ it is given by

$$f_H = |g| \left(\frac{1+b}{1+bg^2} \right)^{1/2} \tag{3.29}$$

or,

$$\omega_H = \Omega_2 \left(\frac{1+b}{1+bg^2} \right)^{1/2} = \left(\frac{\omega_{p1}^2 \Omega_2^2 + \omega_{p2}^2 \Omega_1^2}{\omega_{p1}^2 + \omega_{p2}^2} \right)^{1/2} \tag{3.30}$$

in terms of the plasma frequencies ω_{p1} and ω_{p2} of the two ion species. The ion-ion hybrid resonance has been discussed by Perkins (1977) and Cramer & Yung (1986) in connection with the resonance heating of fusion plasmas with minority ion species, usually positively charged.

The dependence of $\kappa_x^2 = k_x^2/k_z^2$ on the normalized frequency f in the case $\alpha < |g|$, derived from Eq. (3.25) for $g = 0.1$, $k_y = 0$, $\alpha = 0.08$ and $b = 0.2$, is indicated in Figure 3.8. A lower (Alfvén) resonance and an upper (ion-ion hybrid) resonance are evident. The Alfvén resonance is very sharp compared with the ion-ion hybrid resonance, that is, there is a cutoff very close to the Alfvén resonance, as is also shown in Figure 3.7. As b increases, the Alfvén resonance moves down, and the hybrid resonance moves up in frequency in accordance with

Eqs. (3.28) and (3.30), and the three cutoffs move away from the resonances in accordance with Eqs. (2.135)-(2.139).

In the shorter wavelength range $\Omega_2 < v_A k_z \ll \Omega_1$ ($|g| < \alpha \ll 1$) (to the right in Figure 3.7(a) and (b)), the higher resonant frequency is given by

$$f_1 = (\alpha^2 + bg^2)^{1/2} \tag{3.31}$$

or,

$$\omega_1 = (v_A^2 k_z^2 + b\Omega_2^2)^{1/2} \tag{3.32}$$

and the lower by

$$f_2 = \alpha|g|/(\alpha^2 + bg^2)^{1/2} \tag{3.33}$$

or,

$$\omega_2 = v_A k_z \Omega_2 / (v_A^2 k_z^2 + b\Omega_2^2)^{1/2}. \tag{3.34}$$

Which of the two resonance frequencies is to be interpreted as the Alfvén resonance frequency now clearly depends on the size of b. If the secondary ion mass density is small enough that $b < \alpha^2/g^2$, we have

$$\omega_1 \simeq v_A k_z \tag{3.35}$$

which is the Alfvén resonance frequency based only on the plasma ions, while the other (ion-ion hybrid) resonance frequency is $\simeq \Omega_2$. For any value of b there is thus always a value of α high enough that the secondary ion motion is frozen out of the Alfvén resonance. If α becomes comparable to unity and $\alpha^2 \gg bg^2$, primary ion cyclotron effects modify the Alfvén resonance frequency:

$$\omega_A = v_A k_z (1 + v_A^2 k_z^2 / \Omega_i^2)^{-1/2}. \tag{3.36}$$

In the case $b > \alpha^2/g^2$, we find for the higher resonant frequency

$$\omega_2 \simeq \alpha/b^{1/2} \tag{3.37}$$

which is the Alfvén resonance frequency based on the secondary ion mass density, and the hybrid frequency is

$$\omega_1 \simeq b^{1/2}\Omega_2. \tag{3.38}$$

The dependence of κ_x^2 on the frequency in the case $\alpha > g$ is similar to that shown in Figure 3.8, except that, for small secondary ion mass density ($b < \alpha^2/g^2$), the upper resonance is the Alfvén resonance occurring close to the frequency $v_A k_z$, and the lower resonance is the hybrid resonance occurring close to the frequency Ω_2. As in Figure 3.8, as b increases the resonances shift and become broader due to a shift of the cutoffs.

We now consider a wave of fixed frequency and k_z propagating in a plasma with a density stratified in the direction transverse to the equilibrium magnetic field. This corresponds to fixing f and varying α, that is, a horizontal straight line on Figures 3.7(a) or (b). Noting from Eq. (2.140) that the cutoff frequency Ω_m is always above the lower ion cyclotron frequency Ω_2, we see that such a wave can only ever encounter one of the resonances, but not both. Which branch of the dispersion curve the wave is on, and whether the resonance is classified as the Alfvén resonance or the ion-ion hybrid resonance, again depends on the frequency and the other parameters.

Further use of this theory in the discussion of resonances in plasmas with dust grains, treated as secondary ions, is found in Chapter 7. As in the case of single ion species plasmas, the resonances in a multi–ion plasma are removed if kinetic theory and electron inertia effects are included. Wave energy that is propagated into such resonant points is heavily absorbed or mode converted to short–wavelength modes, and may contribute to the heating of fusion plasmas. We briefly mention here some results of calculations for multi–ion species plasmas. At the ion-ion hybrid resonance, in the case of heavy minority species, only electron heating is found to be possible (via mode conversion into ion-Bernstein waves) (Perkins 1977). However, at the Alfvén resonance, roughly equal ion and electron heating is possible (Winglee 1984); mode conversion of the magnetoacoustic wave still occurs through the Alfvén resonance, but minority ions cyclotron-resonance-damp the short–wavelength shear Alfvén wave that arises.

3.2.4   Effects of Collisions

We have seen in Section 3.2.2 that perpendicular resonances in the local approximation of the dispersion relation may be removed by thermal and electron inertia effects. Another way in which resonances are removed is by collisional processes. We investigate here, using the local approximation, the effects of collisional damping on the waves in a zero pressure plasma, with either ion-neutral collisions or resistivity.

Ion-Neutral Collisions

The inclusion of ion–neutral collisions in the Hall model of a single ion species, cold, mildly nonuniform plasma with a cold neutral component gives, from Eq. (2.126) with zero resistivity, the following local value of k_x^2:

$$k_x^2 = -k_y^2 - k_z^2 + \frac{s\omega^2\left(s\omega^2 - v_A^2 k_z^2\right)}{v_A^2\left(s\omega^2 - v_A^2 k_z^2(1 - s^2 f^2)\right)} \tag{3.39}$$

where $f = \omega/\Omega_i$. This equation differs from the collisionless result (Eq. 3.7) by the presence of the complex–valued factor s, defined in Eq. (2.104). In the sense that Eq. (3.39) gives a single value of k_x^2 for given ω and k_z, it describes a single mode, namely the fast magnetoacoustic mode.

It is evident from Eq. (2.103) that the shear Alfvén wave dispersion relation in the low–frequency limit is still independent of k_x in the presence of ion-neutral collisions. However, with finite ion cyclotron corrections, the Alfvén wave becomes the ion cyclotron mode damped by the collisions. The Alfvén resonance of the magnetoacoustic wave is also modified by the

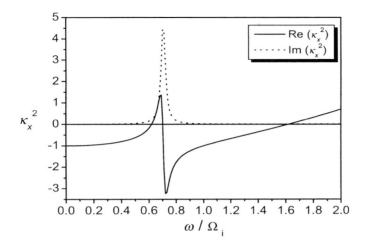

Figure 3.9: The real (solid curve) and imaginary (dotted curve) parts of the normalized square of the perpendicular wavenumber $\kappa_x^2 = k_x^2/k_z^2$ for a plasma with ion-neutral collisions, plotted against normalized frequency $f = \omega/\Omega_i$, for $\alpha = 1$, $k_y = 0$, $\rho_n/\rho_i = 0.5$ and $\nu_{ni}/\Omega_i = 0.05$.

ion-neutral collisions. If we set the denominator in Eq. (3.39) to zero,

$$s\omega^2 - v_A^2 k_z^2 \left(1 - s^2 \frac{\omega^2}{\Omega_i^2}\right) = 0 \tag{3.40}$$

we note that Eq. (3.40) cannot be satisfied, for real frequency and real k_z, at any point in the density profile, so that the real part of k_x^2 is limited in magnitude. The imaginary part of k_x^2 is a maximum near the former Alfvén resonant point, so that large damping occurs there. These features are shown in the plot of the real and imaginary parts of k_x^2 against f in Figure 3.9. Thus energy is absorbed at the former resonance by strong collisional damping of the original magnetoacoustic wave, and not by mode conversion to a second, short–wavelength mode as occurs with thermal effects (and, as is shown in the next section, with resistivity). This is reflected by the fact that the wave in a plasma that is nonuniform transverse to the magnetic field is still described by a second–order differential equation, with its singularities removed by the ion-neutral collisions, rather than a fourth–order differential equation.

The formation of the Alfvén resonance and cutoff fields has been observed in a cylindrical partially ionized plasma in experiments reported by Amagishi & Tsushima (1984). Eq. (3.40) has also been referred to as defining the dispersion relation of the "shear Alfvén wave" for the Hall plasma with ion-neutral collisions (Amagishi & Tanaka 1993); for real k_z it gives a complex ω, indicating a damped wave with frequency close to the Alfvén resonant frequency. The high perpendicular wavenumber components of a localized wave packet in such a plasma will approximately satisfy relation (3.40).

Resistivity

In a resistive plasma, but with no ion-neutral collisions, there is a qualitative difference between the low–frequency ideal Alfvén wave and the resistive Alfvén wave, in that the wave now has structure perpendicular to the magnetic field. The Alfvén resonance is again removed, because of the existence of the short perpendicular wavelength dissipative Alfvén mode discussed in Section 2.5.

We consider a cold plasma in a magnetic field, oriented in the z–direction. The plasma density and magnetic field vary with x. The resistivity is assumed isotropic ($\delta_\perp$ and $\delta_\parallel$ as defined in Eq. (2.127) are set to $\delta = \eta/\mu_0\omega$). Waves are considered with fixed frequency and wavenumbers k_y and k_z. Assuming a local approximation, at low frequency we have for the Alfvén wave, from Eq. (2.108),

$$k_x^2 = -k_y^2 - k_z^2 + i\frac{(\omega^2 - v_A^2 k_z^2)}{\omega^2\delta} \tag{3.41}$$

and for the magnetoacoustic wave,

$$k_x^2 = -k_y^2 - k_z^2 + \frac{\omega^2}{v_A^2 - i\omega^2\delta}. \tag{3.42}$$

The Alfvén wave is therefore, for small resistivity, a short scale–length (in the x–direction) overdamped mode. On either side of the Alfvén resonance position it has small scale–length perpendicular to the magnetic field, just as in the case of small thermal or electron inertia effects leading to the KAW and IAW modes. There is no Alfvén resonance in the magneto-acoustic dispersion equation (3.42).

Inclusion of the Hall term leads to a coupling of the resistive Alfvén and magnetoacoustic modes, described by the following quadratic equation in $k_\perp^2 = k_x^2 + k_y^2$, derived from Eq. (2.126):

$$Ak_\perp^4 + Bk_\perp^2 + C = 0 \tag{3.43}$$

with

$$A = \omega^2(\omega^2\delta + iv_A^2)\delta$$
$$B = [\omega^2(1 + ik_z^2\delta) - v_A^2 k_z^2](v_A^2 - 2i\omega^2\delta) + f^2 v_A^4 k_z^2 \tag{3.44}$$
$$C = f^2 v_A^4 k_z^4 - [\omega^2(1 + ik_z^2\delta) - v_A^2 k_z^2]^2.$$

Just as for the thermal effects, resistivity leads to the removal of the Alfvén resonance in the magnetoacoustic wave dispersion relation (3.7) in the cold Hall plasma model, via the inducement of a second mode. The real and imaginary parts of the normalized square of the perpendicular wavenumber $\kappa_\perp^2 = k_\perp^2/k_z^2$ for the two solutions of Eq. (3.43) are shown in Figure 3.10 plotted against f in the vicinity of the Alfvén resonance point, for the case $\alpha = 1$. A suitable dimensionless parameter involving the resistivity in the Hall model is $r = \mu_0 v_A^2/\eta\Omega_i = \alpha R_m$, if the magnetic Reynolds number R_m (Eq. 2.112) is defined with the scale–length $L = 1/k_z$. In the case of Figure 3.10, $r = 100$. We note that the solution $\kappa_{\perp 1}^2$

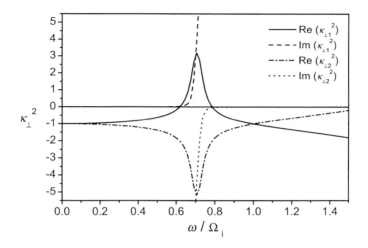

Figure 3.10: The real and imaginary parts of the normalized square of the perpendicular wavenumber $\kappa_\perp^2 = k_\perp^2/k_z^2$ for a cold plasma with resistivity, plotted against normalized frequency $f = \omega/\Omega_i$, for $\alpha = 1$ and $\mu_0 v_A^2/\eta\Omega_i = 100$. The two solutions for $\kappa_\perp^2$ are denoted by $\kappa_{\perp 1}^2$ and $\kappa_{\perp 2}^2$.

connects the magnetoacoustic mode on the left of the resonance to the resistive Alfvén mode on the right of the resonance, while the solution $\kappa_{\perp 2}^2$ connects the two modes in the opposite direction.

The short scale–length ($|k_\perp^2| \gg 0$) resistive Alfvén mode is obtained from Eq. (3.43) by neglecting C, which gives

$$k_\perp^2 = -k_z^2 + i\frac{(\omega^2 - (1 - f^2)v_A^2 k_z^2)}{\omega^2 \delta} \tag{3.45}$$

which is the finite ion cyclotron frequency modified version of the resistive Alfvén wave relation (3.41). At the Alfvén resonance, $B \simeq 0$, giving

$$k_\perp^2 = \pm\frac{v_A k_z^2}{\Omega_i \delta^{1/2}}\frac{1+i}{\sqrt{2}} \tag{3.46}$$

for the two modes at the resonance.

3.3 Waves in Smooth Nonuniformities

We now allow for smoothly varying plasma density or equilibrium magnetic field, by establishing differential equations describing the spatial variation of the waves of given frequency in the stratified plasma. In general, if the wavelengths are comparable to the scale–lengths and the locally uniform approximation cannot be used, these equations cannot be solved analytically, but perturbation methods may be used, or numerical solutions can be found. As in the discussion of the local approximation of the dispersion relations, we take the direction of stratification to be the x–axis, and assume a constant k_y and k_z. In this section we derive the differential equations describing the waves, using the ideal MHD model to describe pressure effects, and the collisionless cold plasma model to describe finite ion cyclotron frequency effects.

3.3.1   Ideal MHD

The first case we consider is waves in a plasma with nonzero plasma β, but at frequencies much lower than the ion cyclotron frequency of the ion species, employing the ideal MHD model. Up to this point we have only considered plasmas in which the zero-order magnetic field is uniform and directed along the z–axis. In this section we allow the equilibrium magnetic field, as well as the plasma density, to vary with the perpendicular x–coordinate. The magnetic field is assumed to lie in the y-z plane, but is otherwise allowed to vary with x in magnitude and direction:

$$\boldsymbol{B}_0(x) = B_{0y}(x)\hat{\boldsymbol{y}} + B_{0z}(x)\hat{\boldsymbol{z}}. \tag{3.47}$$

Such a magnetic field in general has an associated current density also lying in the y-z plane, with a resulting magnetic force given by Eq. (1.17). Thus a current sheet will be present in a region of rapidly varying magnetic field.

The steady–state version of the MHD equation of motion (1.12), with the additional assumption of zero gradient of the plasma convective velocity, then implies that the equilibrium particle pressure p_0 must also vary with x to produce a total pressure balance with the magnetic pressure:

$$p_{T0} = p_0 + \frac{B_0^2}{2\mu_0} = \text{constant}. \tag{3.48}$$

The plasma density ρ_0 variation is independent of the condition (3.48), but it is also allowed to vary with x, such that the zero–order adiabatic gas law (Eq. 1.13) is satisfied.

Magnetoacoustic modes localized on current sheets of nonzero width have been investigated by Hopcraft & Smith (1986), Musielak & Suess (1989) and Seboldt (1990). One study (Hopcraft & Smith 1986) examines surface waves in the vicinity of the Earth's geomagnetic current sheet that are associated with wave activity in geomagnetic substorms. It uses the Harris sheet model (Harris 1962) of the current sheet in which the magnetic field does not rotate but reverses sign by passing through zero in the centre of the sheet, where

$$B_{0z} \propto \tanh(x/h), \quad p_0 \propto \text{sech}^2(x/h). \tag{3.49}$$

The current sheet has the effective width h. To preserve the ideal MHD equilibrium given by Eq. (3.48) in this case, the particle pressure has a maximum at the sheet centre where the magnetic field goes to zero. The normalized magnetic field strength and plasma pressure profiles in the Harris model are shown in Figure 3.11, together with the corresponding plasma density profile assuming the adiabatic gas law with $\gamma = 5/3$. Another study (Musielak &

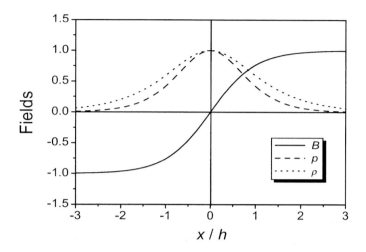

Figure 3.11: The magnetic field strength B, the plasma pressure p, and the plasma density ρ in the Harris current sheet model, all normalized to their maximum values, plotted against normalized distance.

Suess 1989) uses the cold MHD limit, $p = 0$, but considers a finite width current sheet formed by the rotation of the magnetic field vector with constant strength. The cold plasma limit requires that the magnetic field strength must remain constant, otherwise total pressure does not balance across the surface.

In this section we allow for both rotation of the field and variation of the field magnitude, as well as nonuniform density. It is inconvenient to use the characteristic variables given by Eq. (2.8) in the nonuniform plasma case. Instead, we start from the linearized MHD equation of motion (1.12) expressed in terms of the perturbed fluid velocity v, magnetic field B and (particle plus magnetic) pressure p_T:

$$\rho_0 \frac{\partial v}{\partial t} = -\nabla p_T + \frac{1}{\mu_0}(B \cdot \nabla)B_0 + \frac{1}{\mu_0}(B_0 \cdot \nabla)B. \tag{3.50}$$

We also employ Faraday's equation (1.3), combined with Ohm's law with zero resistivity (Eq. 1.8):

$$\frac{\partial B}{\partial t} = \nabla \times (v \times B_0) \tag{3.51}$$

and the linearized version of the adiabatic equation of state (1.13). For a wave of given frequency ω, we can then derive two coupled differential equations in p_T and the velocity component v_x in the stratification direction (Musielak & Suess 1989):

$$\frac{dp_T}{dx} - \tau v_x = 0 \tag{3.52}$$

$$\frac{\mathrm{d}v_x}{\mathrm{d}x} - \frac{q^2}{\tau}p_{\mathrm{T}} = 0 \tag{3.53}$$

where

$$\tau = \mathrm{i}\frac{\rho_0}{\omega}(\omega^2 - (\mathbf{k}\cdot\boldsymbol{v}_{\mathrm{A}})^2) \tag{3.54}$$

and

$$q^2 = k^2 - \frac{\omega^4}{\omega^2(c_{\mathrm{s}}^2 + v_{\mathrm{A}}^2) - c_{\mathrm{s}}^2(\mathbf{k}\cdot\boldsymbol{v}_{\mathrm{A}})^2}. \tag{3.55}$$

Here $\boldsymbol{k} = (0, k_y, k_z)$ is the constant wavevector in the plane perpendicular to the stratification, with $k = |\boldsymbol{k}|$, and $\boldsymbol{v}_{\mathrm{A}} = \mathbf{B}_0/\sqrt{\mu_0\rho_0}$ is the vector Alfvén velocity. The Alfvén wave and fast and slow magnetoacoustic waves described by the ideal MHD model in Section 2.2 are recovered from Eqs. (3.52) and (3.53) in the limit of a uniform plasma density and magnetic field (in the z–direction), in which case $q^2 = -k_x^2$ from Eq. (3.2). If the density or magnetic field varies monotonically with x, $-q^2$ has the same variation as k_x^2 shown in Figure 3.1.

The second–order differential equation for v_x is then

$$\frac{\mathrm{d}}{\mathrm{d}x}\left(\frac{\tau}{q^2}\frac{\mathrm{d}v_x}{\mathrm{d}x}\right) - \tau v_x = 0 \tag{3.56}$$

and that for p_{T} is

$$\tau\frac{\mathrm{d}}{\mathrm{d}x}\left(\frac{1}{\tau}\frac{\mathrm{d}p_{\mathrm{T}}}{\mathrm{d}x}\right) - q^2 p_{\mathrm{T}} = 0. \tag{3.57}$$

There are two possible singularities of Eqs. (3.56) and (3.57). One singularity occurs where $\tau = 0$, that is, at a point where

$$\omega = |\boldsymbol{k}\cdot\boldsymbol{v}_{\mathrm{A}}|. \tag{3.58}$$

This is equivalent to the Alfvén resonance condition (3.6).

The condition (3.58) is also said to define the *continuous shear Alfvén wave spectrum*, or the *Alfvén continuum*, in a nonuniform plasma. For a given frequency and $\boldsymbol{k}$, condition (3.58) is possibly satisfied only at isolated points in the density and magnetic field profile. However, if just $\boldsymbol{k}$ is specified, every point in the plasma profile is associated with a different frequency, given by Eq. (3.58). In a smoothly varying plasma profile we therefore have a continuous spectrum of frequencies, which formally correspond to localized shear Alfvén waves with the same wavevector.

Consider the case where the magnetic field is constant, $\mathbf{B}_0 = B_0\hat{\boldsymbol{z}}$, but the plasma density has a profile $\rho_0(x)$. The differential equations for the wave components v_y and B_y are then

$$\rho_0(x)\frac{\partial v_y}{\partial t} = -\mathrm{i}k_y p_{\mathrm{T}} + \frac{B_0}{\mu_0}\frac{\partial B_y}{\partial z} \tag{3.59}$$

$$\frac{\partial B_y}{\partial t} = B_0\frac{\partial v_y}{\partial z}. \tag{3.60}$$

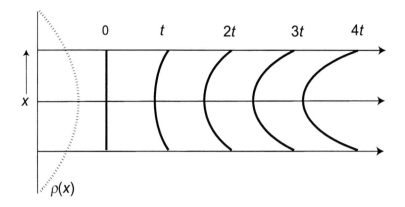

Figure 3.12: The distortion of the wavefront of an Alfvén wave propagating along the magnetic field in the z–direction, in a density profile $\rho(x) \propto 1 - x^2$, at four times after launch.

Thus if $k_y = 0$, we have a shear wave solution with nonzero v_y and B_y, uncoupled from the magnetoacoustic wave variable p_T. The shear wave fields both have arbitrary x–dependence, and upon Fourier transforming are related by

$$v_y = -\frac{\omega}{k_z B_0} B_y. \tag{3.61}$$

The frequency is spatially varying in the x–direction for a fixed wavenumber in the z–direction, and vice versa:

$$\omega = \frac{B_0 k_z}{\sqrt{\mu_0 \rho_0(x)}}. \tag{3.62}$$

Thus a shear wave wavefront that is initially flat in the x-y plane becomes distorted as each point on the front propagates along the magnetic field with its local Alfvén speed. This is shown in Figure 3.12, where the wavefront of an Alfvén wave propagating along the magnetic field in a plasma with a parabolic density profile is shown at four times after its launch. However, if oblique propagation is allowed ($k_y \neq 0$), or the magnetic field is not uniform, the shear wave is coupled to the magnetoacoustic wave. The existence of the continuous spectrum is also related to a *phase mixing* in time, leading to the decay of the wave fields (see Chapter 8).

A second singularity of Eq. (3.56) or Eq. (3.57) occurs where $1/q^2 = 0$, that is, at a position in the density and magnetic field profile where the following condition holds:

$$\omega^2 = \frac{c_s^2 (\mathbf{k} \cdot \mathbf{v}_A)^2}{(v_A^2 + c_s^2)}. \tag{3.63}$$

This condition is equivalent to the cusp or compressive resonance condition (3.5).

We shall use the differential equations (3.52) and (3.53), or (3.57), with the two singularities corresponding to the two types of wave resonances, to investigate waves in the nonuniform plasma, in particular the Alfvén resonance absorption process in the ideal MHD model, and in Chapter 4 in our discussion of surface waves localized on current sheets in that model.

3.3.2 Cold Plasma

Before proceeding to our discussion of Alfvén resonance absorption, we consider the other important model for which we can derive a second–order differential equation for the wave fields, namely the cold plasma model. The wave frequency range in this model is well below the electron cyclotron frequency, but covers the cyclotron frequencies of the different ion species. In this case the magnetic field is assumed uniform, in the z–direction, but the plasma density is a function of x. Again, for the nonuniform plasma, it is found not to be helpful to use the characteristic wave variables given by Eq. (2.8) used for the uniform plasma. Instead, we derive spatial differential equations for the components of the wave electric field.

Starting from the wave equation (1.37) for the electric field, derived from Maxwell's equations, assuming a fixed wave frequency, and employing the cold plasma form of the dielectric tensor given in Eq. (2.75), we obtain the following vector wave equation:

$$\nabla \times (\nabla \times \boldsymbol{E}) = \begin{bmatrix} u_1 & iu_2 & 0 \\ -iu_2 & u_1 & 0 \\ 0 & 0 & u_3 \end{bmatrix} \cdot \boldsymbol{E}. \tag{3.64}$$

Here the scaled dielectric tensor components u_1, u_2 and u_3 defined in Eqs. (2.73) and (2.74) are local functions of x, through their dependence on the plasma density. We note that thermal effects would introduce spatial dispersion, or nonlocality, into the dielectric tensor. The right–hand side of Eq. (3.64) would then effectively contain spatial derivatives, and the spatial order of the differential equations would be raised. However, in the cold plasma case the spatial derivatives occur only on the left–hand side of Eq. (3.64).

If electron inertia, thermal effects and collisions are neglected, the component u_3 is very large in magnitude compared with u_1 and u_2, so E_z can be taken to be zero. Using Faraday's equation (1.3), we can then derive from Eq. (3.64) two differential equations in x for the wave field components E_y and B_z, assuming fixed ω, k_y and k_z:

$$\frac{\mathrm{d}E_y}{\mathrm{d}x} - \frac{k_y H}{G} E_y = i\omega \frac{G - k_y^2}{G} B_z \tag{3.65}$$

$$\frac{\mathrm{d}B_z}{\mathrm{d}x} + \frac{k_y H}{G} B_z = \frac{i}{\omega} \frac{G^2 - H^2}{G} E_y. \tag{3.66}$$

The other electric field component is given by

$$E_x = -i\frac{\omega}{G} \left(\frac{H}{\omega} E_y - i k_y B_z \right). \tag{3.67}$$

In Eqs. (3.65)-(3.67), G and H (defined in Eq. (2.79)) are now functions of x through their dependence on v_A. Substituting the dependence $\exp(ik_x)$ into Eqs. (3.65) and (3.66), we obtain the local dispersion relations for the Alfvén and fast magnetoacoustic waves in a mildly nonuniform cold plasma, modified by ion cyclotron effects, discussed in Sections 3.2.2 and 3.2.3.

The second–order differential equation for B_z, derived from Eqs. (3.65) and (3.66), may be written

$$\frac{\mathrm{d}^2 B_z}{\mathrm{d}x^2} - \frac{L'}{L}\frac{\mathrm{d}B_z}{\mathrm{d}x} + \left[k_y \frac{G^2 L}{H^2}(HL)' - k_y^2 \left(\frac{H^2}{G^2} - L \right) + \frac{L}{G} \right] B_z = 0 \tag{3.68}$$

Figure 3.13: A magnetoacoustic wave propagates in a plasma of varying density, and is absorbed at the Alfvén resonance, leading to local heating of the plasma.

where

$$L = \frac{G^2 - H^2}{G} \tag{3.69}$$

and $L' = \mathrm{d}L/\mathrm{d}x$. A singularity of Eq. (3.68) occurs where $G = 0$. For a plasma with a single ion species, this singularity occurs at the point where the generalized Alfvén resonance condition (3.9) is satisfied.

For a wave of fixed frequency in a two ion species plasma, the singularity occurs where the resonance condition (3.26) is satisfied. There is still a single singularity, occurring at the point where the Alfvén speed, and thus the plasma density, satisfies Eq. (3.26), but the resonance can be interpreted as an Alfvén resonance or an ion-ion hybrid resonance, depending on the conditions, as discussed in Section 3.2.3. Eqs. (3.65) and (3.66), or Eq. (3.68), are used in the following sections to describe waves in a cold plasma of nonuniform density, in a frequency range covering the cyclotron frequencies of the ions.

3.4 Alfvén Resonance Absorption

An important effect of a smooth nonuniformity of the plasma on the propagation of the fast magnetoacoustic wave is Alfvén resonance absorption, shown schematically in Figure 3.13. As mentioned in the Introduction, this process has been proposed as a means of providing supplementary heating of fusion plasmas, and as a possible explanation of the heating of the corona of the Sun and other stars. We shall see that the resonance absorption disappears in the two limits of (i) a very sharp discontinuity of the plasma, and (ii) a uniform plasma.

Several approaches have been used to calculate the damping effect of resonance absorption of fast magnetoacoustic waves, including a relatively simple perturbation approach suitable for sharp transitions, and an analytic method, both of which are considered in this section.

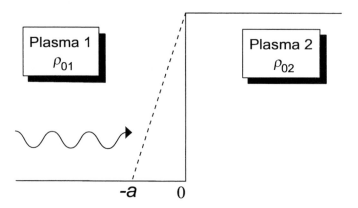

Figure 3.14: A density jump between two regions of uniform plasma density, at the point $x = 0$. The jump is smoothed out by a density ramp starting at $x = -a$. A wave is incident from the left–hand side.

Another approach, the numerical solution of fluid or kinetic differential equations, is discussed in Chapter 6 in the context of the modelling of laboratory experiments.

The damping of the magnetoacoustic wave at the Alfvén resonance can be pictured as arising from coupling of the wave with the continuous spectrum of the shear Alfvén wave. If thermal or electron inertia effects are taken into account, mode conversion into the KAW or IAW modes occurs at the resonance. With resistivity included, mode conversion into the short–wavelength resistive mode occurs. To describe mode conversion, the fourth–order differential equations governing the coupling of the magnetoacoustic mode with the short–wavelength modes must be solved, which usually requires numerical techniques (see Chapter 6).

3.4.1   Narrow Interfaces

Cold Plasma

We first consider the problem of reflection and transmission of a fast magnetoacoustic wave at a discontinuity between two plasmas of different uniform densities at $x = 0$, using the cold plasma model of Section 3.3.2 and assuming a single ion species. The magnetic field is assumed to be uniform and in the z–direction. The wave propagates obliquely to the magnetic field, with fixed wavenumber components k_y and k_z. The plasma occupying the half–space $x < 0$ is labelled as plasma 1, with uniform density ρ_{01} and Alfvén speed v_{A1}, and the plasma occupying the half–space $x > 0$ is labelled as plasma 2, with uniform density ρ_{02} and Alfvén speed v_{A2} (see Figure 3.14).

In plasma 1, a wave with positive x-wavenumber k_{x1} satisfying the dispersion equation (3.25), and with unit amplitude of E_y, is incident on the discontinuity. The wave is partly reflected with amplitude R and negative x–wavenumber, $-k_{x1}$, and partly transmitted with amplitude T and real and positive x–wavenumber k_{x2}, where k_{x2} satisfies Eq. (3.25) in plasma 2. The theory also allows for an imaginary value of k_{x2}, in which case the wave is cut off in plasma 2, with an exponential decay $\exp(-|k_{x2}|)$ of the field amplitudes. Note that there is only a single value of k_x^2 given by Eq. (3.25), in contrast to the two values given by the dispersion equations (3.11) and (3.18) due to a combination of thermal and ion cyclotron effects. This means that there is no mode conversion of the magnetoacoustic wave into reflected and transmitted short–wavelength modes on incidence at the discontinuity, as would be the case if Eq. (3.11) or Eq. (3.18) were used.

The wave fields E_y and B_z are tangential to the discontinuity plane, and so are continuous at $x = 0$. These continuity conditions imply that

$$1 + R = T$$
$$r_{1+} + Rr_{1-} = Tr_2 \tag{3.70}$$

where the ratios $r = B_z/E_y$ for the incident, reflected and transmitted waves can be derived from Eq. (3.65) to give

$$r_{1\pm} = \frac{G_1\,(\pm k_{x1} + ik_y H_1/G_1)}{w(G_1 - k_y^2)}, \quad r_2 = \frac{G_2\,(k_{x2} + ik_y H_2/G_2)}{(G_2 - k_y^2)} \tag{3.71}$$

with

$$G_{1,2} = \frac{\omega^2}{v_{A1,2}^2}\frac{1}{1-f^2} - k_z^2, \quad H = f\frac{\omega^2}{v_{A1,2}^2}\frac{1}{1-f^2} \tag{3.72}$$

and $f = \omega/\Omega_i$.

The resulting transmission coefficient is

$$T = \frac{r_{1+} - r_{1-}}{r_2 - r_{1-}}. \tag{3.73}$$

In the limit $\omega \ll \Omega_i$, this becomes

$$T = \frac{G_1 k_{x2}(k_{x1} + k_{x2})}{G_1 k_{x2}^2 + G_2 k_{x1}^2} \tag{3.74}$$

where in this case

$$k_{x1,2}^2 = \frac{\omega^2}{v_{A1,2}^2} - k_y^2 - k_z^2, \quad G_{1,2} = \frac{\omega^2}{v_{A1,2}^2} - k_z^2. \tag{3.75}$$

We now let the density discontinuity be replaced by a narrow smooth transition between the two constant density plasmas, of width a, as shown in Figure 3.14. The region $x < -a$ is occupied by plasma 1, and the region $x > 0$ by plasma 2. The transition region is assumed to be much narrower than the wavelengths in the y-z plane, that is, such that $|k_y a| \ll 1$ and $|k_z a| \ll 1$. The boundary conditions for the wave fields are that E_y and B_z are continuous across both boundaries $x = 0$ and $x = -a$.

An approximate solution of Eqs. (3.65) and (3.66) can be obtained within the narrow plasma transition region by a perturbation technique (Ginzburg 1964, Stepanov 1965, Cramer

& Donnelly 1983). Thus the wave fields E_y and B_z are expanded in series using the small parameter $\varepsilon = k_z a$;

$$E_y = E_0 + \varepsilon E_1 + \dots \quad \text{and} \quad B_z = k_z (\psi_0 + \varepsilon \psi_1 + \dots) \tag{3.76}$$

where E_0 and $B_{z0} = k_z \psi_0$ are the wave fields at $x = 0$ in the discontinuous case, and the terms of order ε are the corrections to the wave fields due to the nonzero width of the transition region. The reflection and transmission coefficients are then changed to order ε:

$$R = R_0 + \varepsilon R_1 \quad \text{and} \quad T = T_0 + \varepsilon T_1 \tag{3.77}$$

where R_0 and T_0, given by Eqs. (3.70) and (3.73), correspond to the discontinuous case.

Changing to a normalized space variable $\bar{x} = x/a$ (of order unity within the transition), substituting Eq. (3.76) in Eqs. (3.65) and (3.66) and collecting terms of the same order in ε gives, up to first order in ε,

$$\frac{dE_0}{d\bar{x}} = 0, \quad \frac{d\psi_0}{d\bar{x}} = 0 \tag{3.78}$$

$$\frac{dE_1}{d\bar{x}} - \frac{k_y H}{k_z G} E_0 = i\omega \frac{G - k_y^2}{G} \psi_0 \tag{3.79}$$

$$\frac{d\psi_1}{d\bar{x}} + \frac{k_y H}{k_z G} \psi_0 = \frac{i}{\omega} \frac{G^2 - H^2}{k_z^2 G} E_0 \tag{3.80}$$

where now G and H are functions of $\bar{x}$. Integrating Eqs. (3.79) and (3.80) through the transition from plasma 2 to plasma 1, and noting that $E_0 = T_0$ and $\psi_0 = r_2 T_0 / k_z$, we have

$$E_1(\bar{x}) = \frac{1}{k_z} T_0 \int_0^{\bar{x}} \left[\frac{k_y H}{G} + \frac{i\omega(G - k_y^2) r_2}{G} \right] d\bar{x}' \tag{3.81}$$

$$\psi_1(\bar{x}) = \frac{1}{k_z^2} T_0 \int_0^{\bar{x}} \left[\frac{i(G^2 - H^2)}{G} - \frac{k_y H r_2}{G} \right] d\bar{x}'. \tag{3.82}$$

Continuity of E_y and B_z at $x = -a$ then gives, at first order in ε,

$$\begin{aligned} \varepsilon R_1 &= \varepsilon T_1 + \varepsilon E_1(-1) \\ \varepsilon R_1 r_{1-} &= \varepsilon T_1 r_2 + \varepsilon k_z \psi_1(-1). \end{aligned} \tag{3.83}$$

Solving for T_1 gives

$$T_1 = \frac{k_z \psi_1(-1) - r_{1-} E_1(-1)}{r_{1-} - r_2}. \tag{3.84}$$

The integrands in Eqs. (3.81) and (3.82) may contain poles where $G = 0$, that is, where the resonance condition is satisfied. The integrations may be carried out using standard techniques described by Ginzburg (1964) in the context of the similar problem of the resonance absorption of radio waves at the plasma frequency in the ionosphere: a small imaginary part is introduced in G due to a small amount of dissipation, the path of integration is distorted in the complex x-plane to avoid the singularity, and the dissipation is then allowed to approach zero. The result is that E_1 and ψ_1 acquire real and imaginary parts due to the residues of the integrals at the pole $G=0$, which are independent of the size of the small dissipation, and indeed survive in the limit of zero dissipation. These contributions to the fields lead to the damping of the wave at the resonance.

It should be noted that E_y and B_z remain finite through the transition, ensuring the validity of the expansion given by Eq. (3.76), however the fields E_x and B_y become singular at the resonance, as shown by Stepanov (1965) and Stix (1992). In this model there is no indication of how the energy in the wave is finally dissipated. The more realistic models discussed in Chapter 6 show that the energy may be absorbed via mode conversion into a short–wavelength mode, such as the kinetic Alfvén wave or the inertial Alfvén wave, and subsequent collisionless and collisional damping of those waves.

Assuming G has a linear variation at the resonance point $\bar{x}_{\mathrm{r}}$, that is,

$$G \simeq (\bar{x} - \bar{x}_{\mathrm{r}})G' \tag{3.85}$$

we find

$$E_1(-1) = \frac{\pi T_0}{k_z G'}(\omega k_y^2 r_2 + \mathrm{i}f k_y k_z^2) \tag{3.86}$$

$$k_z \psi(-1) = \frac{\pi T_0}{k_z G'}(f^2 k_z^4/\omega - \mathrm{i}f k_y k_z^2 r_2). \tag{3.87}$$

In the low–frequency limit, $f \ll 1$, we see that only E_1 contributes, and we find

$$T_1 = \frac{\pi T_0}{k_z G'}\frac{\omega k_y^2 r_2 r_{1-}}{r_{1-} - r_2}. \tag{3.88}$$

By inspection of the plot of k_x^2 against v_{A}^{-1}, or $\rho^{1/2}$ in Figure 3.4(a) for a single ion species plasma, it is apparent that, if the frequency is very low ($f \ll 1$), when the lower cutoff c_1 almost coincides with the resonance, the Alfvén resonance will only be encountered in the transition if the wave propagates from a higher density plasma 1 (to the right of the cutoff c_2 in the figure) to a lower density plasma 2 where the wave is cut off ($k_{x2}^2 < 0$) (to the left of the cutoff c_1). In that case, if the transition is discontinuous, complete reflection occurs ($|R| = 1$). However, if the transition has nonzero width, it may be shown using Eq. (3.88) that some wave energy has been absorbed at the resonance ($|R| < 1$), that is, the wave has been resonance damped. It can be seen from Figure 3.4 that at higher frequency, the wave may also be propagating in plasma 2 (between the lower cutoff c_1 and the resonance), having experienced damping at the resonance. The wave can also be incident from a lower density plasma 1 and propagate in plasma 2. The perturbation calculation of resonance damping is further employed in the discussion of surface waves in cold plasmas in Chapter 4.

Ideal MHD

The perturbation calculation of the resonance damping of waves that are described by the low–frequency but nonzero β dispersion equation (3.2), and by the differential equations (3.52) and

(3.53), proceeds in an analogous fashion to the cold plasma case. A discontinuous transition in the plasma properties at $x = 0$, as shown in Figure 3.14, is considered first. This may be a discontinuity in the density, or in the magnetic field magnitude and direction. Noting from Eq. (3.52) that

$$ik_x p_T = \tau v_x \tag{3.89}$$

holds in each plasma, we obtain the following transmission coefficient in p_T for a wave that is incident on the discontinuity from plasma 1:

$$T = \frac{2\tau_2 k_{x1}}{\tau_1 k_{x2} + \tau_2 k_{x1}}, \tag{3.90}$$

where τ is defined in Eq. (3.54) and $k_{x1,2} = |iq_{1,2}|$, with q defined in Eq. (3.55). The indices refer to plasmas 1 and 2.

Considering now a narrow transition of width a between the two plasmas (Figure 3.14), we assume that p_T and v_x may be expressed as series in the small parameter $\varepsilon = k_z a$:

$$\begin{aligned} p_T &= p_0 + \varepsilon p_1 + \dots \\ v_x &= v_0 + \varepsilon v_1 + \dots \end{aligned} \tag{3.91}$$

where p_0 and v_0 are the wave fields at the transition in the discontinuous case, and εp_1 and εv_1 are the corrections to the wave fields due to the nonzero width of the transition. Substituting in Eqs. (3.52) and (3.53) and collecting terms of the same order in ε, we have the equations

$$\frac{\mathrm{d}p_0}{\mathrm{d}\bar{x}} = 0, \quad \frac{\mathrm{d}v_0}{\mathrm{d}\bar{x}} = 0 \tag{3.92}$$

$$\frac{\mathrm{d}p_1}{\mathrm{d}\bar{x}} - \frac{\tau}{k_z} v_0 = 0, \quad \frac{\mathrm{d}v_1}{\mathrm{d}\bar{x}} - \frac{q^2}{k_z \tau} p_0 = 0 \tag{3.93}$$

where $\bar{x} = x/a$.

The first–order solutions in the transition region are then found by integrating Eqs. (3.93) from plasma 2:

$$p_1(\bar{x}) = \frac{v_0}{k_z} \int_0^{\bar{x}} \tau \mathrm{d}\bar{x}', \quad v_1(\bar{x}) = \frac{p_0}{k_z} \int_0^{\bar{x}} \frac{q^2}{\tau} \mathrm{d}\bar{x}'. \tag{3.94}$$

Now applying the boundary conditions of continuous p_T and v_x at $x = -a$, we find the correction to the transmission coefficient,

$$\begin{aligned} T_1 &= \frac{i(v_1(-1) + ik_{x1}p_1(-1)/\tau_1)}{k_{x1}/\tau_1 + k_{x2}/\tau_2} \\ &= \frac{iT_0}{k_z(k_{x1}/\tau_1 + k_{x2}/\tau_2)} \int_0^{-1} \left(\frac{q^2}{\tau} - \frac{k_{x1}k_{x2}}{\tau_1 \tau_2} \tau \right) \mathrm{d}\bar{x} \end{aligned} \tag{3.95}$$

where T_0 is the transmission coefficient given by Eq. (3.90) in the discontinuous case.

As in the cold plasma case, T_1 has contributions from the residues at the poles of the integrand in Eq. (3.95). The singularities occur at points where $\tau = 0$, that is, at the Alfvén resonance defined by Eq. (3.6), or where $1/q^2 = 0$, that is, the cusp resonance defined by Eq. (3.5). Inspecting the plot of k_x^2 against v_A^{-1} in Figure 3.1, we see that a propagating wave incident from a higher density plasma 1 (to the right of the cutoff c_2) that is cut off in the lower density plasma 2, will experience Alfvén resonance damping if plasma 2 lies between the cutoffs c_1 and c_2, or both Alfvén and cusp resonance damping if plasma 2 lies to the left of the cusp resonance r_C. On the other hand, a wave that is propagating in plasma 2 (between r_C and c_1) will experience Alfvén resonance damping only. Notice that if $k_y = 0$, then $q^2 \propto \tau$, and the singular contribution in Eq. (3.95) due to the Alfvén resonance disappears. Similarly, T_1 given by Eq. (3.88) is zero if $k_y = 0$. The absence of Alfvén resonance damping in the case $k_y = 0$ is due to the decoupling of the shear and magnetoacoustic waves discussed in Section 3.3.1.

3.4.2   Analytic Derivation

We now reconsider a plasma with a narrow smooth transition in the vicinity of the Alfvén and compressive resonances, and discuss a more rigorous approach to calculating the resonant absorption of wave energy. The ideal MHD model is used, although similar results can be obtained for the cold Hall plasma model. The equilibrium magnetic field is assumed for simplicity to remain in a single direction, $\boldsymbol{B}_0 = B_0(x)\hat{\boldsymbol{z}}$. An analytic solution for resonance absorption can be obtained (Wentzel 1979a) if we assume a linear variation of τ within the transition region of width a, that is,

$$\frac{\mathrm{d}\tau}{\mathrm{d}x} = \tau' = \frac{\tau_2 - \tau_1}{a}. \tag{3.96}$$

τ' is assumed to be a positive constant, and τ_1 and τ_2 are the values of τ on the side of the transition with higher and lower Alfvén speed respectively. A new space variable ζ is introduced:

$$\zeta = k_y \tau / \tau'. \tag{3.97}$$

We note that as ζ is of the order $k_y a$, then $|\zeta| \ll 1$ in the range of interest. Eq. (3.57) is the more convenient differential equation to deal with, and becomes

$$\frac{\mathrm{d}^2 p_T}{\mathrm{d}\zeta^2} - \frac{1}{\zeta}\frac{\mathrm{d}p_T}{\mathrm{d}\zeta} - [1 + C(\zeta)]p_T = 0 \tag{3.98}$$

where

$$C(\zeta) = q^2/k_y^2 - 1 = -\frac{\tau}{k_y^2(v_A^2 + c_s^2)}\frac{\omega^2 - c_s^2 k_z^2}{\tau - \tau_s}. \tag{3.99}$$

Here τ_s, corresponding to ζ_s, gives the point in the transition at which a cusp or compressive resonance, that is, a singularity in q^2, occurs:

$$\tau_s = -\mathrm{i}\rho_0 \omega v_A^2 / c_s^2. \tag{3.100}$$

If c_s is assumed constant, we have

$$C(\zeta) \propto \frac{\zeta}{\zeta - \zeta_s}. \tag{3.101}$$

Equation (3.98) involves either one or two regular singular points within the transition. The Alfvén singularity occurs where $\tau = 0$, so that $C(\zeta) = 0$ there. At that point the two linearly independent solutions of Eq. (3.98) are

$$\xi_1 = \zeta K_1(\zeta) \simeq 1 + \frac{1}{2}\zeta^2 (\ln \zeta - \frac{1}{2}) \tag{3.102}$$

$$\xi_2 = \zeta I_1(\zeta) \simeq \frac{1}{2}\zeta^2 \tag{3.103}$$

where I and K are modified Bessel functions, which have been expanded for small ζ. Note that p_T and v_z remain finite at the Alfvén singularity, but v_x and v_y become proportional to $\ln \zeta$ and ζ^{-1} respectively as $\zeta \to 0$ (from the solution ξ_1).

At the compressive singular point, we have

$$C(\zeta) = \frac{\zeta_s}{\zeta - \zeta_s} C_s^2 \tag{3.104}$$

where

$$C_s^2 = \frac{k_z^2}{k_y^2} \frac{c_s^2}{v_A^2 + c_s^2}. \tag{3.105}$$

The leading terms for the two independent solutions in the vicinity of the compressive singularity are

$$\phi_1 = 1 + C_s^2 \zeta_s (\zeta - \zeta_s)[\ln(\zeta - \zeta_s) - 1] + \frac{1}{2}C_s^2(\zeta - \zeta_s)\left[\ln(\zeta - \zeta_s) - \frac{3}{2}\right] \tag{3.106}$$

and

$$\phi_2 = \zeta - \zeta_s + \frac{1}{2}(\zeta - \zeta_s)^2/\zeta_s. \tag{3.107}$$

At this singularity, v_x becomes proportional to $\ln(\zeta - \zeta_s)$ and v_z to $(\zeta - \zeta_s)^{-1}$, and v_y remains finite. When both singularities are present in the transition, the constant C_s is modified (Wentzel 1979a).

The logarithmic terms such as $\ln \zeta_1$ in ξ_1 make the solution multi–valued. A Riemann surface is used to analytically continue the logarithmic function across the singularity, with the effect of replacing the logarithmic terms in ξ_1 by $\pm i\pi$ (see Ionson (1978) and Wentzel (1979a) for more details). The appearance of i in the solution implies a complex frequency and a temporal damping of the wave if an eigensolution in a bounded plasma is sought, or if surface waves are being considered (see Chapter 4), or a spatial damping if waves of fixed frequency are launched towards the resonances, as we found in the previous section.

Other analytic approaches include calculations of resonant damping in a resistive plasma, using analytic connection formulae at the resonance point to avoid the need to solve the fourth–order resistive equations (Sakurai, Goossens & Hollweg 1991a; Goossens, Ruderman & Hollweg 1995), and this approach has been generalized to include equilibrium fluid flow along the magnetic field in the surface (Goossens, Hollweg & Sakurai 1992).

4 Surface Waves

4.1 Introduction

In this chapter we consider the theory of a particular type of wave in strongly nonuniform plasmas, the Alfvén surface wave, with an emphasis on the inclusion of nonideal effects. The theory of linear surface waves on sharp surfaces is developed from the various fluid plasma models, as well as from the kinetic theory. We review the theory of the waves in ideal plasmas and in plasmas with nonideal but nondissipative effects included, such as finite ion cyclotron frequency, multiple ion species and dust in the plasma. The damping of the surface waves due to fluid dissipative effects such as resistivity and viscosity, and due to collisionless processes such as Landau damping, is then considered, as well as the radiative leakage of wave energy due to coupling into short–wavelength modes. The properties of the waves on nonzero–width surfaces are also considered. In particular, the Alfvén resonance absorption of the waves is analysed using the perturbation theory developed in Chapter 3, and numerical results are discussed. The theory of nonlinear Alfvén surface waves is briefly discussed in Chapter 5.

When there are nonuniformities in the plasma, wave modes can exist with their amplitude a maximum in the nonuniform regions. In the limit where the nonuniformity is sufficiently sharp, thus defining a surface, the possibility arises of low–frequency wave modes with wave fields decaying exponentially in each direction away from the surface. These waves, which are propagating within the surface but are localized in the directions perpendicular to the surface, are known as surface waves and can exist in all highly structured plasma media (Cramer 1995). Given that in reality all plasmas are bounded, we would expect that surface waves may play an important role in a number of naturally occurring and laboratory plasmas. The Alfvén surface wave eigenmodes have been shown by theory and experiment (e.g. Cross 1988) to play an important role in Alfvén wave heating, because they can be easily excited by an antenna in a laboratory fusion plasma that is separated from the vessel walls by a low-density region. Evidence for Alfvén surface waves in the laboratory was first provided by Lehane & Paoloni (1972b), and the idea of Alfvén resonance heating of fusion plasmas using surface wave eigenmodes was suggested by Chen & Hasegawa (1974a).

Alfvén surface waves are also expected to exist in astrophysical plasmas where jumps in density or magnetic field occur, such as the surfaces of magnetic flux tubes in the solar and stellar atmospheres, or the boundaries between plasmas of different properties in the solar wind and the Earth's magnetosphere. The surface waves may be excited by movement of the footpoints of the flux tubes or by instabilities on the plasma boundaries, and may play a role in the heating of the solar corona (Wentzel 1979b). If the surface is not perfectly sharp, but has a nonzero transition width, Alfvén (or compressive) resonance absorption may occur, leading to

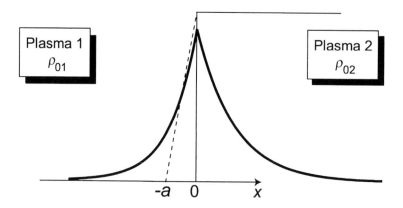

Figure 4.1: The exponentially decreasing fields of a surface wave localized at a density jump between two regions of uniform plasma density, at the point $x = 0$. The jump may be smoothed out by a density ramp starting at $x = -a$.

the ultimate transfer of mechanical energy into heating of the corona. The resonance damping of surface waves at frequencies approaching the ion cyclotron frequency has been discussed by Cramer & Donnelly (1983), and an interesting result is that surface waves may exist with no Alfvén resonance absorption, in contrast to the low–frequency result. Surface waves have been discussed by Wessen & Cramer (1991) for the case of a rotation of the magnetic field across the surface. The role of surface waves in reconnection processes and the resonant absorption of Alfvén waves near a neutral point has been investigated by Uberoi (1994) and Uberoi, Lanzerotti & Wolfe (1996).

We first investigate surface waves at density jumps, assuming the equilibrium magnetic field is uniform, using the cold plasma and ideal MHD models. Then a surface is considered in which the magnetic field varies, forming a current sheet, and again the cold plasma and ideal MHD models are used to derive the surface wave properties. Resonant damping of the waves is considered, as well as collisional and radiative damping. Finally, the kinetic theory of surface waves is briefly reviewed. Only planar geometry is considered in this chapter. Surface waves in cylindrical geometry are considered in Chapters 6 and 7.

4.2   Surface Waves at Density Jumps

Here we derive the dispersion relation for a narrow but nonzero width transition, and obtain the result for a discontinuous density jump in the limit where the width goes to zero. The magnetic field is assumed to be uniform, with $\boldsymbol{B}_0 = B_0 \hat{\boldsymbol{z}}$. The cold plasma model and ideal MHD model are treated in turn. As in Section 3.4.1, a narrow transition of width a between two plasmas of different densities, forming a surface in the y-z plane, is assumed (see Figure 4.1 and Section 3.4.1 for the notation used). The surface is narrow in the sense that $\varepsilon = |k_z a| \ll 1$. In contrast to the transmission and reflection problem of Section 3.4.1, the wave fields are now assumed to have an imaginary k_x in the uniform plasmas on each side of the surface, with an exponential decay away from the surface. The resulting surface wave is thus localized at the transition surface, as shown in Figure 4.1.

4.2.1 Cold Plasma

If there is a cold plasma with a number of ion species on each side of the surface, the dispersion equation for the wave on either side of the surface, with frequency much lower than the electron cyclotron frequency, is given by Eq. (3.25). However, the surface wave has an imaginary k_x on both sides of the surface for real ω, k_z and k_y. We first discuss the case of a cold plasma of uniform density in the half–space $x < -a$, with a narrow interface to a vacuum in $x > 0$.

The general boundary conditions on the electric and magnetic fields are that the tangential components E_y, E_z, B_y and B_z are continuous at each boundary $x = 0$ and $x = -a$. The fields in the plasma correspond to the uniform plasma wave modes discussed in Chapter 2, and use of the boundary conditions at the surface results in a relation between ω, k_z and k_y, which constitutes the surface wave dispersion relation. The properties of surface waves of frequency less than the ion cyclotron frequency are predominantly determined by the fast magnetoacoustic mode discussed in Chapter 2. The short–wavelength mode that arises from nonideal corrections such as thermal corrections or resistivity, is needed for a correct evaluation of the wave frequency and damping and to satisfy the appropriate continuity conditions on $\boldsymbol{B}$ and $\boldsymbol{E}$ at the surface, as is discussed in Section 4.7. As a first approximation however, we assume that thermal effects, electron inertia, and resistivity are neglected. Then just the magnetoacoustic mode described by Eq. (3.25) in the plasma is involved. The boundary conditions that E_y and B_z are continuous at $x = 0$ and $x = -a$, are then sufficient to solve for the single modes on either side of the surface: the component E_z only arises from the short–wavelength mode, so it can be neglected.

The surface wave solutions must have $k_x^2 < 0$ in the uniform plasma region $x < -a$, that is, the wave fields there vary as $\exp(k_p x)$, where $k_p = |k_x|$. It follows from Eq. (3.65) that B_z and E_y in the plasma are related by

$$i\omega B_z = \frac{i k_p G - k_y H}{G - k_y^2} E_y. \tag{4.1}$$

In the vacuum region ($x > 0$) the wave fields can be expressed in terms of plane wave solutions of the vacuum Maxwell's equations, and vary as $\exp(-k_v x)$ where, because the phase velocity is assumed $\ll c$, we have to a good approximation

$$k_v^2 = k_y^2 + k_z^2. \tag{4.2}$$

The vacuum fields have the property

$$i\omega B_z = \frac{k_z}{(k_y^2 + k_y^2)^{1/2}} E_y. \tag{4.3}$$

The dispersion relation for the surface wave is found by requiring that the homogeneous plasma and vacuum solutions for E_y and B_z be matched with the solution inside the nonzero–

width surface. The solution of the wave differential equations (3.65) and (3.66) in the transition region is obtained by the same perturbation technique introduced in Chapter 3 for the transmission and reflection problem. The wavelength in the plane of the surface is assumed to be much larger than the width of the surface transition, so that the wave fields may be written as the series given in Eq. (3.76) in the small parameter ε, resulting in the zero–order and first–order (in ε) differential equations (3.78)–(3.80) in the normalized space variable $\bar{x} = x/a$. The first–order equations (3.79) and (3.80) are integrated from the vacuum at $x = 0$, where (using Eq. (4.3)),

$$E_y = E_0 = E_{\rm v}, \quad \psi_0 = -\frac{i}{\omega k_{\rm v}} E_0. \tag{4.4}$$

The first–order fields at a point $\bar{x}$ in the surface are then given by Eqs. (3.81) and (3.82), but with T_0 replaced by $E_{\rm v}$. The fields in the plasma region $x \le -a$ are given by

$$E_y = E_{\rm p} \exp(k_{\rm p} x) \tag{4.5}$$

$$B_z = B_{\rm p} \exp(k_{\rm p} x) = \frac{i}{\omega} \frac{G_{\rm p} k_{\rm p} - k_y H_{\rm p}}{k_y^2 - G_{\rm p}} E_{\rm p} \exp(k_{\rm p} x) \tag{4.6}$$

where $G_{\rm p}$ and $H_{\rm p}$ are G and H evaluated at $x = -a$.

Continuity at $x = -a$ then gives, to first order in ε,

$$\begin{aligned}
E_{\rm p} \exp(-k_{\rm p} a) &= E_{y0} + \varepsilon E_{y1}(-1) \\
B_{\rm p} \exp(-k_{\rm p} a) &= k_z (\psi_0 + \varepsilon \psi_1(-1)).
\end{aligned} \tag{4.7}$$

Substituting Eqs. (3.81) and (3.82) into Eq. (4.7), and using Eqs. (4.5) and (4.6), gives the dispersion equation expressed as the first two terms of an expansion in ε:

$$\mathcal{D}(\tilde{\omega}) = \mathcal{D}_0(\tilde{\omega}) + \varepsilon \mathcal{D}_1(\tilde{\omega}) = 0 \tag{4.8}$$

with $\tilde{\omega} = \omega - i\gamma$, where ω is the frequency and the damping rate γ is of order $\varepsilon \omega$. Here

$$\mathcal{D}_0(\tilde{\omega}) = \frac{k_{\rm p} G_{\rm p} - k_y H_{\rm p}}{G_{\rm p} - k_y^2} + \frac{k_z^2}{k_{\rm v}} \tag{4.9}$$

and $\mathcal{D}_1$ is the correction:

$$\mathcal{D}_1(\tilde{\omega}) = \frac{k_z}{a} \int_0^{-a} \frac{1}{G} \left[-2 \frac{k_y}{k_{\rm v}} H + \frac{G^2 - H^2}{k_z^2} + \frac{k_z^2}{k_{\rm v}^2}(G - k_y^2) \right] {\rm d}x. \tag{4.10}$$

The dispersion relation for the surface wave on a discontinuous plasma-vacuum boundary, as $\varepsilon \to 0$, is obtained by setting $\mathcal{D}_0(\omega) = 0$ in Eq. (4.9). This leads to an expression for $k_{\rm p} = k_{\rm p}^{(1)}$, which must be consistent with the expression for $k_{\rm p} = k_{\rm p}^{(2)}$ derived from Eq. (3.25), in which $k_x^2 = -k_{\rm p}^2$. Thus we find that

$$(k_{\rm p}^{(2)})^2 - (k_{\rm p}^{(1)})^2 = \frac{(G_{\rm p} - k_y^2)}{G_{\rm p}^2} \left[\frac{k_z^4}{k_{\rm v}^2}(G_{\rm p} - k_y^2) - 2\frac{k_z^2 k_y}{k_{\rm v}} H_{\rm p} + G_{\rm p}^2 - H_{\rm p}^2 \right]. \tag{4.11}$$

Setting Eq. (4.11) to zero gives solutions for the surface wave frequency ω, provided care is taken to exclude spurious solutions introduced by the squaring procedure. It is straightforward to show that the first factor on the right–hand side of Eq. (4.11) must be nonzero. Equating the second factor to zero leads to the dispersion relation for a discontinuous surface.

The damping rate of the surface wave to first order in ε, because of the nonzero width of the surface, is obtained from Eq. (4.8) as

$$\gamma = \varepsilon \frac{\mathrm{Im}\mathcal{D}_1(\omega)}{\partial \mathcal{D}_0/\partial \omega}\bigg|_{\omega=\omega_s} \tag{4.12}$$

where ω_s is the surface wave frequency obtained for the discontinuous density jump. As discussed in Section 3.4.1, the only imaginary contribution to $\mathcal{D}_1$ comes from the pole of the integrand in Eq. (4.10), that is, at that point in the surface transition where $G = 0$, which corresponds to the Alfvén resonance condition (or hybrid resonance in the case of a two ion species plasma). Thus the damping is due to resonance absorption of the surface wave, although in this dissipationless model there is no indication of how the damped energy is finally dissipated. As we have already discussed, more realistic models of resonance absorption show that the energy can be absorbed via mode conversion into a short–wavelength mode that is subsequently damped by collisional or collisionless processes.

The damping rate is found, by evaluating the contribution of the residue at the pole $G = 0$ in the integral in Eq. (4.10), to be

$$\gamma = -\varepsilon\pi \frac{(H + k_z^2 k_y/k_v)^2}{k_z(\partial \mathcal{D}_0/\partial \omega)(\mathrm{d}G/\mathrm{d}\bar{x})} \tag{4.13}$$

where $\partial \mathcal{D}_0/\partial \omega$ is evaluated using $\omega = \omega_s$, and H and $\mathrm{d}G/\mathrm{d}\bar{x}$ are evaluated at $x = x_r$, with x_r the resonant point where $G = 0$.

The surface wave dispersion relation is now investigated in more detail in the low–frequency limit. Ion cyclotron effects in a single ion species plasma, and in a two ion species plasma, are considered in the subsequent two sections.

4.2.2 Low–Frequency Surface Waves

Equation (4.9) for the dispersion function for the discontinuous density jump is valid for frequencies up to and beyond the cyclotron frequency of any ion species. At low frequency $\omega \ll \Omega_i$ in a single ion species plasma, when

$$G = \omega^2/v_A^2 - k_z^2 \quad \text{and} \quad H = 0 \tag{4.14}$$

we obtain

$$\mathcal{D}_0(\omega) = \frac{k_z^2 - \omega^2/v_A^2}{k_p} + \frac{k_z^2}{k_v}. \tag{4.15}$$

From $\mathcal{D}_0(\omega) = 0$, we then obtain the well-known dispersion relation for low–frequency surface waves for a pressureless plasma interfacing with a vacuum, referred to here as the *Alfvén surface wave* (e.g. Cross 1988),

$$\omega_s^2 = v_A^2 k_z^2 \left(\frac{2k_y^2 + k_z^2}{k_y^2 + k_z^2} \right) \tag{4.16}$$

with

$$k_p^2 = k_y^2 + k_z^2 - \frac{\omega^2}{v_A^2} = \frac{k_y^4}{k_y^2 + k_z^2}. \tag{4.17}$$

In the vacuum Eq. (4.2) holds, so that, as required for the surface wave, k_x is imaginary on both sides of the surface. In the limit $k_y \gg k_z$, that is, propagation almost perpendicular to the magnetic field, the dispersion relation is

$$\omega_s = \sqrt{2}v_A k_z. \tag{4.18}$$

This is also the dispersion relation for the surface wave in an incompressible plasma (Hasegawa & Uberoi 1982), as we see in Section 4.5.1.

The resonance damping rate in the limit of low frequency, γ_0, is given from Eq. (4.13) by

$$\gamma_0 = (\pi/8)(k_y a)\omega_s \tag{4.19}$$

if G varies linearly (Wentzel 1979a). Note that the damping rate is independent of the ultimate dissipation mechanism, a result verified when dissipation such as resistivity or mode conversion into a radiating mode is included.

The effects on the dispersion relation of finite ion cyclotron frequency and of electron inertia, electron pressure and resistivity are covered in the following sections.

4.3   Finite Ion Cyclotron Frequency Effects

In large Tokamaks the wave frequency used for Alfvén wave heating may be close to the ion cyclotron frequency, so it is important to understand the behaviour of surface waves, which are easily excited by the antennas at these frequencies. In space plasmas, waves at these frequencies may be excited by instabilities or by energetic proton or relativistic electron beams. For example, surface waves excited at the Earth's magnetopause by the Kelvin-Helmholtz instability driven by the solar wind, feed energy into field line resonances via the Alfvén resonance. The consequent shear Alfvén waves are detected at ground level as the magnetospheric pulsations (Chen & Hasegawa 1974b, Southwood & Hughes 1983), and the Pc-1 pulsations occur at frequencies comparable to the ion cyclotron frequency (see Chapter 7). In addition, MHD waves have been observed at the boundary of the plasma sheet in the Earth's magnetotail with a frequency comparable with the proton cyclotron frequency (Tsurutani *et al.* 1985), and these waves may exist as surface modes in the highly structured plasma there.

4.3.1   Surface Wave Frequency

We consider a cold plasma with a single ion species. The dispersion equation of the surface wave for a plasma-vacuum surface, gained by setting the second factor on the right–hand side

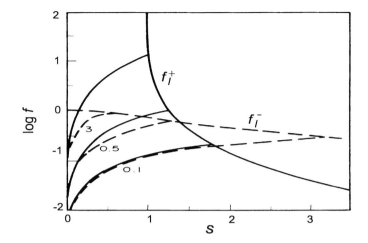

Figure 4.2: The logarithm of the normalized frequency of the surface wave in the cold Hall model, plotted against $s = |k_z/k_y|$, for $\alpha = 0.1, 0.5$ and 3. The limiting values f_l^+ and f_l^- are also shown. The solid lines correspond to $k_y > 0$ and the dashed lines to $k_y < 0$. (Reproduced from Cramer & Donnelly 1983.)

of Eq. (4.11) to zero, can be solved for the surface wave frequency (Cramer & Donnelly 1983), leading to two possible solutions,

$$f^{\pm} = \omega^{\pm}/\Omega_i = \alpha \left[\{ 2 + \alpha^2 + s^2 \}^{1/2} \pm \alpha \right] / (1 + s^2)^{1/2} \tag{4.20}$$

with $s = k_z/|k_y|$ and $\alpha = v_A |k_z|/\Omega_i$, the solution ω^+ being the solution for positive k_y and ω^- the solution for negative k_y. Defining n to be a unit vector perpendicular to the surface and pointing to the vacuum, the positive y–direction corresponds to the $B_0 \times n$ direction. The frequency can also be written, in terms of the angle of propagation θ in the surface and the magnitude of the total wavenumber in the surface k, as

$$f^{\pm} = \frac{v_A k |\cos \theta|}{\Omega_i} \left[\left\{ 1 + \sin^2 \theta + \frac{v_A^2 k^2}{\Omega_i^2} \sin^2 \theta \cos^2 \theta \right\}^{1/2} \pm |\sin \theta \cos \theta| \frac{v_A k}{\Omega_i} \right]. \tag{4.21}$$

The solutions given by Eq. (4.20) are shown in Figure 4.2 plotted against s for representative values of α. Using Eq. (4.20) or Eq. (4.21) we find that the effects of finite ion cyclotron frequency on the surface waves are as follows:

(i) The positive k_y (corresponding to positive azimuthal mode number m in cylindrical geometry) wave (the "+" mode) has a higher frequency than the negative k_y wave and the difference increases as s increases. ω^+ is an increasing function of function of s and α, but s is restricted to the range $0 < s < s_l^+$, where s_l^+ is an upper limit at which $k_x = 0$ and above which k_x^2 is positive, that is, the wave no longer has a surface wave character. The wave is predominantly right–hand polarized. There is otherwise no effect of the ion cyclotron frequency on the wave, which we term the fast surface wave.

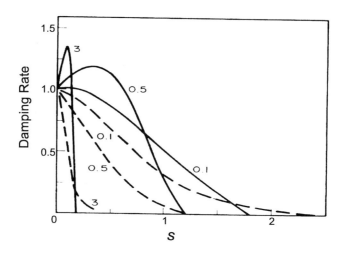

Figure 4.3: The normalized damping rate of the surface wave in the cold Hall model, plotted against s, for $\alpha = 0.1, 0.5$ and 3. The solid lines correspond to $k_y > 0$ and the dashed lines to $k_y < 0$. (Reproduced from Cramer & Donnelly 1983.)

(ii) For the negative k_y wave (the "-" mode), ω_- is again an increasing function of s and α, and s is restricted to the range $0 < s < s_l^-$, where the upper bound $s_l^- = \alpha^{-1}$. At this value of s the Alfvén resonance condition (3.9) is satisfied and $k_x \to i\infty$, and for $s > s_l^-$, k_x^2 is positive. For this mode it is always the case that $\omega < \Omega_i$. The wave is predominantly left–hand polarized. As $s \to s_l^-$ the wave takes on the character of a resonant shear Alfvén (or more generally ion cyclotron) wave localized at the surface. For this reason it has been termed the ion cyclotron surface wave (Cramer & Donnelly 1983).

A discussion of these waves in a cylindrical plasma surrounded by a vacuum with conducting walls, and their application to experimental observations, is presented in Chapter 6, and is also found in detail in Collins, Cramer & Donnelly (1984). The behaviour predicted for the surface waves has also been verified in a number of experiments in a cylindrical plasma device, for example by Amagishi, Saeki & Donnelly (1989).

4.3.2   Resonance Damping

The resonant damping rate for the surface wave in the cold Hall model, on a plasma-vacuum surface of narrow width a, is given by Eq. (4.13), which becomes

$$\gamma = -\varepsilon \pi k_z^3 \frac{[(1+s^2)^{-1/2} + \sigma f]^2}{(\partial \mathcal{D}_0/\partial \omega)(\mathrm{d}G/\mathrm{d}\bar{x})} \tag{4.22}$$

where $\sigma = \mathrm{sign}(k_y)$. Thus the effect of finite frequency is either to increase or decrease the damping rate, depending on the sign of k_y. The low–frequency damping rate (4.19) is given by Eq. (4.22) in the limit $f \to 0$. The damping rates for the full range of allowable s are shown in Figure 4.3, for three values of α. The ratio of the damping rate to the surface wave frequency, γ/ω_s, normalized to the corresponding low–frequency ratio, γ_0/ω_0, is plotted.

We note several features of these curves:

(i) For given α and small values of s, the damping rate ratio increases for the positive k_y mode, and decreases for the negative k_y mode.

(ii) For the positive k_y modes with large value of α, the damping rate goes to zero at a value of s less than the limiting value s_l^+. This behaviour is explained by noting that the frequency f can pass through a value of 1. For $f > 1$, G is always negative and there can be no resonance damping, that is, this mode of the surface wave can propagate undamped by resonant absorption.

(iii) For the negative k_y modes, the damping rate goes to zero as s approaches s_l^-, and the resonance point approaches the high–density edge of the surface.

4.4 Multiple Ion Species

If a minority ion species is introduced into a cold plasma, the components of the tensor u given by Eq. (2.73) are modified, as discussed in Chapters 2 and 3. The main effect of the additional species is to introduce a second resonance frequency into the magnetoacoustic mode wavenumber given by Eq. (3.14), called the ion-ion hybrid resonance frequency. If the minority ion is lighter than the majority ion, the hybrid resonance frequency lies above the Alfvén resonance frequency, but just below the minority ion cyclotron frequency (see Figure 3.7).

A negatively charged minority ion is considered in this section; the case of a negatively charged massive dust grain is treated in Chapter 7. Again, we consider in this section the solutions of Eqs. (3.65) and (3.66) for the case of a narrow surface separating a magnetized plasma and a vacuum.

A solution of the dispersion equation for a plasma-vacuum surface (Cramer & Yung 1986) yields little change, due to the minority species, to the positive and negative k_y surface waves at frequencies less than the majority ion cyclotron frequency, whose properties were discussed in Section 4.3. However, a new negative-k_y surface mode arises at a frequency just above the hybrid resonance frequency, as we show here. The influence of this ion-ion hybrid surface wave on experiments on the Alfvén wave heating of a plasma with impurity ions has been pointed out by Elfimov, Petrzilka & Tataronis (1994a) and Elfimov, tataronis & Hershkowitz (1994b). It may also play a role in the heating of plasmas via the mode conversion of a fast magnetoacoustic wave at the ion-ion hybrid resonance frequency into a short–wavelength electrostatic mode, with subsequent cyclotron damping or electron Landau damping (Klima, Longinov & Stepanov 1975).

Equating the second factor on the right–hand side of Eq. (4.11) to zero, and using Eqs. (2.132) and (2.133), the following quartic equation for f is obtained (Cramer, Yeung & Vladimirov 1998):

$$(1+bg)^2 f^4 - \frac{2\sigma(1+bg)}{(1+s^2)^{1/2}} \alpha^2 f^3 - \left(\frac{2+s^2}{1+s^2}(1+bg^2)\alpha^2 + g^2(1+b)^2 \right) f^2$$

$$+ \frac{2\sigma}{(1+s^2)^{1/2}} g(b+g)\alpha^2 f + \frac{2+s^2}{1+s^2} g^2(1+b)\alpha^2 = 0 \qquad (4.23)$$

where $\sigma = \text{sign}(k_y)$, and we use the notation of Sections 2.6 and 3.2.3. The expression for k_p may be obtained from Eq. (4.9) by setting $\mathcal{D}_0(\omega) = 0$, and for a valid surface wave solution, the condition $k_p > 0$ must be satisfied. This requirement leads to the existence of cutoff frequencies for the surface waves where $k_p = 0$.

It was shown in the previous section that in the case of a single ion species there is one possible positive frequency solution given by Eq. (4.20) for each sign of the wavenumber component in the $\boldsymbol{B}_0 \times \boldsymbol{n}$, or y, direction. A fast wave (f^+) exists for positive k_y, and a slow wave (f^-) for negative k_y. If the secondary ion species has a very large mass or is a dust grain, such that $|g| \ll 1$, the last term in Eq. (4.23) is zero, giving the trivial solution $f = 0$ and a cubic equation in f for the other solutions. There is again a single valid solution of that cubic equation for each sign of k_y and for given s, α and δ (Cramer & Vladimirov 1996b). However, for finite mass and therefore mobile secondary ions there can be two surface wave solutions for each sign of k_y. We consider each sign of k_y separately. The secondary ion is assumed to have the higher mass and to be negatively charged (the case of a negatively charged dust grain as the secondary ion was considered by Cramer, Yeung & Vladimirov (1998), and is mentioned further in Chapter 7).

4.4.1   Surface Wave Solutions

Positive k_y

For propagation of a surface wave in a direction intermediate between the magnetic field ($\boldsymbol{B}_0$) direction and the $\boldsymbol{B}_0 \times \boldsymbol{n}$ direction, where $\boldsymbol{n}$ is the unit vector normal to the surface and pointing to the vacuum (i.e. $k_z > 0$ and $k_y > 0$), two surface modes are found to exist simultaneously over a range of k_z. Figure 4.4 shows the frequencies of the surface waves (the solid curves) obtained from Eq. (4.23) for $g = -0.01$ and for positive k_y and $s = 1$, plotted as a function of $v_A k_z / \Omega_2$ and for two values of b, at frequencies of the order of the lower cyclotron frequency (the upper cyclotron frequency is outside the scale of the diagram). The frequencies of the cutoffs and resonances of k_x are also shown. There are two surface modes over this range of wavenumber, and the larger the relative density of the secondary ion species (the larger b), the greater is the separation in frequency between the two modes.

The lower–frequency surface wave stops at an upper frequency close to the lower ion cyclotron frequency, where a resonance in k_x occurs, and so is referred to here as the lower cyclotron surface wave. This behaviour is analogous to the stopping of the negative k_y surface wave close to the ion cyclotron frequency in a single–ion–species plasma, described in Section 4.3. For small $\alpha/|g|$, the lower cyclotron surface wave has the dispersion relation

$$\omega = \left(\frac{2+s^2}{1+s^2} \right)^{1/2} v_{AT} k_z - \frac{b}{(1+b)^2(1+s^2)^{1/2}} \frac{(v_A k_z)^2}{\Omega_2}. \qquad (4.24)$$

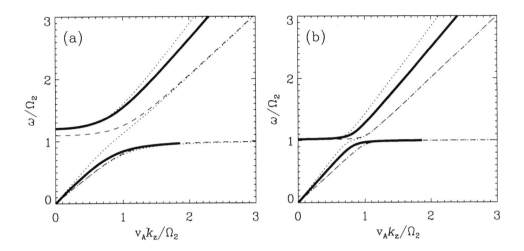

Figure 4.4: The normalized frequency of surface waves for a two ion species plasma, for positive k_y, as a function of normalized wavenumber k_z (solid curves). Cutoffs (dotted curves) and resonances (dashed curves) in the plasma are also shown. The secondary ion is negatively charged, with $g = -0.01$, and $s = 1$. (a) $b = 0.2$, (b) $b = 0.02$. (Reproduced from Cramer, Leung & Vladimirov 1998.)

In the limit $k_z \to 0$ where the quadratic term in k_z can be neglected, this mode is the usual nondispersive Alfvén surface wave (e.g. Wentzel 1979a), but in the plasma of combined mass density of the two ion species.

The higher–frequency surface wave (the upper cyclotron or fast surface wave) has the following dispersion relation for small wavenumber:

$$\omega = \Omega_m \left[1 + \frac{b}{2(1+b)^3} \frac{(1 + (1+s^2)^{1/2})^2}{1+s^2} \left(\frac{v_A k_z}{\Omega_2} \right)^2 \right] \tag{4.25}$$

where Ω_m is defined in Eq. (2.140). The fast surface mode has no cutoff for the range of wavenumber shown, but at much higher frequency (close to the upper ion cyclotron frequency) it stops where a cutoff in k_p (the upper dotted line) is encountered. This mode is the only surface mode that exists for positive k_y when the secondary ions are so massive that they can be considered to be stationary (Cramer & Vladimirov 1996b).

Figure 4.5 shows the same plots as in Figure 4.4, but for $s = 1.3$. The major change is that the fast surface wave has a lower cutoff at a nonzero wavenumber k_z, where the cutoff in k_p is encountered. Using Eq. (4.25) for the frequency in the expression for k_p, obtained from Eq. (4.9) by setting $\mathcal{D}_0(\omega) = 0$, we find

$$\frac{k_p}{k_z} \simeq \frac{1}{s} - \frac{s}{(1+s^2)^{1/2}} \tag{4.26}$$

so that $k_p < 0$ when $s > 1.27$, that is, a fast surface wave cannot exist at $k_z \simeq 0$ and there must exist a cutoff value of k_z as in Figure 4.5.

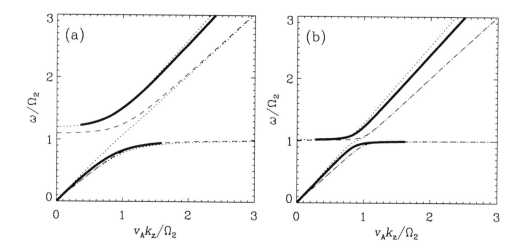

Figure 4.5: The normalized frequency of surface waves for a two ion species plasma, for positive k_y, as a function of normalized wavenumber k_z (solid curves). Cutoffs (dotted curves) and resonances (dashed curves) in the plasma are also shown. The parameters are as for Figure 4.4, except $s = 1.3$. (Reproduced from Cramer, Leung & Vladimirov 1998.)

Negative k_y

For propagation of surface waves in a direction intermediate between the $\boldsymbol{B}_0$ direction and the negative $\boldsymbol{B}_0 \times \boldsymbol{n}$ direction (i.e. $k_z > 0$ and $k_y < 0$), there also exist two surface modes, but with only a very small range of overlap in k_z.

Figure 4.6 shows the surface wave dispersion relations for negative k_y, for $s = 1$ and the same parameters as for Figure 4.4. Again there are two distinct modes; however there is only a small overlap range of wavenumber of the modes. The lower–frequency mode has the following dispersion relation for small k_z:

$$\omega = \left(\frac{2+s^2}{1+s^2}\right)^{1/2} v_{AT}k_z + \frac{b}{(1+b)^2(1+s^2)^{1/2}} \frac{(v_A k_z)^2}{\Omega_2} \tag{4.27}$$

that is, it has the same dispersion relation (4.24), for $k_z \to 0$, as the corresponding positive k_y mode. However the negative k_y mode stops at the second cutoff of k_p, rather than at a resonance.

The higher–frequency surface wave (upper cyclotron surface wave) stops at a lower wavenumber k_z where it encounters the higher resonance. Note that the formal solution of Eq. (4.23) for small k_z is the same as for the positive k_y mode (Eq. 4.25), but k_p is given by

$$\frac{k_p}{k_z} \simeq -\frac{1}{s} - \frac{s}{(1+s^2)^{1/2}} \tag{4.28}$$

so that $k_p < 0$ for all s and the surface wave cannot exist for small k_z. This mode has no upper cutoff for the range of wavenumber shown, however at much higher frequency (close to

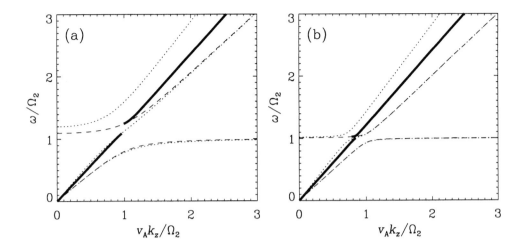

Figure 4.6: The normalized frequency of surface waves for a two ion species plasma, for negative k_y, as a function of normalized wavenumber k_z (solid curves). Cutoffs (dotted curves) and resonances (dashed curves) in the plasma are also shown. The secondary ion is negatively charged, with $g = -0.01$, and $s = 1$. (a) $b = 0.2$, (b) $b = 0.02$. (Reproduced from Cramer, Leung & Vladimirov 1998.)

the upper ion cyclotron frequency) it stops where a resonance (the Alfvén resonance) is again encountered. As s decreases (Figure 4.7 for $s = 0.2$), the two modes are practically continuous, except for a narrow stop-band delineated by the second cutoff and the higher resonance in k_p. The width of the stop–band between the two modes increases as the proportion of charge on the secondary ion species increases. The higher–frequency surface mode also exists in the stationary secondary ion case.

4.4.2 Resonance Damping

Resonance damping of the surface wave may occur if the surface has a nonzero width. An important point to determine is then whether or not the Alfvén resonance is encountered in the surface layer. In the surface the plasma density is assumed to decrease from its value at $x = -a$ to zero at $x = 0$, and the local Alfvén speed correspondingly increases, so that the local value of α increases from its value in the plasma at $x = -a$ to ∞ in the vacuum at $x = 0$. A surface wave might encounter a single value of the ion density, or the local Alfvén speed v_A, where Eq. (3.26) is satisfied, that is, where resonance damping can occur. Thus a surface wave of fixed frequency on a plasma-vacuum interface can experience damping in the transition region between the plasma and the vacuum at either the Alfvén resonance or the hybrid ion-ion resonance, but not both.

A surface wave of given frequency as shown in Figures 4.4-4.7 experiences resonance damping in the surface only if a resonance curve lies to the right of the surface wave solution curve. Whether the resonance encountered is to be interpreted as the Alfvén resonance or

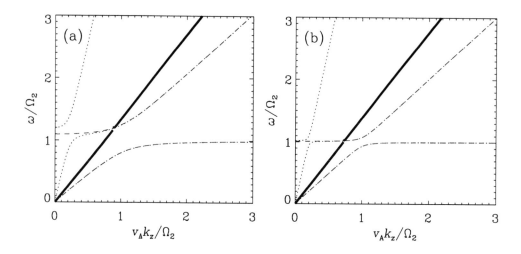

Figure 4.7: The normalized frequency of surface waves for a two ion species plasma, for negative k_y, as a function of normalized wavenumber k_z (solid curves). Cutoffs (dotted curves) and resonances (dashed curves) in the plasma are also shown. The parameters are as for Figure 4.6, except $s = 0.2$. (Reproduced from Cramer, Leung & Vladimirov 1998.)

hybrid resonance depends on the k_z dependence of the local resonance frequency, as discussed in Section 3.2.3. Resonance cannot occur if the right–hand side of Eq. (3.26) is < 0, that is, if the frequency lies in the frequency range separating the lower and upper resonance curves:

$$\Omega_2 < \omega < (1+b)^{1/2}\Omega_2. \tag{4.29}$$

It is evident from Figures 4.4 and 4.5 that the positive k_y surface waves always encounter a resonance in the surface layer. Similarly, Figures 4.6 and 4.7 show that the higher–frequency surface mode for negative k_y always encounters a resonance. However, the lower–frequency mode for negative k_y extends into the frequency range defined by Eq. (4.29) and so does not encounter resonance in that range, and indeed beyond it in the example of Figure 4.7(a). Thus the positive k_y surface waves are always resonance–damped, but the negative k_y, low–frequency mode can propagate free of resonance damping in a small range of frequency above the lower cyclotron frequency. If the resonance is not encountered, no Alfvén resonance damping occurs, and the only damping will be due to global resistive, viscous or other collisional processes (Hollweg 1982), or collisionless Landau damping (Rowe 1991).

If the secondary ion mass is much greater than the primary ion mass, the damping rate for either sign of k_y, using Eq. (4.13), reduces to

$$\gamma = -\varepsilon\pi k_z^3 \frac{\left[(1+s^2)^{-1/2} + \sigma((1+bg)f - bg/f)\right]^2}{(\partial \mathcal{D}_0/\partial\omega)(dG/d\bar{x})}. \tag{4.30}$$

For positive k_y, in the limit $v_A k_z \ll \Omega_2$, this leads to the following damping rate for the fast surface wave:

$$\gamma = \pi(k_z a) s (1 + (1 + s^2)^{-1/2})^2 \Omega_m \tag{4.31}$$

where G is assumed to vary linearly with x in the surface.

4.5   Ideal MHD

We now consider the low–frequency ($\omega \ll \Omega_i$) but nonzero plasma β case, which is of interest for Alfvén surface waves in astrophysical and space plasmas such as isolated flux tubes in the Sun's photosphere and magnetic loops in the solar corona. The analysis of surface modes of frequency much lower than the ion cyclotron frequency in the nonzero β case has been carried out, notably by Roberts (1981) and coworkers, for plane surfaces and cylindrical surfaces. We shall consider the plane surface case here, and the cylindrical surface case in Chapters 6 and 7.

Magnetoacoustic modes localized on current sheets of nonzero width have been investigated by Hopcraft & Smith (1986), Musielak & Suess (1989), Seboldt (1990) and Smith, Roberts & Oliver (1997), either using the Harris sheet model (Harris 1962) of the current sheet introduced in Section 3.3.1, in which the magnetic field does not rotate but reverses sign by passing through zero in the centre of the sheet, or using a finite width current sheet formed by the rotation of the magnetic field vector with constant strength. Here we consider a plane surface separating two plasmas of nonzero β. The magnetic field has constant direction along the z–axis, but may change in magnitude across the surface, with a resulting current sheet in the surface. The case of a rotated magnetic field is considered in the next section.

4.5.1   Dispersion Relation

On each side of a planar surface, where the plasma becomes uniform, the total zero–order particle and magnetic pressure must balance for equilibrium (1 and 2 denoting each side):

$$p_{01} + B_{01}^2/2\mu_0 = p_{02} + B_{02}^2/2\mu_0. \tag{4.32}$$

This can be written as

$$c_{s1}^2/\gamma + v_{A1}^2/2 = (c_{s2}^2/\gamma + v_{A2}^2/2)(\rho_2/\rho_1) \tag{4.33}$$

where γ is the adiabatic index, assumed the same on each side of the surface, and the sound speeds c_s and Alfvén speeds v_A have different values in the plasmas on each side, indicated by the subscripts.

The perturbation calculation of the frequency and damping of surface waves on a narrow surface in this case proceeds in the same way as in the previous sections. Again the small parameter is $\varepsilon = |k_z a|$, and v_x and p are expanded in series in ε as in Eq. (3.91). Again there is only one mode in each plasma, the magnetoacoustic mode with k_x given by Eq. (3.2), so two continuity conditions at each boundary are sufficient to derive the surface wave dispersion

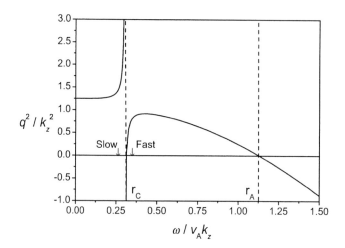

Figure 4.8: q^2/k_z^2 is plotted against $\omega/v_{A1}k_z$ in plasma 1 with $\gamma\beta/2 = 0.1$, for $s = 2$. There is an interface with plasma 2, with $\gamma\beta/2 = 10$, $v_{A2}/v_{A1} = 0.1$ and $\rho_2/\rho_1 = 8.63$. The frequencies of the slow and fast surface waves on this surface are indicated by the arrows. The compressive (r_C) and Alfvén (r_A) resonances are indicated.

relation. The boundary conditions used are not those used in the previous sections for the low-β case, but are that the total pressure (particle plus magnetic) perturbation p_T and the normal velocity component v_x used in Eqs. (3.52) and (3.53) are continuous across any boundary.

Applying the boundary conditions on each side of the surface, as in Section 4.2, yields a zero–order dispersion equation,

$$\mathcal{D}_0 = \tau_1/q_1 + \tau_2/q_2 = 0 \qquad (4.34)$$

and a first–order correction to the dispersion equation,

$$\mathcal{D}_1(\bar{\omega}) = \frac{1}{k_z} \int_0^{-1} (\tau - \tau_1^2 q^2/q_1^2 \tau) d\bar{x}' \qquad (4.35)$$

where $q_1 = ik_{x1}$ and $q_2 = ik_{x2}$ are the two values of q (defined in Eq. (3.55)) on either side of the surface. Eq. (3.54) is used for the definition of τ. The dispersion relation is then obtained by solving Eq. (4.34) for ω, using Eq. (3.2) to find k_{x1} and k_{x2} (which must both be imaginary).

We consider first a sharp surface separating a plasma with small β from a vacuum region. It is found that the usual Alfvén surface wave found in a cold, single ion species plasma occurs close to the frequency given by Eq. (4.16), but with a small modification due to the nonzero β. However, there also occurs a second, slow surface wave with a frequency, correct to lowest order in β,

$$\omega = \left(\frac{\gamma\beta}{2}\right)^{1/2} \frac{v_A k_z^2}{(k_z^2 + 2k_y^2)^{1/2}} = \frac{c_s k_z^2}{(k_z^2 + 2k_y^2)^{1/2}}. \qquad (4.36)$$

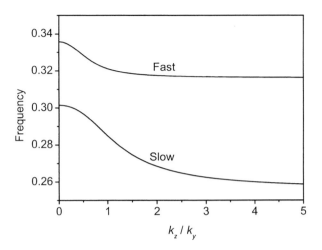

Figure 4.9: Frequency of the slow and fast surface waves, divided by $v_{A1}k_z$, as a function of $s = k_z/k_y$. Here $\gamma\beta_1/2 = 0.1$, $\gamma\beta_2/2 = 10$, $v_{A2}/v_{A1} = 0.1$ and $\rho_2/\rho_1 = 8.63$.

Another interesting case is a surface separating a plasma of low $\beta(=\beta_1)$ from a plasma of high $\beta(=\beta_2)$, with comparable sound speeds c_s in each plasma. This case corresponds to a strong magnetic field region next to an essentially field–free one, as may occur in isolated magnetic flux tubes in the Sun's atmosphere. We then have, from Eq. (4.33),

$$\rho_2/\rho_1 \simeq 1 + 1/\beta_1. \tag{4.37}$$

There are again two surface wave solutions (Wentzel 1979a, Roberts 1981). We may locate these two solutions in a plot of q^2 in the low–β plasma against frequency (Figure 4.8). The frequencies of the two modes are shown in Figure 4.9, plotted against k_z/k_y for typical values of β; the maximum frequency of the fast wave, for given k_z, is $\omega = (2/\gamma)^{1/2}c_s k_z$.

Finally, in the case of a high–γ plasma on each side of the surface, which is equivalent to the incompressible approximation (Uberoi 1994), we have (see Section 3.2.1)

$$q_1 \simeq q_2 \simeq (k_y^2 + k_z^2)^{1/2} \tag{4.38}$$

which gives from Eq. (4.34),

$$\omega^2 = \left(\frac{\rho_1 v_{A1}^2 + \rho_2 v_{A2}^2}{\rho_1 + \rho_2}\right) k_z^2 \tag{4.39}$$

for the Alfvén surface wave frequency, independently of k_y. If the density on one side, say ρ_2, is very small, the dispersion relation is

$$\omega = \sqrt{2}v_{A1}k_z \tag{4.40}$$

which is the same as the low-β result (Eq. 4.18) for propagation almost perpendicular to the magnetic field, in a surface separating a plasma from a vacuum.

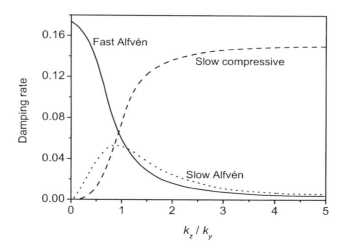

Figure 4.10: Normalized damping rates for the slow and fast surface waves of Figure 4.9, as a function of $s = k_z/k_y$. The fast wave damping rate is normalized to $\pi k_y a w_0$ and the slow wave damping rate is normalized to $\pi k_z a w_0$.

4.5.2   Resonance Damping

The dominant contribution to $\mathcal{D}_1$, leading to damping of the surface wave, is again from the poles of the integrand, one of which is given by $\tau = 0$, or $\omega = v_A k_z$, just as for the low–frequency, low–β case. This corresponds to the Alfvén resonance. However, in addition, a pole arises if there is a singularity of q^2 given by $\omega = v_T k_z$ at a point in the surface transition region: this is the compressive or cusp singularity discussed in Section 3.2.1. Both of these resonances can contribute to the resonance damping of the surface wave. Notice that Alfvén resonance damping of the surface wave disappears if $k_y = 0$ as discussed for the wave transmission problem in Section 3.4.1. Using Eq. (4.12), the Alfvén resonance damping rate is given by

$$\gamma = -\varepsilon\pi \frac{(q_r^2/q_1^2)\tau_1^2}{k_z(\partial \mathcal{D}_0/\partial \omega)(\mathrm{d}\tau/\mathrm{d}\bar{x})} \tag{4.41}$$

where q has the value $q_r = k_y$ at the Alfvén resonance. The damping rate given by Eq. (4.41) reduces to the result (Eq. 4.19) in the low–β limit. The compressive resonance gives rise to a similar damping rate expression. The slow surface wave with frequency given by Eq. (4.36) damps via both the compressive and the Alfvén resonances. The fast surface wave with frequency given by Eq. (4.16) only damps via the Alfvén resonance, because for this wave the compressive singularity is not encountered in the surface. The damping rates for the same conditions as in Figure 4.9 are shown in Figure 4.10. For the slow wave, the compressive resonance damping dominates Alfvén damping at small k_y, and vice-versa for large k_y, with the rates equal at $k_y = k_z$.

A more rigorous derivation of the dispersion relation and the resonant damping uses the theory of Section 3.4.2 (Ionson 1978, Wentzel 1979a). A formal solution of Eq. (3.98) involves the Green's function, in which the following constant appears in the denominator (Sedlacek 1971, Wentzel 1979a):

$$
A = \frac{k_y}{\zeta_1 \zeta_2} \left[\left(\frac{q_1}{k_y} \xi_1 - \frac{d\xi_1}{d\zeta} \right) \bigg|_{\zeta_1} \left(\frac{q_2}{k_y} \xi_2 + \frac{d\xi_2}{d\zeta} \right) \bigg|_{\zeta_2} \right.
$$
$$
\left. - \left(\frac{q_1}{k_y} \xi_2 - \frac{d\xi_2}{d\zeta} \right) \bigg|_{\zeta_1} \left(\frac{q_2}{k_y} \xi_1 + \frac{d\xi_1}{d\zeta} \right) \bigg|_{\zeta_2} \right].
$$

(4.42)

Here ζ_2 and ζ_1 are the values of ζ at $x = 0$ and $x = -a$. The zeroes of A yield the poles of the Green's function integral and thereby determine the dispersion relation of the surface waves. The logarithmic terms such as $\ln \zeta_1$ in ξ_1 make the Green's function multivalued, with the effect of replacing the logarithmic terms in A by $\pm i\pi$. The appearance of i in the dispersion relation implies a complex frequency and a damping of the wave.

We have

$$
A = \frac{q_1}{\zeta_1} + \frac{q_2}{\zeta_2} + k_y \ln \frac{\zeta_2}{\zeta_1} + k_y C^2 \ln \frac{\zeta - \zeta_s}{\zeta_1 - \zeta_s}
$$

(4.43)

so the dispersion equation becomes

$$
\frac{q_1}{\zeta_1} + \frac{q_2}{\zeta_2} - i\pi k_y (1 + C^2) = 0
$$

(4.44)

where the C^2 should be included only if the compressive singularity occurs within the transition.

As in Eq. (4.12), the dispersion equation (4.44) may be expanded around the real component of the frequency to yield a damping rate

$$
\gamma = -\pi k_y (1 + C^2) / \frac{\partial}{\partial \omega} \left(\frac{q_1}{\zeta_1} + \frac{q_2}{\zeta_2} \right)
$$

(4.45)

which reduces to Eq. (4.41) if $C = 0$.

4.6 Magnetic Field Rotation

In this section, the surface of specific interest is a current sheet formed when the magnetic field changes in direction. Current sheets are ubiquitous in space plasmas, such as the terrestrial current sheet, the Jovian current sheet, heliospheric current sheets, filaments of the solar chromosphere and corona, and bow shocks and magnetopauses of the planetary magnetospheres. Current sheets are thought to arise commonly in the solar atmosphere, and hence in the atmospheres of other stars (Wentzel 1979a, Roberts 1981). The surface magnetic field of the Sun is highly inhomogeneous and gives rise to current sheets in the solar wind. These can support surface waves that will contribute to the solar wind wave field. The density and pressure of the plasma inside the sheets are normally much higher than outside, so the sheets are also called plasma sheets (Yamauchi & Lui 1997). Reconnection of magnetic fields can occur in current sheets, and is the mechanism by which the interplanetary magnetic field exerts control over geomagnetic activity. Magnetic flux transported to the magnetotail builds up until reconnection occurs. This process is called a geomagnetic substorm (Hughes 1995).

Up to this point we have only considered surface waves in plasmas in which the zero–order magnetic field is directed along the z-axis. The theory of linear magnetoacoustic surface waves has been investigated for infinitely thin current sheets across which the magnetic field has an arbitrary change of direction, for cold plasmas and low frequency (Hollweg 1982), for zero–β plasmas but frequencies up to the ion cyclotron frequency (Wessen & Cramer 1991), and for low frequency but nonzero plasma β (Cramer 1994). In these works the plasma density was assumed to be the same on either side of the current sheet.

We consider here the ideal MHD and cold plasma cases for a thin surface with a rotating magnetic field and a uniform density (and uniform pressure in the ideal MHD case). The dispersion relation may be found by using the same methods and boundary conditions as in the previous sections, taking into account the rotation of the magnetic field across the surface.

4.6.1   Ideal MHD

In this section, the theory of linear magnetoacoustic surface waves is investigated for current sheets across which the magnetic field has an arbitrary change of direction, in the first place discontinuously, and in the second place via a narrow transition region in which the magnetic field rotates with constant amplitude, so that the gas pressure remains constant. The plasma density and the particle pressure is assumed the same on either side of the current sheet. In contrast to the zero particle pressure, low–frequency case, where a surface wave exists for all angles of rotation of the magnetic field and for all angles of propagation of the wave relative to the magnetic field (Hollweg 1982), it is found that the effect of nonzero pressure is to eliminate the surface wave for certain angles of propagation and to allow the existence of an additional, slower, surface wave for other angles of propagation (Uberoi & Narayan 1986, Cramer 1994).

The equilibrium plasma system consists, in the first place, of two homogeneous equal–density plasmas of equal gas pressures, separated by a sharp interface, and each with a uniform magnetic field imposed in the plane of the interface, but in a different direction in each plasma. In the following section the interface is allowed to have a nonzero width. We label the plasma occupying the half-space $x < 0$ as plasma 1, and the plasma occupying $x > 0$ as plasma 2. We define the z–axis with direction midway between the directions of the magnetic field on either side of the interface, as shown in Figure 4.11. The field $\boldsymbol{B}_2$ is rotated an angle θ with respect to the field $\boldsymbol{B}_1$ in the y-z plane. A current sheet exists at the interface because of the discontinuity of the magnetic field direction. As well as assuming a constant magnitude of $\boldsymbol{B}$ on either side of the interface, that is, $|\boldsymbol{B}_1| = |\boldsymbol{B}_2| = B_0$, we also assume a constant density and constant gas pressure. Thus the only equilibrium parameter that changes across the interface is the magnetic field direction.

Following the analysis of Cramer (1994), the wave fields are assumed to vary as

$$f(x)\exp(ik_y y + ik_z z - i\omega t) \tag{4.46}$$

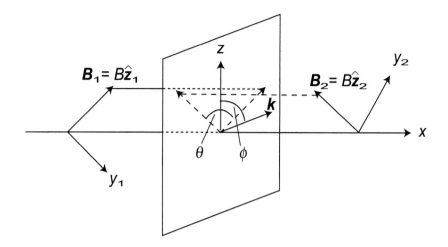

Figure 4.11: The geometry of the surface between two plasmas. The coordinate systems in the two plasmas and in the surface are defined. The magnetic field rotates by the angle θ in the surface, and the wavevector k in the surface plane is at an angle ϕ to the z–axis. (Reproduced from Wessen & Cramer 1991.)

where

$$\boldsymbol{k} = (0, k_y, k_z) = (0, k\sin\phi, k\cos\phi) \tag{4.47}$$

is the wave vector of the propagating wave in the plane of the interface, at an angle ϕ to the z–axis and with a magnitude k. Because the plasmas on either side of the interface have uniform magnetic fields, we can obtain solutions in those plasmas with $f(x) = \exp(\pm\kappa x)$ with $\kappa^2 = q^2$, where q is given by Eq. (3.55). If κ^2 is positive in both regions 1 and 2, a surface mode exists with wave fields localized about the interface.

The surface wave dispersion relation for waves on the sharp interface is calculated by imposing the boundary conditions at the interface. As in the case of a constant direction magnetic field but discontinuous plasma density (Section 4.5.1), the total pressure p_T on each side of the interface must be continuous at $x = 0$, and the velocity component perpendicular to the interface v_x must also be continuous. The governing differential equations for p_T and v_x are still Eqs. (3.52) and (3.53), so the analysis follows as in Section 4.5.1. The result of applying the boundary conditions is a dispersion equation of the same form as Eq. (4.34):

$$\mathcal{D}_0 = \tau_1/\kappa_1 + \tau_2/\kappa_2 = 0. \tag{4.48}$$

Consider mutually rotated local coordinate systems (x, y_1, z_1) and (x, y_2, z_2) in each plasma, with the local z–axis parallel to the local magnetic field. The wavenumbers in the plane of the surface in each plasma, $\boldsymbol{k}_1$ and $\boldsymbol{k}_2$, must be equal because of continuity of the wave fields at the interface, and may be expressed in their local coordinate systems in terms of the components of k as

$$
\begin{aligned}
k_1 &= (0, -k_z \sin(\theta/2) + k_y \cos(\theta/2), k_z \cos(\theta/2) + k_y \sin(\theta/2)) \\
k_2 &= (0, k_z \sin(\theta/2) + k_y \cos(\theta/2), k_z \cos(\theta/2) - k_y \sin(\theta/2)) .
\end{aligned}
\tag{4.49}
$$

We define the normalized phase velocity $V = \omega/kv_{\mathrm{A}}$, where v_{A} is the Alfvén speed in both plasmas, and note that the wavenumber components can be written as

$$
\begin{aligned}
k_{z1} &= k(\cos\phi \cos(\theta/2) + \sin\phi \sin(\theta/2)) \\
k_{z2} &= k(\cos\phi \cos(\theta/2) - \sin\phi \sin(\theta/2)) \\
k_{y1} &= k(-\cos\phi \sin(\theta/2) + \sin\phi \cos(\theta/2)) \\
k_{y2} &= k(\cos\phi \sin(\theta/2) + \sin\phi \cos(\theta/2)).
\end{aligned}
\tag{4.50}
$$

The dispersion equation (4.48) can then be written as

$$
\mathcal{D}_0 = \frac{V^2 - k_{z1}^2/k^2}{q_1} + \frac{V^2 - k_{z2}^2/k^2}{q_2} = 0
\tag{4.51}
$$

where, noting that the vector v_{A} has a different direction in each plasma, we have

$$
\begin{aligned}
q_{1,2}^2 &= k_{y1,2}^2 - \frac{(k^2 V^2 - k_{z1,2}^2)(k^2 V^2 - \beta' k_{z1,2}^2)}{k^2 V^2 (1 + \beta') - \beta' k_{z1,2}^2} \\
&= k^2 \left[\frac{1 - V^2 - \beta'(1 - k_{z1,2}^2/k^2 V^2)}{1 + \beta'(1 - k_{z1,2}^2/k^2 V^2)} \right]
\end{aligned}
\tag{4.52}
$$

with $\beta' = \gamma\beta/2$, and q_1 and q_2 are both assumed to be positive for a surface wave solution. The solution of Eq. (4.51) for V for given θ, ϕ and β then constitutes the dispersion relation.

Zero–Beta Limit

In the zero β limit, a single surface wave is found to exist for all angles of rotation of the magnetic field and for all angles of propagation of the wave relative to the magnetic field. We have for $\beta = 0$,

$$
q_1^2 = q_2^2 = k^2(1 - V^2)
\tag{4.53}
$$

so the dispersion equation is simply

$$
\tau_1 + \tau_2 = 0,
\tag{4.54}
$$

which gives for the dispersion relation

$$
V^2 = \frac{k_{z1}^2 + k_{z2}^2}{2k^2} = \cos^2\phi \, \cos^2(\theta/2) + \sin^2\phi \, \sin^2(\theta/2).
\tag{4.55}
$$

The dimensional version of Eq. (4.55) is

$$
\omega^2 = v_{\mathrm{A}}^2 \overline{k_z^2} = \frac{1}{2}((k \cdot v_{\mathrm{A}1})^2 + (k \cdot v_{\mathrm{A}2})^2)
\tag{4.56}
$$

where $\overline{k_z^2}$ is the average of the square of the wavenumber components parallel to the magnetic field on each side of the interface. In other words, the square of the phase velocity is the average of the squares of the phase speeds of the shear Alfvén waves on either side of the interface. However the waves are compressional, because $p_T \neq 0$. The corresponding attenuation coefficient perpendicular to the surface is the same on each side of the surface and is given by

$$q^2 = k^2 - \overline{k_z^2}. \tag{4.57}$$

The zero–β limit given by Eq. (4.56) was first obtained by Hollweg (1982).

Magnetic field reversal, i.e. $\theta = 180°$, is possibly the most important physical case for current sheets in the solar wind, and in the planetary magnetosphere, where current sheets separate regions of opposite magnetic field polarity in the magnetotail plasma sheet boundary. Indeed, Hopcraft & Smith (1986) and Musielak & Suess (1989) consider this limit only. Note that in the limit $\theta \to 180°$, $q_{1,2} \to k_y$ and $\tau \to 0$, so $p_T \to 0$. Thus the wave is not strictly a surface wave, because one of the boundary conditions (continuous p_T) at the surface cannot be applied. This corresponds to a local shear Alfvén wave excitation. However, for a small angle away from $\theta = 180°$, $p_{T1,2}$ are nonzero although small, and the surface wave exists. Similar behaviour occurs as $\theta \to 0°$: formally the dispersion relation (4.56) implies $\omega \to v_A k_z$ although at $\theta = 0°$ the wave exists only as a shear wave excitation. The phase velocity given by Eq. (4.55) is symmetric in ϕ about $\phi = 90°$, that is, it is independent of the sign of k_z. For $\theta = 90°$, $V = 1/\sqrt{2}$ independently of ϕ.

The polarization of the wave is determined by the following ratios in terms of the wave electric fields $\boldsymbol{E}_1$ and $\boldsymbol{E}_2$ on either side of the surface:

$$\begin{aligned}
\frac{E_{x1}}{E_{y1}} &= -\mathrm{i}\sin(\phi - \theta/2)/(q/k) \\[4pt]
\frac{E_{x2}}{E_{y2}} &= \mathrm{i}\sin(\phi + \theta/2)/(q/k).
\end{aligned} \tag{4.58}$$

This indicates that the waves are generally elliptically polarized, the handedness of the polarization depending solely on the size of ϕ and θ because q is always positive. In plasma 1 the waves are left–hand polarized for $\phi < \theta/2$, that is, the electric field vector rotates in the same direction as the ion cyclotron motion, and the waves are right–hand polarized for $\phi > \theta/2$. In plasma 2 the waves are left–hand polarized for $\phi < 180° - \theta/2$ and otherwise right–hand polarized. Circular polarization occurs for $\phi = 0°, 90°$ and $180°$ (Hollweg 1982).

Note also that for $\phi = 0$ or $\phi = \pi/2$, for any value of β it follows that $q_1 = q_2$ and so Eq. (4.54) is again the dispersion equation. Thus in this case the frequency is the zero–β result, although q_1 and q_2 still depend on β. Note however that in this case $\tau_1 = \tau_2 = 0$, so that $p_T = 0$ and strictly the boundary conditions cannot be satisfied; we have then shear Alfvén waves on each side of the interface. Nevertheless if ϕ is close to but not exactly equal to 0 or $\pi/2$, we can conclude that the frequency of the surface wave is close to the zero-β result.

Nonzero Beta

To interpret the results of solving the dispersion equation (4.51) for $\beta > 0$, it is helpful to consider the behaviour of the perpendicular attenuation coefficient κ, or q, as the frequency is

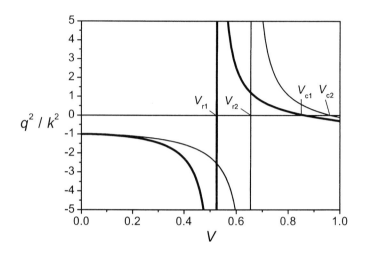

Figure 4.12: The normalized square of the attenuation constant, q^2/k^2, plotted against the normalized phase velocity V. The heavy curve is for the plasma on the left–hand side of the current sheet, and the light curve for the plasma on the right–hand side of the sheet. $\theta = 20°$, $\phi = 32°$ and $\beta' = 1$. The resonance and cutoff phase velocities, V_r and V_c, are shown for each side of the sheet. (Reproduced from Wessen & Cramer 1991.)

changed. When $q^2 > 0$ the waves attenuate in the x–direction and a surface wave may exist, whereas if $q^2 < 0$ on either side of the interface the wave propagates in the x–direction and we no longer have a localized surface wave. A cutoff (where $q^2 = 0$) occurs when

$$V^2 = V_c^2 = \frac{1}{2}\left(1 - \beta' + [(1 - \beta')^2 + 4\beta' k_{z1,2}^2/k^2]^{1/2}\right) \tag{4.59}$$

and a (compressive) resonance (where $1/q^2 = 0$) occurs when

$$V = V_r = \frac{\sqrt{\beta'} k_{z1,2}}{(1 + \beta')^{1/2} k} \tag{4.60}$$

and the surface wave criterion is satisfied, that is, $q^2 > 0$, for phase velocities in the range $V_r < V < V_c$.

A plot of q^2 against V is shown in Figure 4.12 for each side of the interface in a typical case ($\theta = 20°$, $\phi = 32°$, and $\beta' = 1$), with the resonances and cutoffs indicated, and it is evident that the wave will only have surface wave character on each side if the phase velocity is in the range $V_{r2} < V < V_{c1}$. For zero β a single surface wave solution exists for any values of θ and ϕ as was shown above, however for $\beta > 0$ the existence of the resonances and cutoffs leads to a more complicated spectrum of surface waves. Thus in certain ranges of angle of propagation it is found that solutions may not exist, while for other propagation angles there may be two distinct surface modes of different frequency. Uberoi & Narayan (1986) obtained similar results for the case when there is also a density jump across the sheet.

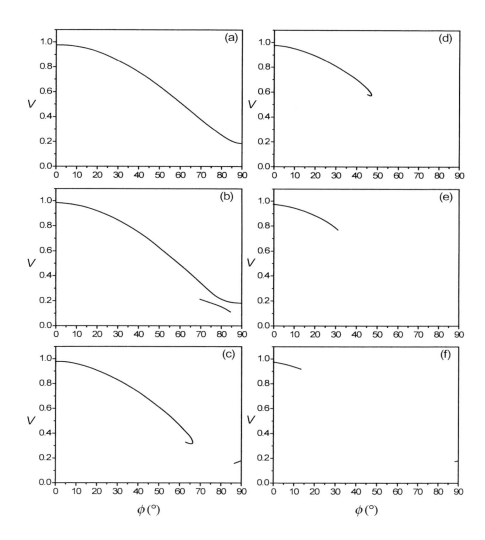

Figure 4.13: The normalized phase velocity of a surface wave on a surface with a rotated magnetic field, plotted against the angle of propagation, for an angle of rotation of the magnetic field of 20°: (a) $\beta' = 0$; (b) $\beta' = 0.2$; (c) $\beta' = 0.4$; (d) $\beta' = 1$; (e) $\beta' = 2$; (f) $\beta' = 5$. (Reproduced from Wessen & Cramer 1991.)

Figure 4.13 shows the dispersion relation, in the form of the phase velocity V plotted against the angle of propagation ϕ, for an angle θ between the magnetic fields of 20°, for six values of β' ranging from $\beta' = 0$ to $\beta' = 5$. The result for $\beta = 0$ (Figure 4.13(a)) is the same as that of Eq. (4.55), but as β increases a second, slower, surface mode appears (Figure 4.13(b)) for a narrow range of ϕ. We refer to the mode of higher phase velocity as the fast mode, while the lower phase velocity mode, where it exists, is referred to as the slow mode.

The end points of the range of ϕ for the slow mode are determined by resonance occurring in the plasma on the negative–x side of the interface. As β increases further (Figure 4.13(c)) the two modes merge at some value of ϕ and a stop–band in ϕ appears where no surface wave exists. Two distinct surface modes still exist for a small range of ϕ just below the stop–band. As β' increases past the value of 1 (Figure 4.13(d)) the extent of the stop–band increases, until at high values of β (Figures 4.13(e) and (f)) surface wave propagation only occurs at angles close to $0°$ and close to $90°$. It is interesting to note that nonzero particle pressure also leads to an additional, slow, surface wave for a surface separating a high–β plasma from a low–β plasma where no rotation of the magnetic field occurs (see Section 4.5.1 and Roberts (1981) and Uberoi & Narayan (1986)). More details of the dispersion relation for a rotating magnetic field can be found in Cramer (1994).

Resonance Damping

As we have seen in previous sections, if the interface between the two plasmas is not sharp, but is allowed to have a small transition width a, there arises the possibility of collisionless damping of the surface wave if the conditions for wave resonance are met in the transition region. In a nonzero–β plasma, this resonance damping can be Alfvén resonance damping or compressive (or "cusp") resonance damping (see Section 4.5.1). The compressive resonance corresponds to a singularity of q, while the Alfvén resonance corresponds to the vanishing of τ. Because both q and τ are functions of x in the transition region, through their dependence on the magnetic field, for a wave of given phase velocity V and wavenumber $\boldsymbol{k}$ the resonance conditions may be satisfied somewhere in the transition region, just as in the case of a plasma density transition considered in the previous sections. The analysis of the resonance damping proceeds using the perturbation approach used in Chapter 3 and Section 4.5.1.

The resulting first–order correction to the dispersion function, $\mathcal{D}_1$, has the same form as in Eq. (4.35). The dominant contribution to $\mathcal{D}_1$ is from the poles of the integrand, that is, from the zero of τ (the Alfvén resonance) and the singularity of q^2 (the compressive resonance). We define $V = V_s$ and $\omega = \omega_s$, for fixed θ and ϕ, to be the phase velocity and frequency solution of the dispersion relation $\mathcal{D}_0 = 0$. Then correct to first order in $\varepsilon = ka$ we have Eq. (4.12) for the damping rate of the surface wave.

To calculate the integral in $\mathcal{D}_1$ it is assumed that the gas pressure is uniform in the transition region, so that the equilibrium magnetic field simply rotates with constant amplitude through the interface transition. The assumed magnetic field in the transition region is

$$\boldsymbol{B}_0 = B_0 \sin(\theta/2 - \theta x/a)\hat{\boldsymbol{y}} + B_0 \cos(\theta/2 - \theta x/a)\hat{\boldsymbol{z}}. \tag{4.61}$$

Then near the Alfvén resonance point x_A:

$$\tau \simeq 2i\rho_0 v_A k(1 - V^2)^{1/2}\theta(x - x_A)/a \tag{4.62}$$

while near the compressive resonance point x_C:

$$q^2 \simeq \frac{V^3 k^2}{2(1 + \beta')^{1/2}(\beta'(1 - V^2) - V^2)^{1/2}} \frac{a}{\theta(x - x_C)}. \tag{4.63}$$

These expressions are used to calculate the residues of the poles in the integrand in Eq. (4.35).

The calculation of the damping rate is straightforward in the $\beta = 0$ case, when only Alfvén resonance damping can occur. In that case it is apparent that Alfvén damping occurs for all values of θ and ϕ because τ must change sign across the interface for the zero–order dispersion relation (4.51) to hold, and so τ vanishes at some point in the transition region. Using the sharp interface $\beta = 0$ solution given by Eq. (4.55) for V_s, the damping rate becomes

$$\gamma = \varepsilon v_A k \frac{\pi}{\theta} \frac{\sin^2 2\phi \sin^2 \theta}{16(1 + \cos 2\phi \cos \theta)}. \tag{4.64}$$

For finite β, the Alfvén resonance is still encountered by the modes shown in Figure 4.13, with small modification to the damping rate. The compressive resonance, however, is encountered for small β only by the slow mode, while for $\beta' > 1$ both slow and fast modes encounter it over the full range of ϕ where they exist, for $\theta = 60°$ and $\theta = 90°$. For $\theta = 20°$ the compressive resonance is not encountered for any of the modes of Figure 4.13. A necessary condition for the compressive resonance to be met, following from Eq. (4.52), is that

$$V^2 < \frac{\beta'}{1 + \beta'} \tag{4.65}$$

which is consistent with the lower likelihood of compressive damping by the low-β modes.

The compressive damping rate, as calculated from Eqs. (4.35) and (4.63), is

$$\gamma = \varepsilon \frac{\pi}{\theta} \frac{\varepsilon_1^2}{q_1^2} \frac{\beta' V}{2(1 + \beta')^{1/2}(\beta'(1 - V^2) - V^2)^{1/2}} \tag{4.66}$$

and is therefore of relative order β for small β.

Thus we have seen that a nonzero plasma β has two main effects: (1) the surface wave cannot propagate in certain ranges of angles of propagation relative to the magnetic field, and (2) a second, slower, surface wave can propagate for certain angles of propagation. Alfvén resonance damping always occurs, as well as, for high β and certain angles of propagation, compressive or cusp resonance damping. The existence and damping of surface waves in a wide current sheet, where the sheet width is greater than the wavelength, and for which the theory presented in this section breaks down, must be investigated by detailed solution of the spatially dependent wave equations, as has been done by Hopcraft & Smith (1986) for the Harris sheet equilibrium. However, the regime of validity of the results discussed here may be estimated for the case of waves near the Earth's geomagnetic current sheet. According to Hopcraft & Smith (1986) the observed wave periods range from 3.7 ms to 35 s, and the measured Alfvén speed ranges from 200 to 1000 km s^{-1}. Assuming that $V \simeq 1$, the wavelength therefore lies in the approximate range 100 m to 35 km, and the surface waves will only exist on current sheets of size less than these values. The geomagnetic current sheet has a considerably greater size, 1000 to 50,000 km, so the observed waves are unlikely to be the surface waves described here, unless they are propagating on localized narrower current sheets. On the other hand, some of the long period waves observed in the solar wind (Barnes 1979), of wavelength 10^3 to 5×10^6 km, could be surface waves propagating on solar wind current sheets of such large sizes.

Other interesting effects arise through the interaction of surface waves on neighbouring tangential discontinuities, as may occur in the Earth's magnetotail. This was considered in the low–frequency limit by Hollweg (1982). Nonzero resistivity of the plasma would result in reconnection and small–scale turbulence of the sheet, which could be responsible for the generation of small–amplitude surface waves at a current sheet. In addition, when there is relative tangential motion between two plasmas separated by a current sheet, the Kelvin-Helmholtz instability will lead to the excitation of small–amplitude surface waves once the relative velocity exceeds some critical value, as we find in Section 5.2.1.

4.6.2   Cyclotron Effects

In the analysis of waves localized on current sheets in the previous section (see also Roberts (1981), Hopcraft & Smith (1986), Musielak & Suess (1989) and Seboldt (1990)), the wave frequency was assumed to be well below the ion cyclotron frequency, that is, the effects of the Hall current in the plasma are neglected. This is certainly a valid assumption for the frequencies of observed MHD fluctuations in the interplanetary solar wind, of periods of the order of minutes, but is not valid for the frequencies of MHD waves observed in the Earth's magnetosphere, in particular the waves observed at the boundary of the plasma sheet in the magnetotail, where the wave period is between 2 and 18 s, compared with the proton cyclotron period of between 5 and 12 s (Tsurutani *et al.* 1985).

To include ion cyclotron effects, we now employ the cold plasma approximation for a single species plasma, in which the Hall term that governs the effects is included in the generalized Ohm's law, but other nonideal effects are neglected. Again we consider, as in the ideal MHD treatment, an interface between two uniform equal density cold plasmas of the same species, singly ionized, and each with a uniform magnetic field in the plane of the interface, but in a different direction in each plasma. The current sheet is force–free because particle pressure is neglected.

Again $k = (0, k_y, k_z)$ is the wavevector of the surface wave in the plane of the sheet, at an angle ϕ to the z-axis and with a magnitude k. Ion cyclotron effects are considered, that is, the frequency can be comparable with the ion cyclotron frequency, but finite electron mass and particle collisions are ignored, ensuring the electric field parallel to the magnetic field in each plasma is zero. Local coordinate systems in each plasma are defined as in the ideal MHD case (see Figure 4.11). Eqs. (3.65) and (3.66) then hold in the two local coordinate systems, and E_x is obtained from Eq. (3.67).

Solving Eqs. (3.65) and (3.66) for the wave field E_y results in the following expressions for E_y in the uniform plasmas on each side of the current sheet:

$$E_{y1} = E_{01} \exp(+\kappa_{p1} x + i k_{y1} y_1 + i k_{z1} z_1 - i\omega t) \tag{4.67}$$

with

$$\kappa_{p1}^2 = k_{y1}^2 - \frac{G_1^2 - H^2}{G_1} \tag{4.68}$$

and

$$E_{y2} = E_{02} \exp(-\kappa_{p2} x + i k_{y2} y_2 + i k_{z2} z_2 - i\omega t) \tag{4.69}$$

with

$$\kappa_{p2}^2 = k_{y2}^2 - \frac{G_2^2 - H^2}{G_2} \tag{4.70}$$

with the subscripts 1 and 2 referring to plasma 1 and plasma 2 respectively. We note that H, defined by Eq. (2.79), has the same value in each plasma. Because we are only interested in modes with their fields localized at the surface, i.e. surface waves, κ_{p1} and κ_{p2} must be real and positive.

Faraday's law gives for the wave magnetic fields:

$$\boldsymbol{B}_1 = \left(-\mathrm{i}k_{z1}, -\frac{k_{z1}(H - k_{y1}\kappa_{p1})}{k_{y1}^2 - G_1}, \frac{k_{y1}H - \kappa_{p1}G_1}{k_{y1}^2 - G_1} \right) \frac{E_{01}}{\mathrm{i}\omega} \tag{4.71}$$

$$\times \exp(\kappa_{p1}x + \mathrm{i}k_{y1}y_1 + \mathrm{i}k_{z1}z_1 - \mathrm{i}\omega t)$$

and

$$\boldsymbol{B}_2 = \left(-\mathrm{i}k_{z2}, -\frac{k_{z2}(H + k_{y2}\kappa_{p2})}{k_{y2}^2 - G_2}, \frac{k_{y2}H + \kappa_{p2}G_2}{k_{y2}^2 - G_2} \right) \frac{E_{02}}{\mathrm{i}\omega} \tag{4.72}$$

$$\times \exp(-\kappa_{p2}x + \mathrm{i}k_{y2}y_2 + \mathrm{i}k_{z2}z_2 - \mathrm{i}\omega t).$$

The wavenumbers in each plasma, $\boldsymbol{k}_1$ and $\boldsymbol{k}_2$, are defined in terms of the wavevector $\boldsymbol{k}$ in the original coordinate system as in Eq. (4.49).

We consider the case of a very sharp transition between the two regions, so to find the surface wave dispersion relation, appropriate boundary conditions at the surface must be used. In the cold plasma model, we usually employ boundary conditions on the electric and magnetic fields. Firstly, the total pressure on each side of the interface must be equal, the pressure being solely magnetic in the cold plasma approximation and given by Eq. (1.18). This reduces to the first–order boundary condition:

$$B_{z1} = B_{z2}. \tag{4.73}$$

Boundary conditions on the electric fields come from integrating Faraday's law across the interface (Stix 1992), and take the form:

$$\hat{\boldsymbol{x}} \times (\boldsymbol{E}_1 - \boldsymbol{E}_2) = v_x(\boldsymbol{B}_{01} - \boldsymbol{B}_{02}) \tag{4.74}$$

where $\boldsymbol{E}$ and v_x are wave fields and hence first–order while $\boldsymbol{B}_{01}$ and $\boldsymbol{B}_{02}$ are the equilibrium fields and therefore zero–order. Here the plasma fluid velocity perpendicular to the surface, v_x, is assumed continuous at the surface. Because we are neglecting effects such as resistivity or electron inertia, the electric field along the equilibrium magnetic field is zero, i.e. $E_{z1} = E_{z2}$. Then Eq. (4.74) reduces to

$$E_{y2}\sin\theta = v_x B_0 \sin\theta$$
$$E_{y1} = (E_{y2} - v_x B_0)\cos\theta + v_x B_0 \tag{4.75}$$

that is,

$$
\begin{aligned}
E_{y1} &= E_{y2} & \text{for} \quad \theta = 0 \\
E_{y1} &= E_{y2} = v_x B_0 & \text{for} \quad \theta \neq 0, \pi \\
E_{y1} + E_{y2} &= 2 v_x B_0 & \text{for} \quad \theta = \pi.
\end{aligned}
\tag{4.76}
$$

A difficulty arises if the E_y's given by the fast mode, satisfying Eqs. (4.67) and (4.69), are used with the continuity condition $E_{y1} = E_{y2}$, because the perpendicular velocity v_x derived from the E_y's will not be continuous across the surface. To ensure continuous v_x, the wave fields must be a combination of the fast mode and the short–wavelength mode induced by corrections such as thermal effects, resistivity, etc. Conversely, continuity of v_x given by a single mode on each side leads to a discontinuity in E_y, which can also be removed by including the second mode.

Note that the problem does not arise in the low–frequency case discussed in Section 4.6.1, because then the Hall term is neglected in the Ohm's law, which becomes $\boldsymbol{E} + \boldsymbol{v} \times \boldsymbol{B}_0 = 0$ (neglecting nonideal effects), and the boundary conditions given by Eq. (4.76) involving v_x are satisfied with a single mode on each side. The effects on surface waves of including the second, short scale–length, mode are discussed in the following section. It was argued by Wessen & Cramer (1991) that when the particle pressure is zero, the plasma can support formally infinite density perturbations and infinitely sharp velocity gradients. In a plasma with finite particle pressure the slow magnetoacoustic wave is present and the velocity perturbation is then due to the sum of two modes. This additional mode has an infinitely sharp gradient in the limit of zero pressure, so there arises the apparent discontinuity in v_x.

It was found by Wessen & Cramer (1991), using the condition $E_{y1} = E_{y2}$, that the inclusion of finite ion cyclotron frequency significantly alters the dispersion relation of the waves from the low–frequency result. The dispersion relation separates into branches through the introduction of Alfvén resonances and cutoffs, and surface modes exist with frequencies both less than and greater than the ion cyclotron frequency. For some angles of rotation of the magnetic field and directions of propagation of the surface wave relative to the magnetic field, two modes can exist for the same wavenumber, and some modes can propagate undamped by Alfvén resonance absorption in a nonzero width surface.

4.7   Radiative and Collisional Damping

For a calculation of the surface wave dispersion relation for a sharp surface, with nonideal corrections such as resistivity, electron inertia and thermal effects discussed in Chapters 2 and 3, the contribution of the short scale–length mode (the QEW; see Section 2.8) on each side of the surface must be included. An extended set of boundary conditions must also be used: the conditions normally employed are that the wave electric and magnetic fields tangential to the surface are continuous there. One approach is to treat the effect of the short–wavelength mode as a perturbation on the dispersion relation gained using the fast magnetoacoustic mode only (Cramer & Donnelly 1992; Collins, Cramer & Donnelly 1984): the frequency is assumed to change by only a small real and/or imaginary part from its value obtained neglecting the QEW.

We may picture the process as a mode conversion or coupling of the original surface wave into the short scale–length mode, although this mode conversion occurs on a sharp surface as opposed to the Alfvén resonance or cusp resonance mode conversion on a smooth surface discussed in earlier sections. The mode conversion takes place upon the reflection and transmission of the magnetoacoustic mode energy flux at the surface, and so may also be interpreted as a scattering process in which the magnetoacoustic mode is scattered into the short scale–length mode by the sudden density jump defining the surface (Rowe 1993). The surface wave can possibly lose energy by radiating energy away from the surface into a short–wavelength mode.

4.7.1 Radiative Damping

To discuss radiative effects, we consider a low–β plasma described by Ohm's law (Eq. 1.31) including electron inertia and electron pressure, but neglecting resistivity. These effects lead to the short (perpendicular) wavelength QEW modes described in Sections 2.8 and 3.2.2, which are further divided into the IAW and KAW modes. The different roles of the IAW and KAW modes in their effect on surface waves can be seen by inspecting Figure 3.5. If electron inertia is dominant over electron pressure, Figure 3.5(b) is applicable and at the surface wave frequency, above the Alfvén resonance frequency, the IAW and MHD modes are both evanescent. Thus no energy is lost from the surface, and the frequency is perturbed by a small (negative) real amount. If electron pressure is dominant, Figure 3.5(a) is applicable and the KAW mode has $k_x^2 > 0$ at the surface wave frequency. The KAW propagates energy away from the surface into the plasma and damping of the surface wave takes place, that is, the frequency acquires a small imaginary part that is proportional to the electron thermal speed. The surface wave has become a *leaky wave*. The radiated KAW mode may subsequently be damped by Landau damping (Hasegawa & Chen 1976).

An analysis can be made in Cartesian geometry for the semi–infinite low–β plasma in $x < 0$ with a plane interface at $x = 0$ to a vacuum in $x > 0$. The equilibrium magnetic field is uniform in the z–direction. In the plasma, the wavenumbers in the x–direction are given by $k_\pm$ for the slow and fast modes respectively (see Section 3.2.2). For the fast compressional mode with $\omega \ll \Omega_i$ and

$$k_-^2 \simeq G - k_y^2 \tag{4.77}$$

it is straightforward to show that the wave magnetic field is

$$\boldsymbol{B}_- = \left(-k_z C_-, -k_y k_z C_-/k_-, (k_-^2 + k_y^2)C_-/k_-\right) \tag{4.78}$$

with the fast mode wavenumber (k_-, k_y, k_z).

For the QEW mode, with $B_z = 0$, the wave field is approximately

$$\boldsymbol{B}_+ = \left(-(k_y + k_z)C_+, k_+(k_y + k_z)C_+/k_y, 0\right) \tag{4.79}$$

with the corresponding wavenumber (k_+, k_y, k_z). The transverse electric (TE) mode vacuum field is

$$\boldsymbol{B}_V = \left(-k_z C_V, -k_y k_z C_V/k_V, -k_z^2 C_V/k_V\right)$$

with the corresponding wavenumber (k_V, k_y, k_z), where $k_V = i(k_y^2 + k_z^2)^{1/2}$. The constants C_-, C_+ and C_V are mode amplitudes, and $k_\pm$ are given by Eqs. (3.14) and (3.15).

We now use the boundary conditions E_y and E_z continuous (which imply that B_x is continuous) and B_y and B_z continuous. The dispersion equation is then given by

$$
\begin{vmatrix}
-k_z & -k_z & -(k_y + k_z) \\
-k_y k_z / k_- & -k_y k_z / k_V & k_+ (k_y + k_z)/k_y \\
(k_-^2 + k_y^2)/k_- & -k_z^2/k_V & 0
\end{vmatrix} = 0.
\tag{4.80}
$$

Separating out the term proportional to k_+, we can write Eq. (4.80) as

$$
\frac{k_+}{k_y} \Delta_0 + \Delta_1 = 0
\tag{4.81}
$$

where Δ_0 is the dispersion function including the fast mode alone, and Δ_1 is the correction due to the short–wavelength mode. We then have the approximate expression for the perturbation in the frequency,

$$
\delta\omega = -\frac{k_y}{k_+} \frac{\Delta_1}{(\partial\Delta_0/\partial\omega)_{\omega_s}}.
\tag{4.82}
$$

The solution of the ideal MHD dispersion equation

$$
\Delta_0(\omega_s) = \frac{k_z^3}{k_V} + \frac{k_z(k_-^2 + k_y^2)}{k_-} = 0
\tag{4.83}
$$

is given by Eq. (4.16).

We also have

$$
k_-^2 = -k_y^4/(k_y^2 + k_z^2)
\tag{4.84}
$$

and

$$
k_+^2 = u_3/2 - k_y^2 \simeq \omega_s^2/2v_A^2\mu
\tag{4.85}
$$

where

$$
\mu = \frac{V_e^2 k_z^2 - \omega^2}{\Omega_i \Omega_e}.
\tag{4.86}
$$

Note that this expression for μ has been derived using a collisionless fluid model of the plasma. This model neglects Landau damping, which is of particular importance when $\omega \simeq V_e k_z$.

The correction to the dispersion function is

$$
\Delta_1 = \frac{k_y k_z}{k_- k_V} \frac{\omega_s^2}{v_A^2}.
\tag{4.87}
$$

We then find for the change in the frequency

$$
\delta\omega/\omega_0 = (\pm)\mathrm{i}\,\frac{k_y^4}{k_z(2k_y^2 + k_z^2)^{3/2}}(2\mu)^{1/2}
\tag{4.88}
$$

where the sign is the same as the sign of k_+. Thus for the KAW ($\mu > 0$), k_+ is negative real and the wave propagates away from the surface in the negative x–direction. Eq. (4.88) gives a damping rate, corresponding to a loss of energy from the surface wave. For the IAW ($\mu < 0$), k_+ is negative imaginary, the wave is evanescent in the plasma, and the frequency is shifted downwards. The frequency shift of the surface wave on a sharp transition due to coupling into the IAW, and damping rate of the wave due to coupling into the KAW, are both proportional to the electron thermal speed V_e.

Considering now the nonzero β, ideal MHD case, if the frequency is low the surface waves on sharp transitions discussed in Section 4.5.1, namely the Alfvén and the slow surface waves, are undamped by mode coupling, since there is a single fast magnetoacoustic mode in each plasma. If a finite ion cyclotron frequency is included, however, a second, short–wavelength mode arises (see Figure 3.6). The dispersion relations of both the Alfvén and the slow surface waves are modified in the same way as was described above for the low-β case. Thus there is mode conversion into the short–wavelength modes from the surface wave and thence a modification of the frequency or damping rate. Consider the Alfvén surface wave on a plasma/vacuum interface. From Figure 3.6, for nonzero small β there is a propagating short–wavelength mode at the (ideal) surface wave frequency which occurs at a frequency given by Eq. (4.16), above the Alfvén resonance frequency. Thus the wave is a leaky wave, that is, energy is lost from the surface wave and ω acquires an imaginary (damping) part. The damping rate can be calculated by a perturbation solution of the full dispersion relation similar to that above, and it is found to be proportional to $\beta^{1/2}$ for a plasma/plasma surface, and proportional to β for a plasma/vacuum surface (Cramer & Donnelly 1992).

4.7.2   Collisional Damping

One dissipative mechanism for surface waves in a partially ionized plasma is ion-neutral friction. It is interesting to note that for a cold plasma, as we have seen in Section 3.2.4, the dissipative ion-neutral collision term does not give rise to an additional short–wavelength mode (e.g. Woods 1962), even if the Hall term is included, so that damping of the surface wave, with frequency up to the range of the ion cyclotron frequency, will occur via direct damping of the MHD mode rather than coupling into another mode.

As an example, we consider a low–frequency surface wave on a cold plasma/vacuum surface. With ion-neutral collisions included, we have

$$G = \frac{s\omega^2}{v_{\mathrm{A}}^2} - k_z^2 \quad \text{and} \quad H = 0 \tag{4.89}$$

with s the neutral mass loading factor defined in Eq. (2.104). We also have

$$k_{\mathrm{p}}^2 = k_y^2 + k_z^2 - \frac{s\omega^2}{v_{\mathrm{A}}^2}. \tag{4.90}$$

Considering the neutral coupling as small, we may write $s = 1 + \varepsilon$, with

$$\varepsilon = \frac{\rho_{\mathrm{n}}}{\rho_0}\frac{1 + i\tau}{1 + \tau^2} \ll 1. \tag{4.91}$$

By the same reasoning as for the resonance damping case (Section 4.2), the approximate surface wave dispersion equation can be written as the first two terms of an expansion in ε:

$$\mathcal{D}(\tilde{\omega}) = \mathcal{D}_0(\tilde{\omega}) + \varepsilon \mathcal{D}_1(\tilde{\omega}) = 0 \tag{4.92}$$

with $\tilde{\omega} = \omega_{\rm s} - i\gamma$, and the damping rate γ given by

$$\gamma = \frac{\mathrm{Im}(\varepsilon \mathcal{D}_1(\omega))}{\partial \mathcal{D}_0/\partial \omega} \Bigg|_{\omega=\omega_{\rm s}} \tag{4.93}$$

where $\omega_{\rm s}$ is the surface wave frequency given by Eq. (4.16).

In this case, the same boundary conditions (E_y and B_z continuous) as used in the collision–free case (Section 4.2.1) may be used, because of the single mode in the plasma. Then $\mathcal{D}(\omega)$ takes the same form as in Eq. (4.9), and becomes

$$\mathcal{D}(\tilde{\omega}) = \frac{k_z^2 - s\omega^2/v_{\rm A}^2}{k_{\rm p}} + \frac{k_z^2}{k_{\rm v}}. \tag{4.94}$$

In the expansion given by Eq. (4.92), $\mathcal{D}_0$ is given by Eq. (4.15) and $k_{\rm p} = k_{\rm p0}$ by Eq. (4.17), and

$$\mathcal{D}_1(\tilde{\omega}) = -\frac{\omega^2}{v_{\rm A}^2} \frac{k_{\rm p0}^2 + k_y^2}{2k_{\rm p0}^3}. \tag{4.95}$$

Thus in this approximation, the damping rate is proportional to the imaginary part of ε. There is also a real frequency shift, proportional to the real part of ε. Alternatively, the exact dispersion equation (4.94) may be numerically solved to find the frequency shift and damping rate. The effects of ion-neutral collisions on Alfvén surface waves in an incompressible plasma with nonzero pressure have been considered by Uberoi & Datta (1998).

Resistive and viscous damping of low–frequency surface waves has been investigated by a number of authors, with the primary aim of explaining the heating of the solar corona, for example Uberoi & Somasundaram (1982), Gordon & Hollweg (1983) and Steinolfson *et al.* (1986). One (nonself–consistent) approach is to use the surface wave solution derived from ideal MHD (Section 4.5) and compute the viscous heating rate due to the wave fluctuations. However, in general these dissipation terms raise the order of the fluid differential equations and so give rise to an additional, short–wavelength mode in the plasma, as discussed in the previous section for thermal and electron inertia effects, with an electric field component parallel to the magnetic field. The difference is that the resistive and viscous short–wavelength modes are highly damped. Again the surface wave couples into these modes with resultant damping of the wave. A perturbation approach, similar to that used for the radiative damping problem discussed above, can be used to calculate the effect of the short–wavelength modes (Collins, Cramer & Donnelly 1984). A self–consistent approach has been used by Steinolfson *et al.* (1986) to calculate the viscous and resistive damping rates in the incompressible limit. Under solar coronal conditions, viscous damping is found to dominate resistive damping by two orders of magnitude. The contribution of viscosity to a second, short–wavelength mode and its effect on the incompressible surface wave has been shown by Uberoi & Somasundaram (1982).

4.8   Kinetic Theory

The collisionless damping of Alfvén surface waves has been postulated as a solar coronal heating mechanism (Assis & Busnardo-Neto 1987; de Assis & Tsui 1991), because the classical processes of resistivity and viscosity are ineffective damping mechanisms in such a hot plasma. A full kinetic theory would be necessary for the proper calculation of the collisionless damping of the waves in this case. A kinetic theory treatment has been attempted for high–frequency ($\omega \simeq \omega_{pe}$) surface waves in unmagnetized plasmas (Shivarova & Zhelyazkov 1982; Guernsey 1969; Alexandrov, Bogdankevich & Rukhadze 1984), in which the collisionless linear damping has been calculated. In a hot magnetized plasma, the magnetoacoustic and KAW modes in the plasma both contribute to the dispersion relation of the surface wave, and energy in the KAW mode is radiated away from the surface with resultant damping of the surface wave, just as in the fluid theory discussed in Section 4.7. Because only the KAW mode has a nonzero electric field component along the background magnetic field, only that mode is Landau damped. The magnetoacoustic mode experiences transit time magnetic damping. However, the damping of the surface wave is independent of whether or not the radiated KAW is Landau damped by the plasma electrons.

The collisionless damping in the hot plasma case can be calculated in one approach by assuming the cold, low–frequency surface wave dispersion relation described in Section 4.2 as a first approximation, and then perturbing this solution, using the kinetic theory dielectric tensor, to obtain the wave electric and magnetic fields responsible for Landau and transit time damping. It was found using this approach (de Assis & Tsui 1991) that the transit time magnetic damping term cancels with a cross term in the dielectric tensor, leaving Landau damping as the only collisionless damping process. A more rigorous calculation of collisionless damping in magnetized plasmas can be made using an image plasma theory described further in this section, analogous to the method of images used extensively in electrostatic theory (e.g. Griffiths 1989). In this approach there is also found to be no contribution (to dominant order) to transit time magnetic damping of the magnetoacoustic mode.

The formal theory of surface waves in a collisionless magnetized plasma, developed in the papers of Rowe (1991, 1992, 1993), and applied to waves of frequency approaching the ion cyclotron frequency, is based on an extension of the plasma response theory used to describe bulk waves in a uniform plasma (e.g. Melrose & McPhedran 1991). This theory has as special cases the fluid results surveyed in the earlier sections, but can also be applied to a kinetic theory description of the plasma, and reduces to the earlier results for unmagnetized plasmas. The fundamental problem in the kinetic theory of a plasma with a nonuniformity such as a surface is that the dielectric tensor of such a medium depends on particle orbits over the entire volume of the medium, that is, it is a nonlocal quantity, and is not easily amenable to Fourier transform techniques. However, a theory of surface waves in an unmagnetized plasma with isotropic particle distribution functions has been described (Guernsey 1969; Barr & Boyd 1972; Alexandrov, Bogdankevich & Rukhadze 1984), which uses an image plasma approach in which the linearized Vlasov equation describing the response to the wave fields of the physical plasma on one side of the surface is mathematically extended into a nonphysical region on the other side of the surface, subject to boundary conditions on the surface such as the assumption of specular reflection of particles, and symmetry conditions on the wave fields. The important feature of this approach is that the semi–infinite physical plasma is assumed to be described by the infinite medium response, that is, by the usual dielectric tensor discussed in Chapter 2.

We briefly summarize the theory here. Consider a sharp surface separating a plasma and a vacuum. Maxwell's equations (1.1)-(1.4) are Fourier transformed in space and time, retaining source terms at the surface, assumed to lie in the y-z plane at $x = 0$, resulting in the following equation for the Fourier transform of the electric field $\boldsymbol{E}(\omega, \boldsymbol{k})$:

$$\Lambda(\omega, \boldsymbol{k}) \boldsymbol{E}(\omega, \boldsymbol{k}) = -\mathrm{i} \frac{\mu_0 c^2}{\omega} \boldsymbol{M}_{\mathrm{s}}(\omega, \boldsymbol{k}_{\mathrm{s}}) \tag{4.96}$$

where the response tensor $\Lambda(\omega, \boldsymbol{k})$ is given by

$$\Lambda(\omega, \boldsymbol{k}) = \frac{c^2}{\omega^2} (\boldsymbol{kk} - k^2 \boldsymbol{\delta}) + \boldsymbol{K}(\omega, \boldsymbol{k}) \tag{4.97}$$

with $\boldsymbol{K}(\omega, \boldsymbol{k})$ the equivalent dielectric tensor. $\boldsymbol{\delta}$ here denotes the unit tensor. We note that the dielectric tensor for a bi–Maxwellian plasma expressed by Eq. (A.10) together with Eq. (A.16) is defined for the case $k_y = 0$. For surface waves we require the more general dielectric tensor with $k_y \neq 0$ defined by Eqs. (A.25) and (A.27).

$\boldsymbol{M}_{\mathrm{s}}(\omega, \boldsymbol{k}_{\mathrm{s}})$ is the surface current on the transition plane within the real-image system, regarded as an extraneous source term, and $\boldsymbol{k}_{\mathrm{s}}$ is the (fixed and real) wavenumber in the y-z plane. The total wavevector is $\boldsymbol{k} = k_n \hat{\boldsymbol{x}} + \boldsymbol{k}_{\mathrm{s}}$, where $\boldsymbol{k}_{\mathrm{s}}$ is complex to allow for a surface wave field dependence. When the source term is zero, Eq. (4.96) is the homogeneous wave equation determining the normal modes of the uniform real–image plasma, and the condition for nontrivial solutions is that the determinant of $\Lambda(\omega, \boldsymbol{k})$ vanishes, that is,

$$\Lambda(\omega, \boldsymbol{k}) = 0. \tag{4.98}$$

For a real system of two uniform plasmas separated by a surface, there are two real-image systems and correspondingly two source terms $\boldsymbol{M}_{\mathrm{s}}^{\pm}(\omega, \boldsymbol{k}_{\mathrm{s}})$. The source terms are determined by imposing the relevant boundary conditions at the physical surface, which are assumed to be those used previously for the low–β fluid model, namely, continuity of the surface components of the wave electric and magnetic fields. Defining the surface electric field as $\boldsymbol{E}_{\mathrm{s}} = E_y \hat{\boldsymbol{y}} + E_z \hat{\boldsymbol{z}}$, the field Fourier transformed in the y and z directions is

$$\boldsymbol{E}_{\mathrm{s}}(\omega, \boldsymbol{k}_{\mathrm{s}}) = \lim_{x \to 0} \int \frac{\mathrm{d}k_x}{2\pi} \mathrm{e}^{\mathrm{i}k_x x} \boldsymbol{E}_{\mathrm{s}}(\omega, \boldsymbol{k}_{\mathrm{s}}). \tag{4.99}$$

The surface electric and magnetic fields on either side of the surface in the real system can be written in terms of $\boldsymbol{M}_{\mathrm{s}}^{\pm}(\omega, \boldsymbol{k}_{\mathrm{s}})$ using Eqs. (4.96) and (4.99) for the two real-image systems. Applying the continuity conditions then leads to the result

$$\boldsymbol{Z}_{\mathrm{s}}^{\pm}(\omega, \boldsymbol{k}_{\mathrm{s}}) \boldsymbol{M}_{\mathrm{s}}^{\pm}(\omega, \boldsymbol{k}_{\mathrm{s}}) = 0 \tag{4.100}$$

where $\boldsymbol{Z}_{\mathrm{s}}^{\pm}(\omega, \boldsymbol{k}_{\mathrm{s}})$ is a matrix,

$$\boldsymbol{Z}_{\mathrm{s}}^{\pm}(\omega, \boldsymbol{k}_{\mathrm{s}}) = \lim_{x \to 0} \int \frac{\mathrm{d}k_x}{2\pi} \mathrm{e}^{\mathrm{i}k_x x} \Lambda(\omega, \boldsymbol{k}) \boldsymbol{Q}_{\mathrm{s}}^{\pm}(\omega, \boldsymbol{k}) \tag{4.101}$$

where $\mathbf{\Delta}(\omega, \boldsymbol{k})$ and $\boldsymbol{Q}_{\mathrm{s}}^{\pm}(\omega, \boldsymbol{k})$ are matrices dependent on the wavenumbers and the dielectric tensor.

The dispersion relation for the surface waves is then determined by setting the determinant of $\boldsymbol{Z}_{\mathrm{s}}^{\pm}(\omega, \boldsymbol{k}_{\mathrm{s}})$ to zero. The essential point to note is that the integrand of Eq. (4.101) is inversely proportional to the determinant $\Lambda(\omega, \boldsymbol{k})$ (Eq. 4.98) describing normal modes in the two real-image plasma systems, that is, it has poles at the values of k_x corresponding to those normal modes. The total electric field is thus determined by the sum of contributions from each of the bulk modes of the plasma. This general theory has been shown to yield the known results for surface waves in a cold plasma (Rowe 1991, 1992). For a hot plasma-vacuum interface, surface wave damping is due to mode conversion to the short–wavelength mode that is either Landau–damped in the plasma or that radiates the surface wave energy away from the interface. In the case of a plasma-plasma interface, the surface wave can lose energy to a radiating short–wavelength mode that propagates into the less dense plasma, provided that the plasma is sufficiently cold. In the case of a plasma slab bounded on both sides by a vacuum, the waves lose energy only via Landau damping, and there is no loss into a radiating vacuum mode, because the vacuum modes are evanescent.

5 Instabilities and Nonlinear Waves

5.1 Introduction

We shall see in Chapters 7 and 8 that Alfvén waves play an important part in the dynamics of a number of space and astrophysical plasmas, such as the solar corona, solar and stellar winds, planetary magnetospheres, and interstellar clouds. Such plasmas are often in a nonequilibrium state and so may provide sources of free energy that can lead to instabilities of Alfvén and magnetoacoustic waves and the development of large amplitudes of the waves. In addition, Alfvén waves in laboratory fusion devices can be destabilized by sources of free energy such as energetic alpha particles. Departures from equilibrium are present either in the phase space distributions of the charged particles, such as beams or anisotropic distributions of particle pitch angles, or in nonequilibrium spatial distributions.

Comprehensive reviews of instabilities, covering the instabilities producing Alfvén waves and magnetoacoustic waves, have already been provided by Melrose (1986) and Gary (1993), so we shall only provide a brief summary of the theory of some instabilities here. Instabilities may derive from normal modes of a system that grow in space or time. Initially, the associated fluctuations are relatively weak, so that linear theory is appropriate to describe the physics in the first approximation. However, linear theory cannot describe the ultimate fate of an instability of a given mode, nor its interactions with other modes. These questions are addressed by nonlinear theory or computer simulations. There has been much effort recently to develop a theory of MHD turbulence, employing insights gained from numerical simulations (Roberts *et al.* 1992), and more analytic approaches (Marsch & Tu 1993).

5.2 Instabilities

We can distinguish macroinstabilities, which occur at relatively long wavelength, and micro-instabilities, which commonly occur at short wavelengths. Macroinstabilities generally depend on configuration space properties, do not involve a resonant interaction with particles, and can be described by fluid theories. Microinstabilities depend on the details of the particle velocity distributions, may be nonresonant or resonant, and may be described by fluid theory or kinetic theory. However, there is of course an overlap between the different categories of instabilities.

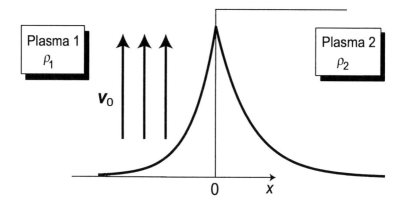

Figure 5.1: The exponentially decreasing fields of a Kelvin-Helmholtz unstable surface wave localized at a density jump between two regions of uniform plasma density, at the point $x = 0$. The plasma in $x < 0$ flows with a velocity v_0 in the y-z plane.

5.2.1   Macroinstabilities

The macroinstabilities that arise because of spatial nonuniformities of the plasma are the instabilities of the surface waves described in Chapter 4. Current–driven macroinstabilities are discussed in Chapter 6, in terms of the instabilities of body and surface Alfvén and magneto-acoustic waves in a cylindrical plasma. Another example of a macroinstability, which we discuss here, is the Kelvin-Helmholtz instability, which occurs when two superposed fluids move with uniform speeds, with a relative drift velocity along the surface of discontinuity. It has great relevance to phenomena such as cometary tails and the magnetospheric boundary, and to astrophysical sources such as the jets in active galactic nuclei.

Kelvin–Helmholtz Instability

Consider a planar discontinuity located in the y-z plane at $x = 0$, between a plasma on side 1 ($x < 0$) moving at a constant velocity v_0 (in the y-z plane) relative to a stationary plasma on side 2 ($x > 0$). The Alfvén speeds on each side are respectively v_{A1} and v_{A2}, and the densities are ρ_1 and ρ_2. The plasmas have nonzero particle pressure, and the frequencies and growth rates are assumed to be much less than the ion cyclotron frequency. The equilibrium satisfies the condition (3.48). The governing wave differential equations for the perturbed fields v and B in a stratified plasma with an x–dependent zero–order velocity $v_0(x)$, are generalizations of Eqs. (3.50) and (3.51) used in Section 3.3.1 to describe waves in stratified stationary plasmas. Thus the equation of motion takes account of the convected fluid element:

$$\rho_0 \frac{\mathrm{d}v}{\mathrm{d}t} = \rho_0 \left(\frac{\partial}{\partial t} + v_0 \cdot \nabla \right) v = -\nabla p_\mathrm{T} + \frac{1}{\mu_0}(B \cdot \nabla)B_0 + \frac{1}{\mu_0}(B_0 \cdot \nabla)B \qquad (5.1)$$

and Faraday's equation uses Ohm's law (Eq. 1.8) with v_0 included:

$$\frac{\partial B}{\partial t} = \nabla \times (v \times B_0 + v_0 \times B).$$ (5.2)

The net effect is that equations of the form (3.52)-(3.55) are obtained, but with the substitution of the Doppler shifted frequency,

$$\omega \rightarrow \omega' = \omega - k \cdot v_0.$$ (5.3)

For the discontinuous transition, the calculation of the surface wave dispersion equation now proceeds as in Section 4.5, but with the important difference that the boundary condition of continuous v_x in the stationary plasma case must be replaced by the requirement of continuous (convected) fluid displacement ξ_x, where $v_x = d\xi_x/dt$. The other boundary condition of continuous p_T remains valid. The dispersion equation analogous to that for the stationary plasma result (4.34) is then derived to be

$$\frac{\omega' \tau_1}{q_1} + \frac{\omega \tau_2}{q_2} = 0$$ (5.4)

where τ_1 and q_1 are as defined in Eqs. (3.54) and (3.55), but with ω replaced by ω'. Equation (5.4) may be written as an algebraic equation of the 10th order in ω; its solutions have been discussed by Pu & Kivelson (1983) with application to instabilities on the Earth's magnetopause interface between the flowing magnetosheath plasma and the magnetospheric plasma.

For an incompressible plasma ($c_s^2 \rightarrow \infty$), with $q_1 = q_2 = k$ (see Eq. (4.38)), the dispersion equation (5.4) simplifies to a quadratic equation in ω (Lee, Albano & Kan 1981):

$$\rho_1(\omega'^2 - (k \cdot v_{A1})^2) + \rho_2(\omega^2 - (k \cdot v_{A2})^2) = 0$$ (5.5)

whose solution may be unstable, that is, ω may have a positive imaginary part. Further simplification occurs if v_0, v_{A1} and v_{A2} are all parallel to the z–axis. The surface wave then becomes unstable if the flow velocity satisfies the condition

$$v_0^2 > \frac{\rho_1 + \rho_2}{\rho_1 \rho_2}(\rho_1 v_{A1}^2 + \rho_2 v_{A2}^2).$$ (5.6)

Finally, we may note that if the transition between the two plasmas has a nonzero width, with smoothly varying flow velocity, the Alfvén resonance condition, $\tau = 0$, is modified by the flow velocity. This can lead to an instability, even if the Kelvin-Helmholtz instability is absent (Ryutova 1988).

5.2.2   Microinstabilities

The classification of a microinstability requires the identification of the source of free energy, and the dispersion properties of the unstable mode. Microinstabilities may be further classified into two categories: hydrodynamic or reactive instabilities, and kinetic instabilities (Melrose 1986).

Temperature Anisotropy: the Firehose Instability

As an example of a hydrodynamic instability involving Alfvén waves, we consider the *firehose instability*, which is caused by an anisotropy of the particle pressures, or temperatures, parallel and perpendicular to the magnetic field. The double–adiabatic fluid model of collisionless plasmas with anisotropic pressures discussed in Chapter 1 can be used to describe this instability. Thus the two equations of motion (1.26) and (1.27) for the parallel and perpendicular components of the fluid velocity, and the double–adiabatic equations of state (1.25), are used. Linear perturbations of the equations are considered, and we assume that the frequency and growth rate of the instability are well below the ion cyclotron frequency. The zero–order and first–order parallel and perpendicular pressures are $p_{0\parallel}(p_{1z})$ and $p_{0\perp}(p_{1\perp})$ respectively.

The characteristic variables introduced in Section 2.2 (Eq. 2.8) can be used, and manipulation of Eqs. (1.26) and (1.27), assuming the zero–order density is uniform and the zero–order magnetic field is uniform in the z–direction, leads to the following set of equations to replace Eqs. (2.10), (2.12) and (2.14):

$$\rho_0 \frac{\partial \zeta_{1z}}{\partial t} - B_0 \left(1 + \frac{\mu_0(p_{0\perp} - p_{0\parallel})}{b_0^2}\right) \frac{\partial J_{1z}}{\partial z} = 0 \tag{5.7}$$

$$\rho_0 \frac{\partial}{\partial t} \nabla \cdot \boldsymbol{v}_1 + \frac{B_0}{\mu_0} \nabla^2 B_{1z} + \nabla_\perp^2 p_{1\perp} + \frac{\partial^2 p_{1z}}{\partial z^2} + \frac{2(p_{0\perp} - p_{0\parallel})}{b_0} \frac{\partial^2 B_{1z}}{\partial z^2} = 0 \tag{5.8}$$

$$\rho_0 \frac{\partial v_{1z}}{\partial t} + \frac{\partial p_{1z}}{\partial z} + \frac{(p_{0\perp} - p_{0\parallel})}{B_0} \frac{\partial B_{1z}}{\partial z} = 0. \tag{5.9}$$

Here the perturbations in the particle pressures are obtained by linearizing the double–adiabatic equations of state (1.25), giving

$$p_{1\perp} = p_{0\perp} \left(\frac{\rho_1}{\rho_0} + \frac{B_{1z}}{B_0}\right) \tag{5.10}$$

$$p_{1z} = p_{0\parallel} \left(\frac{3\rho_1}{\rho_0} - \frac{2B_{1z}}{B_0}\right). \tag{5.11}$$

Equations (2.11), (2.13) and (2.15) are unchanged.

The characteristic variables ζ_{1z} and J_{1z} occur only in Eqs. (2.10) and (2.11), which again describe the shear Alfvén wave, whose dispersion relation is modified by the pressure anisotropy. The resulting dispersion equation is again independent of the perpendicular wavenumber, and is given by

$$\omega^2 = k_z^2 \left(v_A^2 + \frac{p_{0\perp} - p_{0\parallel}}{\rho_0}\right). \tag{5.12}$$

We thus see that the shear Alfvén wave becomes unstable if

$$p_{0\parallel} > p_{0\perp} + \frac{B_0^2}{\mu_0}. \tag{5.13}$$

This *firehose instability* is discussed further by, for example, Clemmow & Dougherty (1969).

The other equations, (2.13), (2.15) and (5.8)-(5.11), can be expressed in terms of the characteristic variables $\nabla \cdot \boldsymbol{v}_1$ and B_{1z} describing the magnetoacoustic mode. The resulting dispersion equation can be found in Clemmow & Dougherty (1969); it is sufficient to say here that the magnetoacoustic wave can become unstable for parallel propagation, in which case it becomes degenerate with the firehose unstable shear Alfvén wave, and for almost perpendicular propagation, the magnetoacoustic wave is unstable if $p_{0\perp}$ *exceeds* $p_{0\|}$ by a sufficient amount, in contrast to the result (5.13) for the firehose instability. This instability of the almost perpendicularly propagating magnetoacoustic wave is known as the *mirror instability*. We may note that the condition $p_{0\perp} > p_{0\|}$, or $T_\perp > T_\|$, is observed more often in space plasmas than $T_\| > T_\perp$, because parallel heated particles move rapidly along the magnetic field, and may leave the region where they are energized more rapidly. The firehose instability is also discussed in Chapter 7 in connection with another hydrodynamic microinstability involving Alfvén waves, the ion ring–beam instability caused by "pick–up" ions in space plasmas.

In order to discuss *kinetic* instabilities driven by a temperature anisotropy, we use the bi–Maxwellian zero–order velocity distribution function with zero drift given by Eq. (A.8) (Gary 1993). For kinetic instabilities, the imaginary part of the frequency is usually small compared with the real part, in contrast to the hydrodynamic instability such as the firehose instability which is purely growing (the real part of ω is zero). For parallel propagation, the imaginary part of the frequency treated as a small perturbation is then, from Eq. (2.149),

$$\gamma = \frac{\pi}{2\omega_\mathrm{r}} \sum_\alpha \frac{\omega_{\mathrm{p}\alpha}^2}{(2\pi V_\alpha^2)^{1/2}} \left[-\frac{T_{\perp\alpha}}{T_{\|\alpha}} \frac{\omega_\mathrm{r}}{k_z} \pm \left(1 - \frac{T_{\perp\alpha}}{T_{\|\alpha}} \right) \frac{\Omega_\alpha}{k_z} \right] \exp \left[\frac{-(\omega_\mathrm{r} \pm \Omega_\alpha)^2}{2k_z^2 V_\alpha^2} \right] \quad (5.14)$$

with V_α the parallel thermal speed, and with the $+$ and $-$ signs corresponding respectively to right–hand circular polarization (RHP) and left–hand circular polarization (LHP) of the wave. For low frequency, i.e. $\omega_\mathrm{r} < |\Omega_\alpha|$, the second term in the first square bracket on the right side of Eq. (5.14) is dominant. Noting that Ω_α has the sign of the particle charge, it then follows from Eq. (5.14) that ions (electrons) with $T_\perp < T_\|$ have the potential to make $\gamma > 0$, that is, drive a RHP (LHP) wave unstable, while for $T_\| < T_\perp$, ions (electrons) can drive a LHP (RHP) wave unstable.

For a nonexponentially small growth rate given by Eq. (5.14), a kinetic microinstability must have a resonance between the particles and the wave, so for anisotropic ions, the LHP mode may become unstable, with a frequency approaching the ion cyclotron frequency. From Eq. (5.14), the growth rate is also sensitive to the parallel ion β. Electromagnetic instabilities due to the condition $T_{\perp \mathrm{i}} > T_{\|\mathrm{i}}$ are often invoked in studies of the magnetosphere, where such anisotropies are often observed. The heating and reflection of solar wind ions at the Earth's bow shock produces strongly anisotropic ion distribution functions immediately downstream of the shock. As the plasma flows into the magnetosheath, the distributions often assume bi–Maxwellian form with $T_{\perp \mathrm{i}} > T_{\|\mathrm{i}}$. In that case, there is the hydrodynamic mirror instability mentioned above, and there is also a kinetic *ion cyclotron anisotropy* (or *Alfvén/ion cyclotron*) instability for $\beta \lesssim 6$ (Gary 1993). For the ion cyclotron anisotropy instability, the wave is LHP in the ion cyclotron mode, and the ions are resonant. Further details of mirror and ion cyclotron anisotropy instabilities, including particle–in–cell simulations of nonlinear effects, may be found in the papers by Davidson & Ogden (1975), Gary *et al.* (1976) and McKean, Winske & Gary (1992).

Drift (Beam) Instabilities

The other type of microinstability that we discuss is the drift or beam instability, where different species of particles have mean velocities relative to each other. This relative streaming provides the free energy driving the instability. The species are described by shifted Maxwellian distributions.   Consider a two–species, three–component plasma, with a tenuous beam of particles of one species, a relatively dense core of the same species, and a background second species to neutralize the plasma. Let the beam and core drift velocities be v_{0b} and v_{0c} respectively, both parallel or antiparallel to the background magnetic field, and a relative drift velocity

$$v_0 = v_{0b} - v_{0c}. \tag{5.15}$$

We assume zero net current in the plasma (current–driven instabilities are discussed in Chapter 6):

$$\sum_\alpha q_\alpha n_\alpha v_{0\alpha} = 0. \tag{5.16}$$

Again, there are hydrodynamic instabilities, with $\omega_r \simeq 0$, and kinetic instabilities. For a kinetic instability of a parallel propagating wave, the imaginary part of the frequency is small compared with the real part, and is given from Eq. (2.149) by

$$\gamma \simeq \frac{\pi}{2\omega_r} \sum \frac{\omega_\alpha^2}{(2\pi V_\alpha^2)^{1/2}} \left(v_{0\alpha} - \frac{\omega_r}{k_z} \right) \exp \left[-\frac{(\omega_r \pm \Omega_\alpha - k_z v_{0\alpha})^2}{2k_z^2 V_\alpha^2} \right]. \tag{5.17}$$

For an instability (positive γ), there is a threshold beam velocity,

$$v_{0b} > \omega_r / k_z \simeq v_A. \tag{5.18}$$

We consider a relatively weak ($n_b \ll n_c$) and energetic ($v_{0b} \gg v_{0c}$) beam.

Plasmas in which the beam and core species are positive ions have instabilities (*electromagnetic ion/ion instabilities*) that have the lowest frequencies and growth rates, of the order of or less than the ion cyclotron frequency, but can grow to the largest fluctuating field energy densities (Gary 1991). These modes are observed in many space plasma regions where ion beams occur, such as the solar wind, planetary bow shocks, the Earth's magnetotail and cometary environments. The parallel propagating modes are found to have the highest growth rates. A cool ion beam ($V_b \ll v_{0b}$) excites a firehose–like nonresonant hydrodynamic instability, which requires $v_{0b} \gg v_A$, and propagates in the direction opposite to the beam direction. The growth rate of this instability is given by

$$\frac{\gamma}{\Omega_i} \simeq \frac{n_b}{2n_e} \frac{v_{0b}}{v_A} \tag{5.19}$$

where Ω_i is the ion cyclotron frequency of the beam and core ions.

A cool ion beam also produces an RHP resonant mode, which at $v_{0b} = 0$ is the RHP fast magnetoacoustic wave. The beam ions are resonant with the RHP mode because of the condition

$$\omega_r + \Omega_i - k_z v_{0b} = 0 \tag{5.20}$$

which can only be satisfied for nonzero beam velocity, $v_{0b} \gtrsim v_A$, and which is known as the *anomalous Doppler effect*. The physical basis of this effect is that the beam is faster than the wave, so in the frame of the beam, the wave slips behind and appears to reverse its polarity. A wave that is RHP in the laboratory frame can then become LHP in the beam frame, and so resonate with the beam ions. The electrons and core ions are nonresonant in this instability, and the unstable wave propagates parallel to the beam direction. The real part of the frequency is $\omega_r \ll \Omega_i$, and the growth rate is given by

$$\frac{\gamma}{\Omega_i} \simeq \left(\frac{n_b}{2n_e} \right)^{1/3} . \tag{5.21}$$

Thus the nonresonant mode can have the higher growth rate, given by Eq. (5.19), if v_{0b}/v_A and n_b/n_e are sufficiently large.

A warm ($V_b \simeq v_{0b}$) or hot ion beam ($V_b \gg v_{0b}$) gives an unstable mode that grows from an LHP Alfvén/ion cyclotron wave, with the beam ions resonant with the LHP mode through the normal Doppler effect with the resonance condition

$$\omega_r - \Omega_i - k_z v_{0b} = 0. \tag{5.22}$$

For $\omega_r < \Omega_i$, the LHP mode will resonate only with ions moving antiparallel to $\boldsymbol{v}_{0b}$, so there has to be a sufficient number of such ions in the warm or hot distribution to drive this instability (Gary 1991).

5.3 Acceleration of Charged Particles

Cosmic rays are believed to be accelerated efficiently at interstellar and intergalactic shock waves, in particular shock waves associated with supernovas, via a *diffusive acceleration* or *Fermi* process involving Alfvén waves (Achterberg & Blandford 1986; Kirk 1994). Energetic particles are detected directly by satellite experiments at the Earth's bow shock. The particles are confined close to the shock front by frequent scattering from low–frequency Alfvén waves. If the shock speed is much larger than the Alfvén speed, the waves are essentially convected at the velocity of the background plasma. The waves on opposite sides of the shock front then approach each other and particles gain energy by being scattered backwards and forwards across the shock front. The scattered waves in the region upstream of the shock front can be generated self–consistently via a beam instability due to the accelerated particles, which try to stream down their density gradient away from the shock. The anisotropy associated with this streaming motion leads to wave generation through a linear instability at the ion cyclotron resonance. In the downstream region there are no density gradients and so no linear wave growth there. However, the waves generated upstream are convected through the shock and so give rise to downstream scattering centres. The downstream plasma must be pressure dominated, that is, of high β, so the waves will be highly damped for propagation oblique to the magnetic field, as was discussed in Section 2.7.3.

Galactic cosmic rays are also trapped by resonant scattering off hydromagnetic waves. We have seen that linear Alfvén waves in a collisionless plasma experience very little Landau damping or transit time magnetic damping. However, a nonlinear Landau damping process can damp the waves, and there may be a steady state with wave excitation by cyclotron resonance of anisotropic energetic particles, being balanced by nonlinear Landau damping by the thermal plasma of the Galaxy (Lee & Völk 1973; Völk & Cesarsky 1982).

Alfvén waves may also play a role in the acceleration of particular species of particles. In the solar wind the dominant ionic component is protons, with an admixture of $\simeq 10\%$ of helium and $\lesssim 1\%$ of other ions. The minor ions have flow speeds, temperatures and temperature anisotropies that differ from protons and from one minor species to another, and which may possibly be explained by their interactions with ion cyclotron waves. Because the magnetic field decreases with radial distance from the Sun, ω/Ω_i increases for outward propagating Alfvén waves. The waves can eventually encounter an ion cyclotron resonance via the "magnetic beach" effect (Stix 1992), and the first resonance encountered is that for the lowest cyclotron frequency. The interaction of the various species with the waves can lead to strongly species–dependent acceleration, with protons being the least affected.

5.4 Nonlinear Waves

The effects of nonlinear amplitudes of Alfvén and magnetoacoustic waves have been vigorously studied in recent years, but in order to make progress it has often been necessary to make various simplifications, such as the assumption of a uniform plasma medium. In a plasma of uniform density and with uniform background magnetic field, it has been found that nonlinear Alfvén wave solitons can propagate parallel to the background field, provided the nonlinearities are balanced by a dispersive term in the wave equations (Mjølhus 1976; Spangler & Sheerin 1982; Sakai & Sonnerup 1983). The dispersive term is due to the finite ion cyclotron frequency effect (sometimes referred to as the finite ion inertia effect), arising explicitly from the Hall term in the generalized Ohm's law. The nonlinear waves are circularly polarized, and the wave magnetic field satisfies the "Derivative Nonlinear Schrödinger" (DNLS) equation. The soliton solutions of this equation for the parallel case have been used as a basis for the description of MHD turbulence in the solar wind (Spangler & Sheerin 1982; Ghosh & Papadopoulos 1987). On the other hand, for propagation at sufficiently large angles to the background field, the dispersive fast and slow MHD waves obey the Korteweg-de Vries equation (KdV), and the Alfvén wave obeys the modified KdV equation (MKdV), which has a weaker nonlinearity.

We proceed in this section to first establish the nonlinear differential equations from the fluid model including the Hall term in the Ohm's law, and then consider low–frequency ($\omega \ll \Omega_i$) and higher frequency ($\omega \lesssim \Omega_i$) nonlinear wave solutions, for parallel and almost parallel propagation.

5.4.1 Wave Equations

We consider a plasma containing a single ion species, and allow nonlinear oscillations of the density ρ, magnetic field $\boldsymbol{B}$ and velocity $\boldsymbol{v}$. Normalizing ρ and $\boldsymbol{B}$ by reference values ρ_0 and B_0, the fluid equations (1.3), (1.11) and (1.12), including the Hall term in Ohm's law (Eq. 1.28), may be written:

$$\frac{\partial \rho}{\partial t} + \nabla \cdot (\rho \boldsymbol{v}) = 0 \tag{5.23}$$

$$\rho \frac{\mathrm{d}\boldsymbol{v}}{\mathrm{d}t} = -\nabla p + v_{\mathrm{A}}^2 (\nabla \times \boldsymbol{B}) \times \boldsymbol{B} \tag{5.24}$$

$$\frac{\partial \boldsymbol{B}}{\partial t} = \nabla \times (\boldsymbol{v} \times \boldsymbol{B}) - \frac{v_{\mathrm{A}}^2}{\Omega_{\mathrm{i}}} \nabla \times \left(\frac{1}{\rho} (\nabla \times \boldsymbol{B}) \times \boldsymbol{B} \right). \tag{5.25}$$

where v_{A} is the Alfvén speed and Ω_{i} is the ion cyclotron frequency, using B_0 and ρ_0. The adiabatic gas law (Eq. 1.13) is also assumed.

5.4.2 Low Frequency

We first consider one–dimensional propagation (in the z–direction), with variations of time–scale much longer than the ion cyclotron period, so that the Hall term contribution in Eq. (5.25) (the term in $1/\Omega_{\mathrm{i}}$) can be neglected. Variation only in z implies from Eq. (5.25) that B_z is a constant. If v_z is also a constant, we can move to a frame of reference in which $v_z = 0$, and ρ is then constant in time from Eq. (5.23). We then also assume that ρ is uniform in z. The pressure p is then also constant. The z–component of Eq. (5.25) gives

$$\rho \frac{\mathrm{d}v_z}{\mathrm{d}t} = -\frac{\partial}{\partial z} \left(p + \frac{1}{2\mu_0} (B_x^2 + B_y^2) \right) \tag{5.26}$$

so that the square of the transverse magnetic field $\boldsymbol{B}_\perp$ is constant:

$$B_\perp^2 = B_x^2 + B_y^2 = \mathrm{constant}. \tag{5.27}$$

We find from Eqs. (5.24) and (5.25) the following equation for the transverse magnetic field components:

$$\frac{\partial^2 B_{x,y}}{\partial t^2} = v_{\mathrm{A}}^2 \frac{\partial^2 B_{x,y}}{\partial z^2} \tag{5.28}$$

where we have set $B_z = 1$ and $\rho = 1$. A possible solution of Eqs. (5.27) and (5.28) is

$$B_x = B_{\mathrm{a}} \cos[k(z - v_{\mathrm{A}} t) - \phi], \quad B_y = \pm B_{\mathrm{a}} \sin[k(z - v_{\mathrm{A}} t) - \phi] \tag{5.29}$$

where B_{a} is a constant real amplitude, k is an arbitrary wavenumber, and ϕ is an arbitrary phase angle.

Thus a circularly polarized incompressible wave with phase velocity v_{A}, of either sense of polarization and arbitrary amplitude, is an exact solution of the nonlinear equations, provided the frequency is much less than the ion cyclotron frequency (Sagdeev & Galeev 1969). The transverse magnetic field and plasma fluid velocity for this wave are related by

$$\boldsymbol{v} = \boldsymbol{B}_\perp / (\mu_0 \rho)^{1/2} \tag{5.30}$$

independently of the size of the wave amplitude. It is interesting to note that the circularly polarized wave is also an exact solution of the finite ion Larmor radius modified MHD equations, with dispersion relation (2.191) (Hamabata 1993). The exact nonlinear solution given by Eq. (5.29) can however be nonlinearly unstable, and can decay due to mode-mode coupling, as we will see in Section 5.5.

Note that if we attempt to construct a linearly polarized, parallel propagating, nonlinear shear Alfvén wave, by combining the oppositely polarized solutions (5.29), we find that $B_\perp^2$ is no longer a constant. This implies from Eqs. (5.23) and (5.26) that v_z and ρ are no longer constants, so that a simple linearly polarized solution of the nonlinear equations is not a possible exact solution. Such a solution can however be employed as an approximation for moderately nonlinear fields, because the resulting density perturbation may be regarded as being of second order in the wave amplitude (Hollweg 1971).

5.4.3 Higher Frequency

We may note that large–amplitude MHD fluctuations have been observed in the Earth's magnetosphere, in particular at the boundary of the plasma sheet in the magnetotail, with a wave period of 2 to 18 s, compared with the local proton cyclotron period of 5 to 12 s (Tsurutani *et al.* 1985). It is evident that finite ion cyclotron effects should be included in the description of the nonlinear waves.

If we now retain the dispersive Hall term in Eq. (5.25), we find that the circularly polarized wave given by Eq. (5.29) is no longer an exact solution. Again considering propagation of the wave solely in the z–direction, and acknowledging the important role of circularly polarized waves, it is convenient to work with the complex transverse fields

$$B_\pm = B_x \pm iB_y. \tag{5.31}$$

Any physical transverse magnetic field can be expressed as the real part of a suitable combination of B_+ and B_-. If $B_- = 0$, the resulting phase difference between B_x and B_y implies that the wave, described purely by the $+$ sign field, has right–hand circular polarization. Similarly, a wave described by the $-$ sign field has left–hand circular polarization.

Noting that B_z is again constant, we obtain from Eqs. (5.24) and (5.25) (Ovenden, Shah & Schwartz 1983),

$$\frac{\partial^2 B_\pm}{\partial t^2} - \frac{\partial}{\partial z}\left(\frac{v_A^2}{\rho}\frac{\partial B_\pm}{\partial z}\right) + \frac{\partial}{\partial z}\left(v_z\frac{\partial B_\pm}{\partial t} + \frac{d}{dt}(v_z B_\pm)\right)$$
$$\pm \frac{iv_A^2}{\Omega_i}\frac{\partial}{\partial z}\left[\frac{d}{dt}\left(\frac{1}{\rho}\frac{\partial B_\pm}{\partial z}\right)\right] = 0. \tag{5.32}$$

For an isothermal equation of state, with $p \propto \rho$, Eqs. (5.23) and (5.25) lead to

$$\frac{\partial^2 \rho}{\partial t^2} - c_s^2\frac{\partial^2 \rho}{\partial z^2} = \frac{\partial^2}{\partial z^2}\left(\frac{v_A^2}{2}|B_\pm|^2 + \rho v_z^2\right) \tag{5.33}$$

with the speed of sound $c_s = (p/\rho)^{1/2}$.

The set of nonlinear equations (5.32)-(5.33), or (5.23)-(5.25), can be reduced to a simpler nonlinear equation that has well–known solutions, provided the dispersive Hall term is relatively small, that is, if the frequency is a small but nonzero proportion of the ion cyclotron frequency. It is necessary to define a new set of "stretched coordinates", using a small parameter that measures the importance of the dispersive terms. A suitable parameter is

$$\varepsilon = v_A k / \Omega_i = k L_0 \tag{5.34}$$

where

$$L_0 = v_A / \Omega_i = c / \omega_{pi} \tag{5.35}$$

is a characteristic length, the ion inertial length, and k is the wavenumber along z of the linear wave. A characteristic time is Ω_i^{-1}, so that the frequency, on this time–scale, is of order ε.

The small size of ε means we are effectively using a *long—wavelength* approximation in terms of the length L_0. The dispersive Hall term is of order ε, and a nonlinear partial differential equation with soliton–type solutions can be obtained by requiring the nonlinear terms to also be of order ε. The linear dispersion relation in the low–frequency limit defines a phase velocity in the z-direction, which we denote by c (= v_A in this case). If we move to a frame of reference moving at c, to lowest order in ε the wave is stationary, and dispersion introduces a frequency shift of order ε^2, because we have from Eq. (2.45) that

$$\frac{\omega}{\Omega_i} \simeq \varepsilon \pm \varepsilon^2 / 2. \tag{5.36}$$

This means that a suitable "stretched" time coordinate is $\tau = \varepsilon^2 t$. In this frame, the wavenumber is of order ε, and a suitable "stretched" spatial coordinate is $Z = \varepsilon(z - ct)$.

The fields are assumed to be predominantly due to Alfvénic fluctuations, with density and parallel velocity perturbations smaller (i.e. of higher order in ε) than the transverse magnetic fields and velocities. Thus the fields ρ, $\boldsymbol{B}_\perp = (B_x, B_y)$ and $\boldsymbol{v} = (u, v, w)$ are expanded as the following power series in ε (Mio *et al.* 1976):

$$\rho = 1 + \varepsilon \rho_1 + \varepsilon^2 \rho_2 + \dots \tag{5.37a}$$
$$B_x = \varepsilon^{1/2} B_{x1} + \varepsilon^{3/2} B_{x2} + \dots \tag{5.37b}$$
$$B_y = \varepsilon^{1/2} B_{y1} + \varepsilon^{3/2} B_{y2} + \dots \tag{5.37c}$$
$$w = \varepsilon w_1 + \varepsilon^2 w_2 + \dots \tag{5.37d}$$
$$u = \varepsilon^{1/2} u_1 + \varepsilon^{3/2} u_2 + \dots \tag{5.37e}$$
$$v = \varepsilon^{1/2} v_1 + \varepsilon^{3/2} v_2 + \dots \tag{5.37f}$$

Equation (5.33) then becomes, at order ε,

$$\left(\frac{\partial^2}{\partial t^2} - c_s^2 \frac{\partial^2}{\partial z^2} \right) \rho_1 = \frac{\partial^2}{\partial z^2} \left(\frac{v_A^2}{2} |B_\pm|^2 \right) \tag{5.38}$$

which indicates that density perturbations are driven by the ponderomotive force due to the transverse magnetic pressure gradient.

Substituting Eqs. (5.37) into the nonlinear equations (5.23)-(5.25), using the scaled time and space variables, and collecting terms of the same order in ε, yields the following lowest–order equations:

$$-c\frac{\partial u_1}{\partial Z} = v_A^2 \frac{\partial B_{x1}}{\partial Z} \tag{5.39a}$$

$$-c\frac{\partial v_1}{\partial Z} = v_A^2 \frac{\partial B_{y1}}{\partial Z} \tag{5.39b}$$

$$-c\frac{\partial B_{x1}}{\partial Z} = \frac{\partial u_1}{\partial Z} \tag{5.39c}$$

$$-c\frac{\partial B_{y1}}{\partial Z} = \frac{\partial v_1}{\partial Z} \tag{5.39d}$$

$$-c\frac{\partial w_1}{\partial Z} = -c_s^2\frac{\partial \rho_1}{\partial Z} - \frac{v_A^2}{2}\frac{\partial(B_{x1}^2 + B_{y1}^2)}{\partial Z} \tag{5.39e}$$

$$c\frac{\partial \rho_1}{\partial Z} = \frac{\partial w_1}{\partial Z}. \tag{5.39f}$$

The first four equations of Eqs. (5.39) imply that $c = v_A$. This is the result to lowest order in ε from the linear dispersion relation given by Eq. (2.45) or Eq. (5.36) for parallel propagation with the Hall term, if we note that the Ω_i-dependent term is of higher order in ε. The last two equations of Eqs. (5.39) serve to define the first–order parallel velocity and the density perturbation, in terms of the first–order transverse fields.

The equations corresponding to the next order in ε are, with the first–order fields collected on the right–hand side,

$$-v_A\frac{\partial u_2}{\partial Z} - v_A^2 \frac{\partial B_{x2}}{\partial Z} = G$$

$$= -\left(\frac{\partial u_1}{\partial \tau} + w_1\frac{\partial u_1}{\partial Z}\right) + v_A\rho_1\frac{\partial u_1}{\partial Z} \tag{5.40a}$$

$$-v_A\frac{\partial v_2}{\partial Z} - v_A^2 \frac{\partial B_{y2}}{\partial Z} = H$$

$$= -\left(\frac{\partial v_1}{\partial \tau} + w_1\frac{\partial v_1}{\partial Z}\right) + v_A\rho_1\frac{\partial v_1}{\partial Z} \tag{5.40b}$$

$$-v_A\frac{\partial B_{x2}}{\partial Z} - \frac{\partial u_2}{\partial Z} = I$$

$$= -\frac{\partial B_{x1}}{\partial \tau} - \frac{\partial}{\partial Z}(w_1 B_{x1}) - \frac{v_A}{\Omega_i}\frac{\partial^2 v_1}{\partial Z^2} \tag{5.40c}$$

$$-v_A\frac{\partial B_{y2}}{\partial Z} - \frac{\partial v_2}{\partial Z} = J$$

$$= -\frac{\partial B_{y1}}{\partial \tau} - \frac{\partial}{\partial Z}(w_1 B_{y1}) + \frac{v_A}{\Omega_i}\frac{\partial^2 u_1}{\partial Z^2}. \tag{5.40d}$$

For consistency of Eqs. (5.40) it is required that $G/v_A = I$ and $H/v_A = J$, which leads to the following two equations for the first–order fields:

$$\frac{\partial B_{x1}}{\partial \tau} + \frac{v_A}{4(1 - \beta')}\frac{\partial}{\partial Z}\left(B_{x1}(B_{x1}^2 + B_{y1}^2)\right) - \frac{v_A^2}{2\Omega_i}\frac{\partial^2 B_{y1}}{\partial Z^2} = 0 \tag{5.41a}$$

$$\frac{\partial B_{y1}}{\partial \tau} + \frac{v_A}{4(1-\beta')}\frac{\partial}{\partial Z}\left(B_{y1}(B_{x1}^2 + B_{y1}^2)\right) + \frac{v_A^2}{2\Omega_i}\frac{\partial^2 B_{x1}}{\partial Z^2} = 0 \qquad (5.41b)$$

where $\beta' = \gamma\beta/2 = c_s^2/v_A^2$.

Thus nonlinear partial differential equations linking the slow time and space variation of the first–order wave fields have been derived from consistency relations on the second–order wave fields. Changing back to the original time and space variables, and noting that

$$\varepsilon^2 \frac{\partial}{\partial \tau} = \frac{\partial}{\partial t} + c\frac{\partial}{\partial z} \qquad (5.42)$$

the following equation is obtained, coupling the variables $B_\pm$ defined in Eq. (5.31) to the density perturbation:

$$\frac{\partial B_\pm}{\partial t} + v_A\frac{\partial B_\pm}{\partial z} \pm i\frac{v_A^2}{2\Omega_i}\frac{\partial^2 B_\pm}{\partial z^2} + \frac{v_A}{2}\frac{\partial}{\partial z}(\rho_1 B_\pm) = 0. \qquad (5.43)$$

If the fourth (coupling) term in Eq. (5.43) is neglected, the first three terms give the linear dispersion relation (5.36).

The density perturbation forced by the transverse magnetic field pressure is, from Eq. (5.39),

$$\rho_1 = \frac{1}{2(1-\beta')}|B_\pm|^2 + \text{constant}. \qquad (5.44)$$

Thus for a low–β plasma the density is enhanced in a region of strong transverse magnetic field, while for $\beta' > 1$ a plasma "hole" accompanies regions of high field strength, as for Langmuir solitons in unmagnetized plasmas (Spangler & Sheerin 1982).

We note that if the third (dispersive) term in Eq. (5.43) is neglected, a solution of Eqs. (5.43) and (5.44) is given by Eq. (5.29) with

$$|B_\pm|^2 = \text{constant} \quad \text{and} \quad \rho_1 = 0 \qquad (5.45)$$

that is, a constant amplitude, transverse, circularly polarized plane wave of infinite extent as in Section 5.4.2. However, if the dispersive term is included, B_+ and B_- no longer satisfy the same differential equation, and the solution (5.45) no longer applies.

Using Eq. (5.44) with the constant set to zero, Eq. (5.43) becomes

$$\frac{\partial B_\pm}{\partial t} + v_A\frac{\partial B_\pm}{\partial z} + \frac{v_A}{4(1-\beta')}\frac{\partial}{\partial z}(|B_\pm|^2 B_\pm) \pm i\frac{v_A^2}{2\Omega_i}\frac{\partial^2 B_\pm}{\partial z^2} = 0. \qquad (5.46)$$

This is the "Derivative Nonlinear Schrödinger" (DNLS) equation describing nonlinear one–dimensional waves, derived (for $\beta = 0$) by Mio *et al.* (1976) and Mjølhus (1976), and (for $\beta > 0$) by Spangler & Sheerin (1982) and Sakai & Sonnerup (1983).

A solution to Eq. (5.46) is sought in the form of an envelope–modulated carrier wave (Spangler & Sheerin 1982; Spangler, Sheerin & Payne 1985; Spangler & Plapp 1992):

$$B_\pm = b(z,t)\exp i\theta(z,t) \qquad (5.47)$$

with a real phase θ, and where the envelope amplitude b is real and is stationary in a frame moving with the envelope velocity V_E, that is, $b = b(\zeta)$, where

$$\zeta = z - V_E t. \tag{5.48}$$

After substituting Eq. (5.47), the real part of Eq. (5.46) may be integrated over ζ to yield the following equation describing a nonlinear phase modulation:

$$\frac{\partial \theta}{\partial \zeta} = \mp \frac{\Omega_i}{v_A^2} \left[V_E - v_A - \frac{3 v_A}{8(1 - \beta')} b^2 + C_0 b^{-2} \right] \tag{5.49}$$

where C_0 is an arbitrary constant of integration, which determines the nature of the solution.

Setting $C_0 = 0$, envelope soliton solutions may be obtained. Substituting Eq. (5.49) into the imaginary part of Eq. (5.46) yields a pseudo equation of motion,

$$\frac{\partial^2 b}{\partial \zeta^2} = -\frac{dV(b)}{db} \tag{5.50}$$

where b can be considered as a pseudo–particle spatial coordinate and ζ plays the role of the pseudo–time (Spangler & Sheerin 1982). The pseudo–potential is

$$V(b) = A b^2 + B b^4 + C b^6 \tag{5.51}$$

with coefficients

$$A = \frac{1}{2} \left(\frac{\Omega_i}{v_A^2} \right)^2 (V_E - v_A)^2 \tag{5.52}$$

$$B = -\frac{1}{8(1 - \beta')} \left(\frac{\Omega_i}{v_A^2} \right)^2 (V_E - v_A) \tag{5.53}$$

$$C = \frac{1}{128(1 - \beta')^2} \left(\frac{\Omega_i}{v_A^2} \right)^2 v_A^2. \tag{5.54}$$

For a soliton solution to exist, the pseudo–potential must form a potential well, and properties of the soliton are determined by the characteristics of the potential well.

Solitons exist for both $\beta' < 1$ and $\beta' > 1$, and for both senses of circular polarization (Spangler, Sheerin & Payne 1985). Two types of soliton exist for the same envelope speed V_E. If B is positive, that is, if $V_E > v_A$ (super–Alfvénic) and $\beta' < 1$, or $V_E < v_A$ (sub–Alfvénic) and $\beta' > 1$, a relatively high–amplitude "Hi" soliton exists, with an envelope described by

$$b^2(\xi) = b^2_{\max} \left[\frac{\sqrt{2} - 1}{\sqrt{2} \cosh \xi - 1} \right] \tag{5.55}$$

and

$$b^2_{\max} = 8(V_E - v_A)(1 - \beta')(\sqrt{2} + 1). \tag{5.56}$$

The normalized spatial coordinate appearing in Eq. (5.55) is

$$\xi = \frac{2\Omega_i}{v_A^2}(V_E - v_A)\zeta. \tag{5.57}$$

If B is negative, a lower–amplitude "Lo" soliton exists, with

$$b^2(\xi) = b^2_{\max} \left[\frac{\sqrt{2} + 1}{\sqrt{2}\cosh \xi + 1} \right] \tag{5.58}$$

and

$$b^2_{\max} = -8(V_E - v_A)(1 - \beta')(\sqrt{2} - 1). \tag{5.59}$$

The solitons have a property commonly associated with soliton solutions, namely, that a larger amplitude soliton is of smaller spatial extent.

We can gain some physical insights into the formation of the solitons (Ovenden, Shah & Schwartz 1983), by noticing that if the amplitude b is small, the wavenumber of the carrier wave is

$$k \simeq \frac{\partial \theta}{\partial \zeta} \simeq \mp \frac{\Omega_i}{v_A^2}(V_E - v_A) \tag{5.60}$$

so that

$$V_E \simeq v_A \left(1 - \frac{k v_A}{\Omega_i} \right) = v_g \tag{5.61}$$

where v_g is the group velocity of the wave with linear dispersion relation (5.36).

Consider the case $\beta' < 1$. The increased mass density in the region of high magnetic field strength given by Eq. (5.44) leads to a lower local Alfvén speed, causing the associated build–up in wave energy in this region. The transport of energy to this region occurs at a rate given by the group velocity, that is, the envelope speed given by Eq. (5.61), resulting in an unchanged soliton envelope. The wave is thus modified continuously, with a balance maintained between energy entering and leaving the region. If dissipation is included, it is found that the model describes intermediate (Alfvénic) shock waves. It has been shown by Wu & Kennel (1992) that a modified set of equations results, the "DNLS-Burgers" equations.

5.4.4 Oblique Propagation

In this section we derive simplified weakly nonlinear wave equations in three space dimensions describing propagation of a wave obliquely to the ambient magnetic field in a uniform cold plasma, assuming that the dispersion is "small" in the same sense as in the parallel propagation case (Mjølhus & Wyller 1986; Cramer 1991). The obliqueness, measured by $|\sin \theta|$, is also assumed to be small, so that the wave is almost parallel propagating. The resulting equations are not standard equations with known solutions, although in the case of propagation parallel to the magnetic field they reduce to the DNLS equation. It is found from these equations that the obliqueness of propagation gives rise to the generation of the second harmonic of the linear wave, which is absent in the case of parallel propagation. The equations may form a basis for discussion of surface waves, or of the three–dimensional stability of the parallel propagating solitons.

The wave equations are scaled in t and z as in Section 5.4.3, with the scaling parameter now defined as $\varepsilon = v_A k_z / \Omega_i$, where k_z is the wavenumber along the ambient magnetic field in the linear limit. A scaling for the x and y variables is also necessary: a suitable scaling with the x and y variation weaker, that is, of higher order in ε, than the z variation is found to be $X = \varepsilon^{3/2} x$ and $Y = \varepsilon^{3/2} y$.

The fields ρ, $\boldsymbol{B} = (B_x, B_y, B_z)$, and $\boldsymbol{v} = (u, v, w)$ are expanded as the power series given in Eqs. (5.37), with the addition of

$$B_z = 1 + \varepsilon B_{z1} + \varepsilon^2 B_{z2} + \dots . \tag{5.62}$$

The lowest–order equations are the same as in Eqs. (5.39), except for:

$$c\frac{\partial B_{z1}}{\partial Z} = \frac{\partial u_1}{\partial X} + \frac{\partial v_1}{\partial Y} \tag{5.63a}$$

$$c\frac{\partial \rho_1}{\partial Z} = \frac{\partial u_1}{\partial X} + \frac{\partial v_1}{\partial Y} + \frac{\partial w_1}{\partial Z}. \tag{5.63b}$$

The equations corresponding to the next order in ε are in the same form as Eqs. (5.40), but with the following new expressions for the functions G, H, I and J:

$$G = v_A^2 \left(B_{z1} \frac{\partial B_{x1}}{\partial Z} - \frac{\partial B_{z1}}{\partial X} - B_{y1} \left(\frac{\partial B_{y1}}{\partial X} - \frac{\partial B_{x1}}{\partial Y} \right) \right)$$
$$- \left(\frac{\partial u_1}{\partial \tau} + u_1 \frac{\partial u_1}{\partial X} + v_1 \frac{\partial u_1}{\partial Y} + w_1 \frac{\partial u_1}{\partial Z} \right) + v_A \rho_1 \frac{\partial u_1}{\partial Z} \tag{5.64a}$$

$$H = v_A^2 \left(B_{x2} \left(\frac{\partial B_{y1}}{\partial X} - \frac{\partial B_{x1}}{\partial Y} \right) - \frac{\partial B_{z1}}{\partial Y} + B_{z1} \frac{\partial B_{y1}}{\partial Z} \right)$$
$$- \left(\frac{\partial v_1}{\partial \tau} + u_1 \frac{\partial v_1}{\partial X} + v_1 \frac{\partial v_1}{\partial Y} + w_1 \frac{\partial v_1}{\partial Z} \right) + v_A \rho_1 \frac{\partial v_1}{\partial Z} \tag{5.64b}$$

$$I = -\frac{\partial B_{x1}}{\partial \tau} + \frac{\partial}{\partial Y}(u_1 B_{y1} - v_1 B_{x1})$$
$$- \frac{\partial}{\partial Z}(w_1 B_{x1} - u_1 B_{z1}) - \frac{v_A}{\Omega_i}\frac{\partial^2 v_1}{\partial Z^2} \tag{5.64c}$$

$$J = -\frac{\partial B_{y1}}{\partial \tau} + \frac{\partial}{\partial Z}(v_1 B_{z1} - w_1 B_{y1})$$
$$- \frac{\partial}{\partial X}(u_1 B_{y1} - v_1 B_{x1}) + \frac{v_A}{\Omega_i}\frac{\partial^2 u_1}{\partial Z^2}. \tag{5.64d}$$

Again, for consistency of Eqs. (5.64) it is required that $G/v_A = I$ and $H/v_A = J$, which leads to the following two equations for the first–order magnetic fields:

$$\frac{\partial B_{x1}}{\partial \tau} + \frac{v_A}{4}\frac{\partial}{\partial Z}\left(B_{x1}(B_{x1}^2 + B_{y1}^2 + 2B_{z1}) \right)$$
$$- \frac{v_A}{4}\frac{\partial}{\partial X}(B_{x1}^2 + B_{y1}^2 + 2B_{z1}) - \frac{v_A^2}{2\Omega_i}\frac{\partial^2 B_{y1}}{\partial Z^2} = 0 \tag{5.65a}$$

$$\frac{\partial B_{y1}}{\partial \tau} + \frac{v_A}{4}\frac{\partial}{\partial Z}\left(B_{y1}(B_{x1}^2 + B_{y1}^2 + 2B_{z1})\right)$$

$$-\frac{v_A}{4}\frac{\partial}{\partial Y}(B_{x1}^2 + B_{y1}^2 + 2B_{z1}) + \frac{v_A^2}{2\Omega_i}\frac{\partial^2 B_{x1}}{\partial Z^2} = 0. \tag{5.65b}$$

Equations (5.65) must of course be completed by the $\nabla \cdot \boldsymbol{B} = 0$ equation:

$$\frac{\partial B_{x1}}{\partial X} + \frac{\partial B_{y1}}{\partial Y} + \frac{\partial B_{z1}}{\partial Z} = 0. \tag{5.66}$$

Changing back to the original time and space variables, Eqs. (5.65)–(5.66) may be combined in the form (Mjølhus & Wyller 1986)

$$\frac{\partial B_\pm}{\partial t} + v_A\frac{\partial B_\pm}{\partial z} + \frac{v_A}{4}\frac{\partial}{\partial z}\left((|B_\pm|^2 + 2B_z)B_\pm\right)$$

$$-\frac{v_A}{4}\tilde{\nabla}_{\perp\pm}(|B_\pm|^2 + 2B_z) \pm i\frac{v_A^2}{2\Omega_i}\frac{\partial^2 B_\pm}{\partial z^2} = 0 \tag{5.67}$$

$$\frac{\partial B_{z1}}{\partial z} + \nabla_\perp \cdot \boldsymbol{B}_\perp = 0 \tag{5.68}$$

where

$$B_\pm = B_{x1} \pm iB_{y1}, \quad \boldsymbol{B}_\perp = (B_{x1}, B_{y1}) \tag{5.69}$$

$$\tilde{\nabla}_{\perp\pm} = \frac{\partial}{\partial x} \pm i\frac{\partial}{\partial y}, \quad \nabla_\perp = \left(\frac{\partial}{\partial x}, \frac{\partial}{\partial y}\right). \tag{5.70}$$

It should be noted that for parallel propagation, that is, setting $\partial/\partial x = \partial/\partial y = 0$, we have $B_{z1} = 0$, and Eq. (5.67) reduces to the DNLS equation (5.46) (with $\beta' = 0$). It can also be shown that if a transformation is made to a new coordinate system with the new z–axis at the angle θ to the background magnetic field, and if $\sin\theta$ is of order $\varepsilon^{1/2}$, then the wave fields satisfy the DNLS equation, but with elliptical polarization (Kennel *et al.* 1988; Mjølhus 1989). Comparing Eq. (5.46) with Eqs. (5.65), it is evident that the x and y dependence, i.e. the obliqueness of propagation, gives rise to terms that are quadratic in the wave fields. Using a description of the nonlinear wave in terms of interacting harmonics of the linear wave, such quadratic terms mean second–harmonic excitation, which is absent in the parallel propagation case given by Eq. (5.46). The second–harmonic generation is discussed in connection with nonlinear surface waves in Section 5.7.

5.5   Parametric and Modulational Instabilities

We have seen in the previous sections that nonlinear Alfvén waves can form steady–state propagating structures. We must however be concerned about the stability of such structures: this leads us to the next important topic in nonlinear wave theory, that of *parametric instabilities*.

A general resonant system can be set into oscillation by periodically varying the parameters of the system (Pippard 1978): this process is known as parametric excitation, or parametric instability. An example is a pendulum whose length is slightly varied at twice the natural frequency. In our case it may be the density or the magnetic field in the plasma that is periodically varied, or pumped, giving rise to parametric excitation of modes of the plasma, in particular Alfvén and magnetoacoustic waves. Parametric instabilities driven by an external pump electromagnetic field have been intensively studied from the point of view of applications to laser fusion. *Modulational instabilities* are a type of parametric instability in which an amplitude modulation of a pump wave, giving rise to side–band daughter waves, grows in time.

Parametric instabilities can also be viewed as a type of nonlinear wave-wave interaction: the varied parameter itself corresponds to a field variable of a large–amplitude wave (or pump wave), which couples nonlinearly with smaller–amplitude fluctuations in the plasma, according to the rules of conservation of energy (frequency) and momentum (wavenumber). If the fluctuations are themselves normal wave modes of the plasma, and the pump wave damps, with an accompanying growth of the fluctuations, we have a parametric decay instability of the pump wave. Parametric instabilities involving Alfvén waves may be categorized as those where Alfvén waves are nonlinearly excited by an external pump, and those where Alfvén waves themselves undergo nonlinear parametric decay, each of which we now consider.

5.5.1   Excitation by a Magnetoacoustic Pump

We first consider the parametric excitation of Alfvén waves by a pump magnetic field that is, in the first approximation, uniform and fixed in one direction, but is modulated in time at some frequency. Such a pump may sometimes be regarded as corresponding to a large–amplitude, infinite–wavelength, magnetoacoustic wave. Pumps in the form of forced magnetoacoustic oscillations generated by external solenoidal windings have been employed for plasma heating and diagnostics of laboratory devices. If the forced oscillation is resonant with a natural mode of the device (taking into account the boundary conditions on the fields), the field amplitude can reach relatively large values, and so nonlinear effects, such as parametric instabilities of the pump, can be important.

The possibility of parametric decay of magnetoacoustic oscillations into Alfvén waves was predicted theoretically by Montgomery & Harding (1966) and Vahala & Montgomery (1971), and verified experimentally by Lehane & Paoloni (1972a), where the amplification of a torsional Alfvén wave launched into a region of magnetic field modulation was observed. Maximum amplification occurred when the pump or modulation frequency was twice the wave frequency, indicating that the (standing wave) magnetic field pump was decaying into two oppositely propagating Alfvén waves of the same frequency and wavelength. Associated with the pumped magnetic field is a pumped plasma density, which can parametrically excite two oppositely propagating acoustic waves (Lashmore-Davies & Ong 1974). The general parametric coupling of obliquely propagating Alfvén waves, and slow and fast magnetoacoustic waves, due to a magnetic pump was treated by Cramer (1976, 1977). The effects of boundary conditions on parametric excitation in a cylindrical plasma were considered by Elfimov & Nekrasov (1973) and Cramer & Sy (1979).

We now proceed to consider the case of parametric instability of a magnetic pump in a plasma with zero particle pressure. The analysis proceeds by expanding the total fields in terms of quantities associated with the pump, associated with the expansion parameter $\bar{\varepsilon}$, and quantities associated with the excited waves, described by the parameter ε':

$$
\begin{aligned}
\boldsymbol{B} &= \boldsymbol{B}_0 + \bar{\varepsilon}\bar{\boldsymbol{B}} + \varepsilon'\boldsymbol{B}' \\
\boldsymbol{v} &= \bar{\varepsilon}\bar{\boldsymbol{v}} + \varepsilon'\boldsymbol{v}' \\
\rho &= \rho_0 + \bar{\varepsilon}\bar{\rho} + \varepsilon'\rho'
\end{aligned}
\tag{5.71}
$$

where the subscript 0 refers to the steady background quantities, and $\boldsymbol{B}_0 = B_0\hat{\boldsymbol{z}}$.

Substitution of Eq. (5.71) into the dimensional versions of the nonlinear equations (5.23)-(5.25) (neglecting the particle pressure but keeping the Hall term), and retaining terms of order less than or equal to $\bar{\varepsilon}\varepsilon'$, but excluding terms of order $(\varepsilon')^2$ or $(\bar{\varepsilon})^2$, yields a set of expanded equations for the excited wave fields (Cramer & Sy 1979):

$$
\rho_0\frac{\partial\boldsymbol{v}'}{\partial t} - \frac{1}{\mu_0}(\nabla\times\boldsymbol{B}')\times\boldsymbol{B}_0 = \bar{\varepsilon}\boldsymbol{F}
\tag{5.72}
$$

$$
\frac{\partial\boldsymbol{B}'}{\partial t} - \nabla\times(\boldsymbol{v}'\times\boldsymbol{B}_0) + \frac{v_{\text{A}}^2}{\Omega_{\text{i}}B_0}\nabla\times\left((\nabla\times\boldsymbol{B}')\times\boldsymbol{B}_0\right) = \bar{\varepsilon}\nabla\times\boldsymbol{G}
\tag{5.73}
$$

$$
\frac{\partial\rho'}{\partial t} + \nabla\cdot(\rho\boldsymbol{v}') = -\bar{\varepsilon}\nabla\cdot(\bar{\rho}\boldsymbol{v}' + \rho'\bar{\boldsymbol{v}})
\tag{5.74}
$$

where the terms $\boldsymbol{F}$ and $\boldsymbol{G}$ appearing on the right–hand sides of the equations, which couple the pump fields and the excited wave fields, are defined by

$$
\begin{aligned}
\boldsymbol{F} = -\bar{\rho}\frac{\partial\boldsymbol{v}'}{\partial t} - \rho'\frac{\partial\bar{\boldsymbol{v}}}{\partial t} - \rho_0\boldsymbol{v}'\cdot\nabla\bar{\boldsymbol{v}} - \rho_0\bar{\boldsymbol{v}}\cdot\nabla\boldsymbol{v}' \\
+ \frac{1}{\mu_0}(\nabla\times\boldsymbol{B}')\times\bar{\boldsymbol{B}} + \frac{1}{\mu_0}(\nabla\times\bar{\boldsymbol{B}})\times\boldsymbol{B}'
\end{aligned}
\tag{5.75}
$$

and

$$
\begin{aligned}
\boldsymbol{G} = \bar{\boldsymbol{v}}\times\boldsymbol{B}' + \boldsymbol{v}'\times\bar{\boldsymbol{B}} + \frac{v_{\text{A}}^2}{\Omega_{\text{i}}B_0}\Bigg[(\nabla\times\bar{\boldsymbol{B}})\times\boldsymbol{B}_0\frac{\rho'}{\rho_0} + (\nabla\times\boldsymbol{B}')\times\boldsymbol{B}_0\frac{\bar{\rho}}{\rho_0} \\
- (\nabla\times\boldsymbol{B}')\times\bar{\boldsymbol{B}} - (\nabla\times\bar{\boldsymbol{B}})\times\boldsymbol{B}'\Bigg].
\end{aligned}
\tag{5.76}
$$

Equations (5.72)-(5.74) can be used to describe the parametric excitation of obliquely propagating plane waves (Cramer 1975) or waves in a cylindrical laboratory plasma (Cramer & Sy 1979). However, here we consider the simpler case of excited waves propagating parallel or antiparallel to the magnetic field direction (Cramer 1975). The pump fields are due to a large–amplitude standing magnetoacoustic wave of frequency ω_0, with wavenumber k_0 in the x–direction (perpendicular to the magnetic field direction):

$$
\begin{aligned}
\boldsymbol{B}^{(0)} &= B_0(1 + \bar{\varepsilon}\cos k_0 x\cos\omega_0 t)\hat{\boldsymbol{z}} \\
\rho^{(0)} &= \rho_0(1 + \bar{\varepsilon}\cos k_0 x\cos\omega_0 t) \\
\boldsymbol{v}^{(0)} &= \bar{\varepsilon}v_{\text{A}}\sin k_0 x\sin\omega_0 t\hat{\boldsymbol{x}}
\end{aligned}
\tag{5.77}
$$

with $\omega_0 = v_A k_0$. Considering distances x such that $k_0 x =$ order $\bar{\epsilon}$ (e.g. a cylindrical plasma with radius $R \ll 1/k_0$), we have spatially uniform pump fields to order $\bar{\epsilon}$:

$$
\begin{aligned}
\boldsymbol{B}^{(0)} &= B_0(1 + \bar{\epsilon} \cos \omega_0 t)\hat{\boldsymbol{z}} \\
\rho^{(0)} &= \rho_0(1 + \bar{\epsilon} \cos \omega_0 t) \\
\boldsymbol{v}^{(0)} &= 0.
\end{aligned}
\tag{5.78}
$$

The functions $\boldsymbol{F}$ and $\boldsymbol{G}$ then simplify to

$$
\begin{aligned}
\boldsymbol{F} &= 0 \tag{5.79}\\
\boldsymbol{G} &= \boldsymbol{v}' \times \bar{\boldsymbol{B}} \tag{5.80}
\end{aligned}
$$

and $\rho' = 0$.

The complex excited wave fields have a time and space dependence $\exp(ikz - i\omega t)$, and satisfy one of the two dispersion equations (from Eq. (2.45))

$$
\omega^2 = v_A^2 k^2 \left(1 \mp \frac{\omega}{\Omega_i}\right).
\tag{5.81}
$$

The waves are characterized by complex velocity amplitudes $v_+ = v_x' + iv_y'$ corresponding to the slow ion cyclotron mode for $Real(\omega) = \omega_s > 0$, and $v_- = v_x' - iv_y'$ corresponding to the fast mode for $Real(\omega) = \omega_f > 0$, respectively. Remembering that the sense of circular polarization is defined in reference to the screw sense of the fields in the direction of propagation of the wave, the waves propagating in the positive z-direction, that is, with $Real(\omega) > 0$ for $k > 0$, are left–hand circularly polarized if given by v_+, right–hand circularly polarized if given by v_-. Waves propagating in the negative z-direction, that is, with $Real(\omega) < 0$ for $k > 0$, are also left–hand polarized for v_+, but correspond to the fast wave, and are right–hand polarized for v_-, but correspond to the slow ion cyclotron wave.

Equations (5.72)-(5.73) with (5.79)-(5.80) become

$$
\frac{\partial^2 v_\pm}{\partial t^2} \mp \frac{iv_A k^2}{\Omega_i} \frac{\partial v_\pm}{\partial t} + v_A^2 k^2 v_\pm = -\bar{\epsilon} \cos \omega_0 t v_A^2 k^2 v_\pm.
\tag{5.82}
$$

These equations are solved by taking the Fourier transform (Nishikawa 1968). For example, the equation in v_+ becomes the following algebraic equation in $V_+(\omega)$, the Fourier transform of $v_+(t)$:

$$
\left(-\omega^2 + v_A^2 k^2 \left(1 - \frac{\omega}{\Omega_i}\right)\right) V_+(\omega) = -\frac{\bar{\epsilon}}{2} v_A^2 k^2 (V_+(\omega + \omega_0) + V_+(\omega - \omega_0)).
\tag{5.83}
$$

We assume that the frequency ω is close to the natural slow mode frequency of the system:

$$
\omega = \omega_1 = \omega_s + \mathcal{O}(\bar{\epsilon})
\tag{5.84}
$$

and that $\omega_s < \omega_0$.

There is a strong parametric interaction of the wave with the pump if one of the Fourier modes on the right–hand side of Eq. (5.83) also corresponds to a natural mode frequency. Because the interaction described by Eq. (5.83) is only between modes described by V_+, the

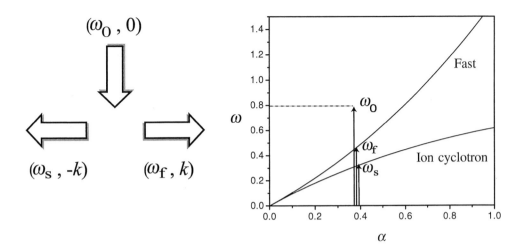

Figure 5.2: Schematic representation of the parametric decay of a pump magnetic field into a slow (ion cyclotron) wave and a fast wave, propagating in opposite directions along the magnetic field. The dispersion relation for the two waves is also shown, with an example of the frequency sum rule for the decay indicated. Here ω is measured in units of Ω_i, and $\alpha = v_A |k|/\Omega_i$.

mode $V_+ (\omega - \omega_0)$ can be resonant if $\omega - \omega_0 \simeq -\omega_f$, that is, the second wave is propagating anti–parallel to the magnetic field in the fast mode. The mode $V_+ (\omega + \omega_0)$ does not correspond to a natural mode, that is, is nonresonant, and so may be neglected. For the standing wave pump considered here, we can therefore say that there are *resonance conditions* for the (positive) frequencies ω_s and ω_f, and the wavenumbers k_s and k_f, of the excited waves:

$$\omega_0 = \omega_s + \omega_f, \quad 0 = k_s + k_f. \tag{5.85}$$

The wavenumbers are opposite in sign, indicating the opposite directions of propagation of the two waves. The pump wave can also be said to *decay* into the two excited waves. From the resonance conditions and the linear dispersion relations (5.81) we find that

$$k^2 = 2\frac{\Omega_i^2}{v_A^2}\left(\left(1 + \frac{1}{4}\frac{\omega_0^2}{\Omega_i^2}\right)^{1/2} - 1\right) \tag{5.86}$$

and

$$\omega_{s,f} = \frac{\omega_0}{2} \mp \frac{v_A^2 k^2}{2\Omega_i}. \tag{5.87}$$

This parametric excitation of the slow and fast waves by the pump magnetic field is shown schematically in Figure 5.2, which also shows an illustration of the frequency sum rule given by Eq. (5.85) on the dispersion relation diagram for the two modes.

A second equation may be obtained from Eq. (5.83) by making the substitution $\omega \rightarrow \omega - \omega_0$, and neglecting the nonresonant response $V_+ (\omega - 2\omega_0)$. The resulting two equations coupling the resonant responses are

$$\left(-\omega^2 + v_A^2 k^2 \left(1 - \frac{\omega}{\Omega_i}\right)\right) V_+(\omega) = -\frac{\bar{\varepsilon}}{2} v_A^2 k^2 V_+(\omega - \omega_0) \tag{5.88a}$$

$$\left(-(\omega - \omega_0)^2 + v_A^2 k^2 \left(1 - \frac{(\omega - \omega_0)}{\Omega_i}\right)\right) V_+(\omega - \omega_0) = -\frac{\bar{\varepsilon}}{2} v_A^2 k^2 V_+(\omega). \tag{5.88b}$$

From consistency of these two equations there results a dispersion equation:

$$\left(\omega^2 - v_A^2 k^2 \left(1 - \frac{\omega}{\Omega_i}\right)\right) \left((\omega - \omega_0)^2 - v_A^2 k^2 \left(1 - \frac{(\omega - \omega_0)}{\Omega_i}\right)\right)$$

$$= \frac{\bar{\varepsilon}^2}{4} v_A^4 k^4 = \frac{\bar{\varepsilon}^2}{4} \omega_s^2 \omega_f^2. \tag{5.89}$$

Writing $\omega = \omega_r + i\gamma$, with ω_r and γ real, we find a growth rate

$$\gamma = \frac{\bar{\varepsilon}}{2} \frac{v_A^2 k^2}{\omega_0} = \frac{\bar{\varepsilon}}{2} \frac{\omega_s \omega_f}{\omega_0}. \tag{5.90}$$

The same analysis may be applied to excitation of the v_- mode, and it is found that again a slow and a fast wave are excited, only in the opposite directions to the v_+ modes, and with the opposite polarizations.

In the low–frequency, long–wavelength limit, $\omega_0 \ll \Omega_i$ and $v_A^2 k^2 \ll \Omega_i^2$, both excited waves have the same frequency, $\omega_{s,f} = \omega_0/2$, and the growth rate is

$$\gamma = \bar{\varepsilon} \omega_0 / 8. \tag{5.91}$$

Thus in the low–frequency limit the two oppositely polarized waves can be combined to give a linearly polarized shear Alfvén wave propagating parallel or antiparallel to the magnetic field, in other words the pump can excite two oppositely propagating shear Alfvén waves of half the pump frequency (Vahala & Montgomery 1971).

5.5.2   Instability of Alfvén waves

We now consider the nonlinear instabilities of propagating Alfvén waves. The coupling of a large–amplitude linearly polarized Alfvén wave to density and magnetic field fluctuations was analysed by Lashmore-Davies (1976). However, the circularly polarized wave, being an exact nonlinear solution given by Eq. (5.29) for low frequency, is a more natural starting point. The instabilities of the exact nonlinear circularly polarized Alfvén wave solution have been investigated by Goldstein (1978) and Derby (1978) for a low-β plasma, using MHD theory at low frequency.

The instability is an example of a *modulational interaction* (Vladimirov *et al.* 1995). For nonzero β, a *four–wave* coupling occurs in which the initial forward propagating Alfvén wave excites a forward propagating density wave and magnetic wave, as well as a backward propagating magnetic wave. The driven waves are not necessarily normal modes of the plasma, but may be forced or "virtual" waves. We note that the derivation of Eq. (5.29) depended on the assumption of a constant ρ. If ρ is allowed to fluctuate, the wave is unstable. The density perturbation produces a modulation of the Alfvén wave, such that daughter magnetic waves are produced at the side–bands. In the limit of very low β, a mode–coupling parametric "decay" instability is recovered in which the density wave is an ion-acoustic wave and the backward propagating magnetic wave is an Alfvén wave (Sagdeev & Galeev 1969). The amplitude of the other transverse magnetic wave becomes negligible, so the decay instability is a *three–wave* coupling.

We adopt the analysis of Goldstein (1978) and Derby (1978) to derive the growth rate of the modulational instability, at frequencies much lower than the ion cyclotron frequency. There is a steady magnetic field in the z–direction. Consider the pump as a monochromatic, circularly polarized, large–amplitude Alfvén wave with frequency ω_0 and wavenumber k_0, propagating along the steady magnetic field direction. The pump has a magnetic field $\bar{\varepsilon}\bar{B}_\perp$ perpendicular to the z–axis, with components given by Eq. (5.29). We define a rotating unit vector to describe left and right–hand circular polarizations;

$$\hat{e}_\pm = (\hat{x} \pm i\hat{y})/\sqrt{2}. \tag{5.92}$$

The pump field amplitude $\bar{B}_\perp$ can then be written as the real part of a complex vector:

$$\bar{B}_\perp = Real(B \exp[-i(k_0 z - \omega_0 t)]\hat{e}_\pm) \tag{5.93}$$

with $B = \sqrt{2}B_a \exp(i\phi)$, where B_a and ϕ are as defined in Eq. (5.29). If B_a is of order B_0, the size of the pump magnetic field relative to the steady field is measured by the parameter $\bar{\varepsilon}$, as for the magnetoacoustic pump considered in the previous section.

The fields, including the pump wave fields characterized by $\bar{\varepsilon}\bar{B}_\perp$ and $\bar{\varepsilon}\bar{v}_\perp$, and the small fluctuations assumed to vary in the z–direction only and characterized by $\varepsilon'B'_\perp$, $\varepsilon'v'_\perp$ and $\varepsilon'\rho'$, are then written in the same series form as in Eq. (5.71):

$$v(z,t) = \bar{\varepsilon}\bar{v}_\perp(z,t) + \varepsilon'v'_\perp(z,t) + \varepsilon'v'_\parallel(z,t)$$

$$B(z,t) = B_0\hat{z} + \bar{\varepsilon}\bar{B}_\perp(z,t) + \varepsilon'B'_\perp(z,t) \tag{5.94}$$

$$\rho(z,t) = \rho_0 + \varepsilon'\rho'(z,t).$$

Note that there is no density oscillation in the pump wave. Only fluctuations of the same sense of polarization, or helicity, as the pump wave can be excited, and the fluctuation fields can be written in the form

$$B'_\perp(z,t) = Real(B'(z,t)\hat{e}_\pm). \tag{5.95}$$

We start from the general nonlinear equations (5.23)-(5.25), retaining the particle pressure gradient term but dropping the Hall term. Substituting Eq. (5.94), eliminating v', and retaining terms of order $\bar{\varepsilon}\varepsilon'$, we find the following differential equations describing one–dimensional propagation of the excited waves:

$$\left(\frac{\partial^2}{\partial t^2} - v_A^2\frac{\partial^2}{\partial z^2}\right)B'_\perp = -\bar{\varepsilon}\left[\frac{B_0}{\rho_0}\frac{\partial}{\partial z}\left(\rho'\frac{\partial}{\partial t}\bar{v}_\perp\right) + \frac{\partial^2}{\partial z\partial t}\left(v'_\parallel\bar{B}_\perp\right)\right.$$

$$\left. + B_0\frac{\partial}{\partial z}\left(v'_\parallel\frac{\partial}{\partial z}\bar{v}_\perp\right)\right] \tag{5.96}$$

$$\left(\frac{\partial^2}{\partial t^2} - c_s^2 \frac{\partial^2}{\partial z^2}\right)\rho' = \bar{\varepsilon}\frac{1}{2\mu_0}\frac{\partial^2}{\partial z^2}(\bar{\boldsymbol{B}}_\perp \cdot \boldsymbol{B}'_\perp) \tag{5.97}$$

$$\frac{\partial\rho'}{\partial t} + \rho_0\frac{\partial v'_\parallel}{\partial z} = 0. \tag{5.98}$$

A physical interpretation of the nonlinear interactions may be gained by noting that the right–hand side of Eq. (5.96) derives from the nonlinear current density arising from the beating of the pump Alfvén wave and the density fluctuations. The right–hand side of Eq. (5.97) derives from the ponderomotive force resulting from the beating of the pump and the induced magnetic fluctuations.

Taking the space-time Fourier transform of Eqs. (5.96)-(5.98) leads to three equations coupling a Fourier mode of the density fluctuation, of frequency ω and wavenumber k, to two Fourier modes of the magnetic field fluctuations, with frequencies $\omega_\pm = \omega \pm \omega_0$ and wavenumbers $k_\pm = k \pm k_0$:

$$(\omega^2 - c_s^2 k^2)\rho'(k,\omega) = \bar{\varepsilon}\frac{k^2}{2\mu_0}\left[BB'^*(-k_-, -\omega_-) + B^* B'(k_+, \omega_+)\right] \tag{5.99}$$

$$(\omega^2 - v_A^2 k_+^2)B'(k_+, \omega) = -\bar{\varepsilon}\frac{k_+\omega_0^2 B}{2k_0\rho_0}\left[1 - \frac{\omega}{\omega_0}\frac{k_0}{k} - \frac{\omega_+\omega}{\omega_0^2}\frac{k_0}{k}\right]\rho'(k,\omega) \tag{5.100}$$

$$(\omega^2 - v_A^2 k_-^2)B'(-k_-, -\omega) = \bar{\varepsilon}\frac{k_-\omega_0^2 B^*}{2k_0\rho_0}\left[1 - \frac{\omega}{\omega_0}\frac{k_0}{k} + \frac{\omega_-\omega}{\omega_0^2}\frac{k_0}{k}\right]\rho'(k,\omega). \tag{5.101}$$

If the pump wave is not circularly polarized (Lashmore-Davies 1976), coupling to higher–order side–bands occurs.

The dispersion equation resulting from Eqs. (5.99)-(5.101) is (Goldstein 1978, Derby 1978)

$$(\omega^2 - c_s^2 k^2)(\omega - v_A k)[(\omega + v_A k)^2 - 4\omega_0^2]$$
$$= \frac{\eta v_A^2 k^2}{2}[\omega^3 + \omega^2 v_A k - 3\omega\omega_0^2 + \omega_0^2 v_A k)] \tag{5.102}$$

where $\eta = \bar{\varepsilon}^2|B|^2/B_0^2$. Eq. (5.102) is fifth–order in ω, with complex solutions $\omega_r + i\gamma$, where $|\gamma|$ is the decay rate of the pump wave and the growth rate of the daughter waves. Unstable solutions are found for a left–hand polarized pump wave, for $\beta' \lesssim 1$, in a broad range of k, with $|k| \lesssim k_0$. Note that $\omega_\pm$ may not be equal to $\pm v_A k_\pm$, so the daughter magnetic waves are not necessarily normal modes of the plasma, i.e. they may be forced oscillations.

However, consider the limit of a very low–β plasma, $c_s^2 \to 0$. Then there is a solution of Eq. (5.102) given by $\omega_r \simeq c_s k \ll \omega_0$, and the growth rate γ then has a maximum near $k \simeq 2k_0$. The frequency of one daughter magnetic wave is $\omega_- \simeq -\omega_0$, and its wavenumber is $k_- \simeq k_0$, that is, for this wave $\omega_- \simeq -v_A k_-$. Thus the forward moving pump Alfvén wave decays into a forward moving acoustic wave and a backward moving Alfvén wave, and we have the parametric decay instability of the pump Alfvén wave into those two natural modes of the system (Sagdeev & Galeev 1969). The decay is shown schematically in Figure 5.3. The upper side–band magnetic wave is also excited, but with small amplitude, because it is not a normal mode.

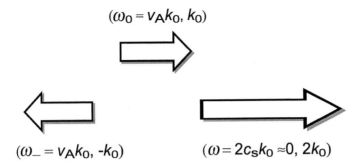

Figure 5.3: Schematic representation of the decay of an Alfvén wave in a low–β plasma into an acoustic wave and a backward propagating daughter Alfvén wave. Each wave has the indicated frequency-wavenumber pair (ω, k), and the length of each arrow indicates the size of the corresponding wavenumber k.

The relations between the frequencies and wavenumbers for the decay instability can be written as the resonance conditions

$$\omega_0 = \omega + |\omega_-|, \quad k_0 = k - k_- \tag{5.103}$$

which show explicitly the conservation of frequency (energy) and wavenumber (momentum) in the break–up of the pump Alfvén wave into the two daughter waves, analogous to the resonance conditions for the decay of a magnetoacoustic pump, given by Eq. (5.85). The solution of Eq. (5.102) for the decay instability is

$$\omega^2 \simeq c_s^2 k^2 - \frac{\omega_0^2}{4} \eta \frac{v_A}{c_s} \tag{5.104}$$

so in the very low–β limit and for large enough pump amplitude (large η), when the last term in Eq. (5.104) dominates, the growth rate of the daughter waves is (Galeev & Oraevskii 1963)

$$\gamma \simeq \frac{\omega_0}{2} \frac{\eta^{1/2}}{\beta'^{1/4}} \tag{5.105}$$

with $\beta' = c_s^2/v_A^2$. As β increases, the daughter waves become less resonant with the normal modes, there is a broad band of unstable k values, and the modulational instability loses its parametric decay nature. One application of this theory is to Alfvén waves in the solar wind (Goldstein 1978). The decay instability is not expected to be of importance in the solar wind ($\beta \simeq 1/2$), or the vicinity of the Sun down to about twice the solar radius ($\beta \simeq 0.1$).

Various extensions of these results have been reported in the literature. The effects of nonzero β were investigated by Jayanti & Hollweg (1993), where it was pointed out that the result (5.105), although correct for $0 < \beta' \lesssim 1$, is not in fact valid for $\beta' \to 0$. By a careful analysis for the different ranges of values of β', the maximum growth rates were found to be, in addition to the result (5.105):

$$\gamma_{\max} \propto \begin{cases} \eta^{1/3} & (\beta' \simeq 0) \\ \eta^{1/3}/(\beta' - 1)^{3/2} & (\beta' \gtrsim 1) \\ \eta^{3/4} & (\beta' \simeq 1). \end{cases} \tag{5.106}$$

The effects of finite ion cyclotron frequency can be included by keeping the Hall term in Eq. (5.25), and have been investigated for $\beta' \lesssim 1$ (Sakai & Sonnerup 1983; Terasawa *et al.* 1986; Wong & Goldstein 1986; Longtin & Sonnerup 1986). In this case, there is a modulational instability with a broad band of unstable k with $|k| < k_0$, just as for the low–frequency case. In addition, there is an instability at $|k| > k_0$ with a smaller growth rate and a very narrow bandwidth in k, called the "beat wave" instability (Wong & Goldstein 1986). The modulational instability of parallel and almost parallel propagating, periodic and solitary solutions of the DNLS equation have been considered by Kennel *et al.* (1988). A high–β plasma with anisotropic particle pressure can also be treated, using the CGL system of equations (1.26) and (1.27), including finite ion Larmor radius effects (Hamabata 1993). Daughter waves that propagate obliquely to the steady magnetic field direction have been considered (Kuo, Whang & Lee 1988; Kuo, Whang & Schmidt 1989; Champeaux *et al.* 1997a; Champeaux, Passot & Sulem 1997b). In this case the plane wave front of the pump wave interacts with a density corrugation in the plane, and breaks up into filaments, due to the *filamentation instability*.

The instabilities of standing, rather than propagating, pump waves have been studied by Hung (1974) and Lashmore-Davies & Ong (1974), with a generalization to the strongly driven regime, where nonresonant acoustic modes are generated, by Chian & Oliveira (1994) and Chian (1995). Standing structures of Alfvén waves in the Earth's environment may be generated as the waves are guided along field lines in the magnetosphere-ionosphere system and reflected by the lower ionosphere. Such standing waves have been observed by the Viking satellite in the 0.1–1 Hz frequency range trapped within a density cavity of high v_A between a half and a few Earth radii. Another example of a standing wave that may be subject to modulational and decay instabilities is that of the Io plasma torus.

The parametric and modulational instabilities may lead to a turbulent dissipation process, leading to conversion of the Alfvén wave energy to particle kinetic energy, for example by the acoustic waves being Landau damped. The density perturbation that accompanies the modulational instability gives rise to an electrostatic, longitudinal electric field, and a longitudinal gradient in magnetic field strength, which leads to Landau and transit time magnetic damping. A density perturbation also arises for two arbitrarily circularly polarized Alfvén waves, that nonlinearly interact to give a low–frequency longitudinal beat wave that also suffers these collisionless damping mechanisms via wave-particle interaction with the thermal particles: this process is known as nonlinear Landau damping of the Alfvén waves (Lee & Völk 1973; Völk & Cesarsky 1982). Alfvén waves may also be involved in parametric instabilities with higher frequency and electrostatic modes. There is evidence of close correlations between whistler waves, Langmuir waves and Alfvén waves propagating along the auroral magnetic field lines, in association with low energy (100 eV to 10 keV) field–aligned beams of electrons (Chian 1995). Thus nonlinear wave-wave interactions involving these modes may be occurring.

5.6 Nonlinear Kinetic and Inertial Alfvén Waves

The dispersive character of KAWs and IAWs, as shown by their dispersion relations (2.226) and (2.229), combined with nonlinear steepening, can give rise to soliton solutions, known as Solitary Kinetic Alfvén Waves (SKAW). A compressive solitary nonlinear kinetic Alfvén wave was shown to exist by Hasegawa & Mima (1976) for $\beta > m_e/m_i$. As in the linear case (Section 2.8.3), the low–β assumption allows the use of the two–potential fluid method, with only shear perturbations in the magnetic field. Charge neutrality is again assumed. As in the theory of parallel propagating waves described by the DNLS equation (Section 5.4.3), an equation of motion for a pseudo–particle is derived, with a pseudo or "Sagdeev" potential, which has solitary wave solutions. In the small amplitude limit, the soliton is described by the KdV equation. The structure of the nonlinear solitary KAW is scaled by the ion Larmor radius in the perpendicular direction, but the wave propagates almost in the parallel direction with a sub–Alfvénic velocity, accompanied by a hump in the plasma density. A high–β correction was included by Yu & Shukla (1978).

A nonlinear inertial Alfvén wave (for $\beta < m_e/m_i$) was considered by Shukla, Rahman & Sharma (1982). A rarefactive soliton was found to exist, with perpendicular structure scaled by the electron skin depth, a super–Alfvénic velocity along the magnetic field direction and a dip in the plasma density. If ion inertia and the displacement current are included, both super– and sub–Alfvénic rarefactive solitons are found to exist for $\beta < m_e/m_i$ (Kalita & Kalita 1986). The case where $\beta \simeq m_e/m_i$ was considered by Wu *et al.* (1996), and it was found that both hump and dip solitons are possible. Linear and nonlinear KAW and IAW modes in two–ion plasmas have been considered by Faria *et al.* (1998).

Nonlinear mode couplings between finite amplitude KAWs or IAWs provide the possibility of the formation of solitary vortices. Drift-Alfvén vortices are localized structures with spatial scales determined by the ion Larmor radius and the electron inertial length (Kuvshinov *et al.* 1999). Further discussion of nonlinear KAW and IAW modes in space plasmas is found in Chapter 7, where observational evidence for the waves is also discussed.

The process of Alfvén resonance absorption of waves is influenced by nonlinear effects. The wave fields at the resonant point may grow up to large amplitude. Thus the short–wavelength KAW and IAW modes that are excited at the resonance may have the nonlinear properties described above. The effects of nonlinearity were shown to accelerate the Alfvén resonant absorption of surface waves on a thin smooth surface in a viscous and resistive plasma by Ruderman & Goossens (1993). Nonlinear interactions cause shorter–wavelength fields to arise, which are more effectively damped by the dissipative processes. Nonlinear wave-wave coupling, such as parametric interactions, were shown by Lundberg (1994) to effectively damp the sharply peaked wave fields within the thin surface due to resonant absorption. The nonlinear processes within the thin surfaces of a magnetized slab were found to be more important than those from the main slab. The ponderomotive force produced by the fields about the Alfvén resonance were derived for a resistive plasma model by Tataronis & Petrzilka (1996). If the Hall term is retained, and the nonlinear equations are time–averaged, wave momentum is absorbed. The absorbed energy and momentum can be used in a laboratory plasma to drive current, and control the local density profile and the plasma transport.

5.7 Nonlinear Surface Waves

We have seen in Chapter 4 that linear MHD or Alfvén surface waves can exist in highly structured plasmas with nonuniform densities and magnetic fields, with fields concentrated about more or less well–defined surfaces such as density discontinuities or current sheets. The theory of nonlinear surface waves is less well developed than the theory of nonlinear waves in uniform media. A major difference of surface waves to waves in uniform media is that their description is inherently two–dimensional, that is, variation of the wave fields perpendicular to the surface as well as in the surface must be allowed for, and so must also be taken into account for nonlinear waves. The nonlinear theory of high–frequency surface waves in unmagnetized plasmas has been reviewed by Vladimirov, Yu & Tsytovich (1994). A small number of papers has been written on nonlinear surface waves of frequency less than the ion cyclotron frequency in magnetized plasmas, which we review in this section.

If stable nonlinear wave structures such as soliton modes are to arise, there must be dispersive terms in the wave equations to balance the nonlinear terms. One way that dispersion can be introduced into the linear dispersion relation of surface waves, even in the case of a frequency much less than the ion cyclotron frequency, is through geometric effects. If a cylindrical plasma is surrounded by a vacuum, or by another plasma of different density or pressure, or if a finite width slab of plasma is considered, a surface wave that is dispersive can propagate on the interfaces. In these cases the dispersive term results from the finite width of the cylinder or the slab.

For low–frequency waves in a slab or cylindrical plasma, the nonlinear wave equation that arises is the Benjamin-Ono equation (Roberts 1985), or the Korteweg-de Vries equation (Hollweg & Roberts 1984; Roberts & Mangeney 1982). Even more exotic equations can arise in other circumstances (Ruderman 1988). A cubic nonlinear Schrödinger equation has been derived by Sahyouni, Zhelyazkov & Nenovski (1988) by calculating the ponderomotive force due to a finite–amplitude fast symmetric magnetoacoustic surface wave, and by Zhelyazkov, Murawski & Goossens (1996) in a treatment of modulations of slow surface waves on a plasma slab. Geometrical dispersion effects on MHD surface waves on beamed galactic jets, and the resulting soliton wave structures, have been invoked as an explanation of observed radio knots in the collimated portion of jetlike radio sources (Fiedler 1986).

Another way of introducing dispersion is to include the Hall or ion cyclotron term in the generalized Ohm's law, just as in the treatment of nonlinear waves in uniform plasmas. We now consider this case in more detail.

5.7.1 Nonlinear Surface Waves with Hall Dispersion

As we have seen in Chapter 4, surface waves can propagate in a low–β plasma on density discontinuities at frequencies up to and beyond Ω_i. The dispersion relations (4.20) and (4.21) for linear surface waves in a cold plasma show that for a fixed direction of propagation in the surface, there is a single nondispersive mode, independent of the sign of k_y, if the terms in $1/\Omega_i$ are neglected, that is at low frequency. This mode is related to the fast compressional Alfvén wave and is called an Alfvén surface wave. As the frequency increases, a two–mode structure of the surface wave dispersion relation develops, depending on the sign of k_y or $\sin\theta$, where θ is the angle of propagation in the surface, which becomes degenerate for $\omega \ll \Omega_i$, but leads to distinctive properties as the frequency approaches Ω_i. The "+" mode, which has electric field in the plasma predominantly right–hand polarized, is hardly affected by the ion cylotron resonance, whereas the "-" mode, with left–hand polarized electric field in the plasma, has a limiting upper frequency at which the surface wave takes on the character of the shear Alfvén wave. As mentioned in Section 4.6.2, MHD fluctuations have been observed in the Earth's magnetosphere, in particular at the boundary of the plasma sheet in the magnetotail, with a wave period of 2 to 18 s, compared with the local proton gyroperiod of 5 to 12 s (Tsurutani *et al.* 1985). These fluctuations have large amplitude (wave magnetic field comparable with the ambient field), the plasma is highly structured, and the frequencies are not very small compared to the ion cyclotron frequency. It is therefore possible that nonlinear surface Alfvén waves at frequencies approaching the ion cyclotron frequency may contribute to the spectrum of the fluctuations.

To analyse the corresponding nonlinear surface waves, we require equations describing the three–dimensional variation of the fields. The nonlinear wave equations derived in Chapter 5 for a uniform plasma, which lead to the DNLS equation, are applicable to variation of the wave fields along the steady magnetic field direction, or at small angles ($\sin \theta = $ order $\varepsilon^{1/2}$) to it. Solutions of the weakly nonlinear oblique wave equations (5.65) and (5.66) would have to be linked to the (linear) vacuum fields by means of the boundary conditions at the surface, in order to obtain a surface wave solution. Rather than following this path of solution, we proceed to obtain an alternate solution to the problem of the nonlinear surface wave in terms of an interacting harmonics description, utilizing a theory originally developed for nonlinear surface waves in solid state plasmas (Agranovich & Chernyak 1982).

This method uses a description in terms of interacting harmonics to derive a nonlinear dispersion relation, which is then used to derive a nonlinear Schrödinger equation (NLS) describing the weakly nonlinear wave (Cramer 1991). In this approach, the second–harmonic response due to the nonlinearity is calculated in second order in the wave amplitude, and then the second harmonic interacts with the first harmonic to give a third–order first–harmonic response and so a third–order (nonlinear) modification to the dispersion relation. The solution of the nonlinear equations (5.23)-(5.25), with the particle pressure neglected but the Hall term retained, is assumed to be weakly nonlinear, so that the magnetic field may be written as a power series in the wave amplitude:

$$B_x = b_x^{(1)} + b_x^{(2)} + b_x^{(3)} + \dots \qquad (5.107)$$

where the nth term is of the nth order in the wave amplitude. The second–order term is a second harmonic of the linear wave, arising from the nonlinear terms in the wave equations. Similar expansions hold for B_{y1} and B_{z1}, as well as the electric fields. The surface wave dispersion relation depends on the boundary conditions assumed at the surface. As in Section 4.3 the plasma is assumed to fill the semi–infinite space $x < 0$, with a vacuum for $x > 0$, and the boundary conditions for the wave magnetic field are that

$$
\begin{aligned}
&b_x^{(1)}(x = 0^-) + b_x^{(2)}(x = 0^-) = B_x(x = 0^+) \\
&b_x^{(n)}(x = 0^-) = 0 \text{ for } n \geq 3 \\
&B_z(x = 0^-) = B_z(x = 0^+)
\end{aligned}
\qquad (5.108)
$$

and $\boldsymbol{B}$ in the vacuum is the sum of a first harmonic and a second harmonic.

Equations (5.23)-(5.25) are expanded to third order in the field amplitudes, and the first and second harmonics of the fields are calculated. Further details are given by Cramer (1991). The dispersion relation with its nonlinear correction is found to be

$$
\begin{aligned}
f &= f_{\mathrm{L}} - \frac{9k_z^6 + 14k_y^2 k_z^4 - 12k_y^4 k_z^2 - 20k_y^6}{64 k_y k_z^2 (k_z^2 + k_y^2)^{3/2}} |a|^2 + \mathcal{O}(\alpha) \\
&= f_{\mathrm{L}} - \frac{9 - 13 \sin^2 \theta - 13 \sin^4 \theta - 3 \sin^6 \theta}{64 \sin \theta \cos^2 \theta} |a|^2 + \mathcal{O}(\alpha)
\end{aligned}
\tag{5.109}
$$

where a is the amplitude of the x–component of the magnetic field, normalized to the equilibrium field, $\alpha = v_{\mathrm{A}} k_z / \Omega_{\mathrm{i}}$, and f_{L} is the linear frequency, normalized to Ω_{i}. Note that the dominant term in the nonlinear correction is independent of α, that is, is independent of the frequency. The sign of the nonlinear correction depends on the sign of $\sin \theta$, so is opposite for the LHP and RHP modes.

5.7.2   Surface Alfvén Wave Solitons

The nonlinear dispersion relation (5.109) may be used to derive a nonlinear Schrödinger equation (NLS), and thence soliton properties. The frequency can be written generally, for small dispersion and nonlinear term, as

$$
f \simeq \left(\frac{\partial f_{\mathrm{L}}}{\partial k} \right)_{k=0} k + \frac{1}{2} \left(\frac{\partial^2 f_{\mathrm{L}}}{\partial k^2} \right)_{k=0} k^2 + f_2 |a|^2.
\tag{5.110}
$$

The following NLS equation follows from Eq. (5.110) (Grozev, Shivarova & Boardman 1987):

$$
i \left[\frac{\partial a}{\partial t} + \left(\frac{\partial f_{\mathrm{L}}}{\partial k} \right)_{k=0} \frac{\partial a}{\partial x} \right] + \frac{1}{2} \left(\frac{\partial^2 f_{\mathrm{L}}}{\partial k^2} \right)_{k=0} \frac{\partial^2 a}{\partial x^2} - f_2 |a|^2 a = 0
\tag{5.111}
$$

with a soliton solution

$$
a = \left[-\frac{1}{f_2} \left(\frac{\partial^2 f_{\mathrm{L}}}{\partial k^2} \right)_{k=0} \right]^{1/2} \frac{e^{i \delta t}}{v_{\mathrm{g}} t_1} \operatorname{sech} \left[\frac{t - x/v_{\mathrm{g}}}{t_1} \right]
\tag{5.112}
$$

where

$$
\delta = (\partial^2 f_{\mathrm{L}} / \partial k^2)_{k=0} / (2 v_{\mathrm{g}}^2 t_1^2)
\tag{5.113}
$$

and the group velocity is

$$
v_{\mathrm{g}} = (\partial f_{\mathrm{L}} / \partial k)_{k=0}
\tag{5.114}
$$

and t_1 is the pulse halfwidth. From Eqs. (4.21) and (5.109),

$$
-\frac{1}{f_2} \left(\frac{\partial^2 f_{\mathrm{L}}}{\partial k^2} \right)_{k=0} = \frac{v_{\mathrm{A}}^2 \sin^2 \theta \cos^4 \theta}{\Omega_{\mathrm{i}}^2 (9 - 13 \sin^2 \theta - 13 \sin^4 \theta - 3 \sin^6 \theta)}.
\tag{5.115}
$$

Note that the sign of the nonlinear correction compensates for the sign of the dispersion, so the sign of the right–hand side of Eq. (5.115) is independent of the sign of $\sin\theta$, that is, it is the same for the LHP and RHP modes. If the quantity defined in Eq. (5.115) is positive, the solution (Eq. 5.112) describes a "bright" soliton, that is, a localized pulse propagating in the equilibrium plasma; otherwise the soliton is "dark", consisting of a decrease in intensity of a continuous background wave.   The bright soliton is found to exist for $0° < |\theta| < 42.7°$ and $137.3° < |\theta| < 180°$, independent of whether the linear wave is the LHP or RHP mode.

If the soliton solution (Eq. 5.112) is written as

$$|a| = A\operatorname{sech}((v_gt - x)/\lambda_E) \tag{5.116}$$

the width of the soliton envelope λ_E is related to the amplitude A by

$$\lambda_E = \left[-\frac{1}{f_2}\left(\frac{\partial^2 f_L}{\partial k^2}\right)_{k=0}\right]^{1/2}/A. \tag{5.117}$$

Thus the ratio of the envelope width to the carrier wavelength is of the order of $k\lambda_E$, which is approximately f_L/A, neglecting a directional factor. If the soliton magnetic field amplitude is comparable to the ambient magnetic field, as is the case for solar wind magnetic perturbations, $A \simeq 1$ and $k\lambda_E \simeq f_L \simeq \alpha$, so the soliton width is small compared with the carrier wavelength, which is of the order of 4000 km in the Earth's magnetotail.

6 Laboratory Plasmas

6.1 Introduction

In the quest for controlled nuclear fusion utilizing magnetically confined plasmas, two areas where Alfvén and magnetoacoustic waves may make a strong contribution is to the heating of the plasma to fusion temperatures, and to the driving of currents that produce a magnetic field that hopefully will confine the charged particles. The waves also play roles in the various instabilities and turbulence that may arise in the controlled fusion devices. In addition, Alfvén and magnetoacoustic waves have been studied from a fundamental viewpoint in many laboratory experiments, with the aim of understanding the basic properties of the waves in the context of both fusion and space plasmas. The Alfvén wave heating scheme and Alfvén wave current drive, to be described in this chapter, are attractive for Tokamaks and Stellarators because of the availability of high–power radio frequency generators and the simplicity of the wave launching antenna structures. The plasmas in such machines are low–β, and the wave properties are strongly influenced by the boundary conditions at the walls of the device, which can act like a waveguide or resonant cavity. Thus eigenmodes of the waves may exist, with spectra of discrete eigenfrequencies. This contrasts to waves in most space and astrophysical plasmas, where boundary conditions are usually not important. An exception is found in solar magnetic flux tubes, where, even though the spatial scale is vastly different to laboratory plasmas, the waves can form eigenmodes through their interaction with the relatively sharp "walls" of the flux tubes, and the tube can act like a waveguide.

An extensive survey of Alfvén waves in laboratory plasmas can be found in the book by Cross (1988), and the early experiments were described in the book by Hasegawa & Uberoi (1982). In this chapter, we first introduce general concepts associated with the theory of wave modes in bounded plasmas. Then we briefly discuss the theory of waves in the specific geometries relevant to laboratory devices, and their relation to the theory developed in the earlier chapters. A straight circular cylindrical geometry is normally used as a good first approximation in describing waves in laboratory plasmas, and we make much use of this geometry in this chapter. However, the Tokamak geometry is in fact toroidal, which leads to modifications of the expected wave spectrum, which we also discuss. The driving of plasma currents by Alfvén waves is then considered. Another important aspect of waves in laboratory devices is the launching of the waves by antennas. In the case of Alfvén waves, antennas, or indeed naturally occurring processes such as the instabilities of particle beams in the Earth's magnetosphere, can launch waves that are very localized to particular field lines, as was discussed in Section 2.8.5. The interplay of localized waves with eigenmodes of the bounded plasma is also considered here.

6.2   Modes of Bounded Plasmas

The theory of waves in bounded plasmas is important for the description of laboratory plasmas. Such plasmas, in particular those with cylindrical symmetry, are discussed in depth in this chapter. Here we briefly discuss the modes that arise in a nonuniform plasma with a rectangular slab geometry. Thus the plasma is taken to be of finite extent in the x–direction, with boundaries at $x = a$ and $x = b$, with $b > a$, but is infinite in extent in the y and z directions. We consider first the example of a plasma described by the ideal MHD model, with an equilibrium x-dependent magnetic field that lies in the y-z plane, described by Eq. (3.47), and pressure described by Eq. (3.48). Dissipative effects are initially neglected. The small–amplitude waves are assumed to have given wavenumber components k_y and k_z, and have low frequency compared with the ion cyclotron frequency, and so the perturbation x–component of plasma velocity for a wave of given frequency obeys Eq. (3.56). The boundary conditions chosen, corresponding to rigid conducting walls confining the plasma, are that $v_x = 0$ at $x = a$ and $x = b$.

For a given plasma density and magnetic field dependence on x, there will be two solutions of the second–order differential equation (3.56), which may be interpreted as the spatial parts of two magnetoacoustic waves propagating (or evanescent) in the $\pm x$–directions. The application of the boundary conditions leads to a set of eigenvalues for the frequency ω, with a corresponding set of spatial eigenfunctions for the field v_x. If the density and magnetic field are slowly varying, such that their spatial scale–length is much greater than the local wavelength of the wave, an approximate solution may be obtained using a WKB approach, where the solution of Eq. (3.56) can be assumed to have the form

$$v_x = \hat{v}_x \exp(i\Phi) \tag{6.1}$$

where $\hat{v}_x$ is slowly varying in x. Substituting this into Eq. (3.56) and neglecting the second–order derivative of the phase Φ, we then have

$$\Phi = \pm \int (-q^2)^{1/2} dx. \tag{6.2}$$

The boundary conditions then lead to the quantization condition

$$\int_a^b (-q^2)^{1/2} dx = n\pi \tag{6.3}$$

which defines the set of allowed eigenfrequencies, through the dependence of q on ω, corresponding to the integers n. This procedure is valid provided that there are no singularities in the integrand in Eq. (6.3), that is, the cusp resonance defined by Eq. (3.5), where $1/q^2 = 0$, is not encountered in $a < x < b$. In the uniform plasma case, the exact solution is a standing wave with sinusoidal eigenfunctions, with the quantization condition

$$k_x = n\pi/l \tag{6.4}$$

where $k_x = (-q^2)^{1/2}$ and $l = b - a$. For a uniform plasma with a cylindrical boundary, as discussed in the next section, the eigenfunctions are Bessel functions or combinations of them, and the set of eigenfrequencies are determined by the zeroes of the eigenfunctions.

We may note that the Alfvén resonance, $\tau = 0$, is not encountered by the magnetoacoustic wave described by Eq. (3.56) in the WKB approach. As we have seen in Section 3.3.1, one way of interpreting an Alfvén resonance point in a nonuniform plasma is to say that there is a continuous spectrum of frequencies of shear Alfvén waves, each localized at a point in the plasma profile where $\tau = 0$. Another way of expressing this is to state that the eigenfunctions of the Alfvén wave are discontinuous improper eigenfunctions and the spectrum forms a continuous band along the real axis of the complex frequency plane. For the continuous spectrum, the boundary conditions have no effect on the eigenfunction unless the resonance point lies on the boundary. However, one should remember that gradients of the equilibrium plasma parameters, or nonideal effects such as finite ion cyclotron frequency, lead to a coupling between the magnetoacoustic mode and the shear Alfvén wave.

6.2.1 Resistive Plasmas

Turning now to the effects of resistivity of the plasma, we note that the resistive modes of a bounded plasma, such as a Tokamak fusion device, are important because of their connection to instabilities of the plasma, such as the resistive tearing modes. In contrast to the ideal infinitely conducting plasma case, in the resistive plasma the Alfvén wave has a dispersion relation (3.41) dependent on k_x (i.e. the full dispersion relation is fourth order in k_x, or the wave differential equation is fourth order in x), and the resistive Alfvén wave spectrum has continuous, proper eigenfunctions, but with complex frequencies reflecting the damped nature of the waves. The magnetoacoustic wave also has proper eigenfunctions, again with complex frequencies: the damping in this case may be interpreted as resonance damping if the Alfvén resonance is present in the plasma profile (or compressive resonance if the pressure is not zero). We now derive differential equations describing the waves in a cold resistive plasma, stratified in the x–direction, and use them to discuss the bounded plasma modes.

Faraday's equation (1.3), together with the resistive Ohm's law (Eq. 1.8), gives

$$\frac{\partial \boldsymbol{B}}{\partial t} = \nabla \times (\boldsymbol{v} \times \boldsymbol{B}_0 - \eta \boldsymbol{J}) \tag{6.5}$$

from which we obtain

$$-\mathrm{i}\omega B_x = \mathrm{i}(\boldsymbol{k} \cdot \boldsymbol{B}_0)v_x + \frac{\eta}{\mu_0}\nabla^2 B_x \tag{6.6}$$

where $\boldsymbol{k} = (0, k_y, k_z)$. Considering now a uniform plasma density ρ_0, but allowing for a rotating x–dependent magnetic field in the y-z plane of the form given by Eq. (3.47), with a corresponding Alfvén velocity vector $\boldsymbol{v}_A$, we have from Eq. (3.50)

$$\omega\mu_0\rho_0((\nabla \cdot \boldsymbol{v}_1)' - \nabla^2 v_x) = (\boldsymbol{k} \cdot \boldsymbol{B}_0)'' B_x - (\boldsymbol{k} \cdot \boldsymbol{B}_0)\nabla^2 B_x \tag{6.7}$$

where the dash denotes a derivative with respect to x. At the boundaries $x = a$ and $x = b$ we assume a rigid conducting wall, so $B_x = v_x = 0$.

As a first approximation, a WKB solution is again sought, that is, we assume that the wavelength is short compared with the characteristic length–scales of the plasma, so we seek a solution of the form

$$B_x = \hat{B}_x \exp(\mathrm{i}\Phi) \tag{6.8}$$

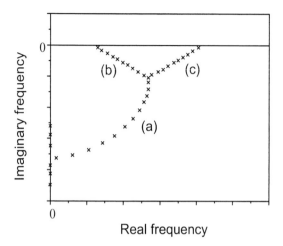

Figure 6.1: Schematic resistive mode spectrum in the complex frequency plane, for a plasma slab. See text for explanation of (a), (b) and (c). (Adapted from Storer & Schellhase 1995).

where Φ is proportional to an inverse positive power of the (assumed small) resistivity (Storer & Schellhase 1995). In this approximation we can eliminate the magnetoacoustic wave and select the Alfvén wave by setting $\nabla \cdot \boldsymbol{v}_1 = 0$ (if the plasma is incompressible this is true in any case). This turns Eqs. (6.6) and (6.7) into two coupled second–order differential equations in v_x and B_x. Keeping only the leading terms in an expansion in η leads to the result

$$(\Phi')^2(-i\omega(\eta/\mu_0)(\Phi')^2 + \omega_A^2(x) - \omega^2) = 0 \tag{6.9}$$

where

$$\omega_A(x) = \boldsymbol{k} \cdot \boldsymbol{v}_A. \tag{6.10}$$

Then

$$\Phi = \pm \int_C \sqrt{\frac{\omega_A^2(x) - \omega^2}{i\omega(\eta/\mu_0)}} dx \tag{6.11}$$

where C denotes a suitable contour in the complex plane. If the magnetic field is uniform, we obtain the result $\Phi = \exp(\pm ik_x x)$ with k_x given by Eq. (3.41) (for $|k_x^2| \gg k_y^2 + k_z^2$).

Imposing the boundary conditions on the fields leads to a quantization condition similar to Eq. (6.3). The resulting spectral points, that is the complex frequency eigenvalues, for the *resistive spectrum*, either lie along curves in the lower half complex frequency plane or are purely damped modes (on the negative imaginary axis), as is shown in Figure 6.1 for the case where $\omega_A(x)$ is linearly varying with x (Storer & Schellhase (1995) and references therein). The eigenfunctions for the spectral points on the lines (b) and (c) are resistively modified Alfvén modes which are of oscillatory form near the edges of the slab associated with either maximum or minimum ω_A. These modes correspond to shear Alfvén waves that are localized about the Alfvén resonance points defined by $Real(\omega) = \omega_A$ (see Section 3.2.4). The eigenfunctions associated with the points on the arc (a) are zero near the edges and have a peaked oscillatory behaviour near the centre of the slab, and the modes are called *resistive modes*. The angles of $30°$ made by the lines near the real axis are characteristic of a linear profile for $\omega_A(x)$. For a cylindrical plasma with a quadratic profile near the axis, the lines make an angle of $45°$ (Dewar & Davies 1984).

The spectrum can be understood if we first consider the spectrum for a uniform plasma and magnetic field, with $\omega_A = $ constant. From Eq. (3.41) and the quantization condition (6.4) we find

$$\omega = \left(\omega_A^2 - \frac{n^4\pi^4\eta^2}{4l^4\mu_0^2}\right)^{1/2} - \frac{in^2\pi^2\eta}{2l^2\mu_0}. \tag{6.12}$$

For low n, there is a single lightly damped wave, with frequency $\simeq \omega_A$. Thus the lines (b) and (c) in Figure 6.1 have collapsed into a single point. As n increases, the damping increases and the spectral points lie along an arc similar to the arc (a), until the first factor on the right–hand side of Eq. (6.12) vanishes, and the points thereafter lie on the negative imaginary axis.

6.3   Cylindrical Geometry

Many early laboratory experiments on waves in plasmas were performed in machines with a cylindrical geometry. More recent devices such as Tokamaks and Stellarators can also, in some circumstances, be approximated by a straight cylindrical geometry, with periodic boundary conditions at the ends of the cylinder. In this and the following sections we consider the bounded Alfvén and magnetoacoustic modes of a plasma with a cylindrical geometry. In addition to laboratory plasmas, this theory is also important for the description of waves in naturally occurring almost cylindrical plasmas, such as plasmas in magnetic flux tubes in the solar atmosphere (see Chapter 7).

In this section we employ the low (but nonzero) β two–fluid model of a single ion species plasma, as used in Sections 2.8.2 and 3.2.2, which enables us to retain ion cyclotron effects and also describe KAW and IAW modes. Collisional effects such as resistivity, ion-neutral collisions and viscosity are not considered here, but they have been included by Woods (1962) in his treatment of modes in a uniform cylindrical plasma. We consider an equilibrium plasma in the form of a circular cylinder with a uniform axial magnetic field. The plasma has axial symmetry, but has a density dependent on the radius, $\rho_0(r)$. The cylindrical coordinates (r, θ, z) are used. Small–amplitude waves are considered, with wave fields of the form

$$g(r)\exp(im\theta + ikz - i\omega t) \tag{6.13}$$

where m (an integer) is the azimuthal mode number, and k is the axial wavenumber (assumed positive). The wave equation takes the form

$$\nabla \times (\nabla \times \boldsymbol{E}) = \boldsymbol{u} \cdot \boldsymbol{E} \tag{6.14}$$

where the tensor u has the same form in cylindrical coordinates as the Cartesian coordinate form used in Eq. (2.75), with components u_1, u_2 and u_3 defined in Eqs. (2.73) and (2.74), with k replacing k_z.

Using Eq. (6.14) and Faraday's law (Eq. 1.3), defining $p = -\partial B/\partial t$, and eliminating the wave fields E_r and B_r, results in the following set of four first–order coupled differential equations in E_θ, E_z, p_θ and p_z:

$$\frac{u_1}{r}\frac{d(rE_\theta)}{dr} = \frac{m}{r}u_2 E_\theta + \frac{m}{r}kp_\theta + \left(u_1 - \frac{m^2}{r^2}\right)p_z \qquad (6.15)$$

$$u_1\frac{dE_z}{dr} = ku_2 E_\theta - (u_1 - k^2)p_\theta - \frac{m}{r}kp_z \qquad (6.16)$$

$$\frac{1}{r}\frac{d(rp_\theta)}{dr} = \frac{m}{r}kE_\theta + \left(u_3 - \frac{m^2}{r^2}\right)E_z \qquad (6.17)$$

$$u_1\frac{dp_z}{dr} = \left(u_2^2 - u_1(u_1 - k^2)\right)E_\theta - \frac{m}{r}ku_1 E_z + ku_2 p_\theta - \frac{m}{r}u_2 p_z. \qquad (6.18)$$

6.3.1   Uniform Plasma

Radial nonuniformity in the equilibrium density is allowed for in Eqs. (6.15)-(6.18), and their solutions are used in subsequent sections to describe modes such as surface waves in a plasma of general radial density profile. However, we consider first a uniform density plasma with uniform magnetic field. The analysis in Chapter 2 of wave modes in a uniform plasma with cartesian coordinates is not affected qualitatively if cylindrical coordinates are employed; the only difference arises in the basis functions used for the wave fields.

By the manipulation of Eqs. (6.15)-(6.18) with constant coefficients to obtain a fourth–order differential equation for $B_z(r)$ (Woods 1962), it can be shown that two body wave modes exist in a plasma of infinite radial extent, for given ω, k and m, with wave magnetic fields (bounded at the origin $r = 0$)

$$B_r(r, k_c) = iCm\frac{J_m(k_c r)}{k_c r} + iAk\frac{dJ_m(k_c r)}{d(k_c r)} \qquad (6.19)$$

$$B_\theta(r, k_c) = -C\frac{dJ_m(k_c r)}{d(k_c r)} - Amk\frac{J_m(k_c r)}{k_c r} \qquad (6.20)$$

$$B_z(r, k_c) = Ak_c J_m(k_c r) \qquad (6.21)$$

where $J_m(k_c r)$ is a Bessel function of the first kind, k_c^2 is a root of

$$u_1 k_c^4 - \left\{(u_3 + u_1)(u_1 - k^2) - u_2^2\right\} k_c^2 + u_3\left\{(u_1 - k^2)^2 - u_2^2\right\} = 0 \qquad (6.22)$$

and C and A are constants in the ratio

$$C/A = \left\{ u_1(u_1 - k_c^2 - k^2) - u_2^2 \right\} / u_2 k = \left\{ \omega^2 / v_A^2 - (k_c^2 + k^2) \right\} / fk \qquad (6.23)$$

with $f = \omega / \Omega_i$. Eq. (6.22) is just Eq. (3.11) with k_c^2 replacing $k_x^2 + k_y^2$.

In general,

$$|u_3 / u_1| \simeq m_i / m_e \gg 1 \qquad (6.24)$$

and so, as long as the Alfvén resonance condition

$$u_1 - k^2 = 0 \qquad (6.25)$$

is not approximately satisfied, the two solutions of Eq. (6.22) for k_c^2 are, as for Eqs. (3.14) and (3.15) in the Cartesian geometry case,

$$k_{c-}^2 = \left\{ (u_1 - k^2)^2 - u_2^2 \right\} / (u_1 - k^2) \qquad (6.26)$$

$$k_{c+}^2 = u_3(u_1 - k^2) / u_1. \qquad (6.27)$$

It is apparent that, under the above conditions, $|k_{c+}| \gg |k_{c-}|$. The waves are the same as discussed for the Cartesian geometry case (Section 3.2.2); the wave associated with wavenumber k_{c-} is called the fast compressional mode, and the wave with wavenumber k_{c+} is a short–wavelength QEW mode. If k_{c+} and k_{c-} are both real, both of these modes are body waves, that is, they are radially oscillating modes in a uniform plasma of infinite extent. The forms given in Eqs. (6.19)-(6.21) for the wave fields imply a standing wave in the radial direction. If k_{c+} or k_{c-} are imaginary, the J_m Bessel functions in Eqs. (6.19)-(6.21) are replaced by the modified Bessel functions I_m or K_m, depending on the boundary conditions, corresponding to radially evanescent solutions.

When the Alfvén resonance condition (6.25) is satisfied the wavenumbers are of similar magnitude, as in Eq. (3.17):

$$|k_-|^2 \simeq |k_+|^2 \simeq f\, |u_1 u_3|^{1/2} . \qquad (6.28)$$

It is apparent from Eqs. (6.19)-(6.21) that B_θ is the dominant field component for both the fast compressional and short–wavelength modes in this case.

6.3.2 Bounded Plasma

We now consider a cylindrical column of uniform density plasma with a sharp interface to a vacuum. The plasma column, of radius a, is surrounded by a vacuum layer and a conducting wall at radius d, as shown in Figure 6.2. In the vacuum region the wave fields can be expressed in terms of four independent solutions of Maxwell's equations that are combinations of the modified Bessel functions $I_m(kr)$ and $K_m(kr)$ and their derivatives. The boundary conditions are that E_θ, E_z, B_θ and B_z are continuous at the plasma-vacuum boundary and that E_θ and E_z are zero at the wall. The fields in the plasma correspond to the uniform plasma wave modes discussed above, and use of the boundary conditions at the surface results in a dispersion

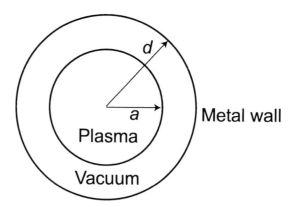

Figure 6.2: A cylindrical plasma column, of radius a, is surrounded by a vacuum layer and a conducting wall at radius d.

relation, that is, a relation between ω, k and m. The solutions so obtained can be interpreted as body waves or surface waves.

The presence of a surface between two regions of plasma of differing properties (or a plasma and a vacuum) means that, after application of the matching conditions on the wave fields across the surface, wave modes may exist that are localized about the surface, that is, the wave fields decay away from the surface. Such a wave mode is a surface wave, which has been analysed for a planar surface in Chapter 4. If the planar surface is parallel to the y-z plane, the surface wave has k_x imaginary on both sides of the surface for real ω, k_z and k_y. In the cylindrical geometry case, a cylindrical surface may represent a sudden change in plasma density, or an interface of the plasma with a vacuum.

If the surface is a circular cylinder about the z-axis, a surface wave's fields must go to zero as $r \to \infty$. Note however that within the cylindrical surface, the wave fields will not generally go to zero as $r \to 0$, and it is not easy to distinguish surface wave from body wave properties in this inner region. However it is shown later in this section that surface waves in the cylindrical case may still be identified by their localization about the surface. Body waves may be characterized by peaks in the wave field amplitude at radial positions less than the surface radius. An additional complication is the introduction of a second boundary, such as a metal wall, surrounding the plasma at a greater radius than the first surface. The electric field must vanish at the metal wall, which is achieved by superposing a spatially growing mode and a spatially decaying mode in the vacuum.

The properties of waves of frequency less than the ion cyclotron frequency are predominantly determined by the fast compressional mode, provided that the wavelength of the QEW mode is short compared with the plasma column size, that is, $|k_{+}a| \ll 1$. The short–wavelength mode is needed for a correct evaluation of the wave frequency and damping and to satisfy the appropriate continuity conditions on $\mathbf{B}$ and $\mathbf{E}$ at $r = a$. If we assume that thermal and electron inertial effects can be neglected, the wave properties can be obtained using the ideal MHD equations including the Hall term, so just the fast compressional mode is involved. The electric field component E_z is zero in the plasma, and Eqs. (6.15)-(6.18) with constant coefficients reduce to the following Bessel equation:

$$\frac{1}{r}\frac{\mathrm{d}}{\mathrm{d}r}\left(r\frac{\mathrm{d}B_z}{\mathrm{d}r}\right) + \left(k_c^2 - \frac{m^2}{r^2}\right)B_z = 0 \tag{6.29}$$

with a Bessel function solution of the form given by Eq. (6.21), where

$$k_c^2 = k_{c-}^2 = -k^2 + \frac{\omega^2(\omega^2 - v_A^2 k^2)}{v_A^2(\omega^2 - v_A^2 k^2(1 - f^2))}. \tag{6.30}$$

The expression for k_c^2 is identical to the corresponding result given by Eq. (3.7) for $k_x^2 + k_y^2$ for Cartesian geometry, as plotted in Figure 3.2 against f for given $\alpha = v_A k/\Omega_i$.

The sufficient boundary conditions now are that E_θ and B_z are continuous at $r = a$, and that $E_\theta = 0$ at $r = d$. The displacement current is ignored, so only the transverse electric (TE) mode in the vacuum annulus has magnetic fields, which can be expressed as

$$B_r^v = -\mathrm{i}GkK_m'(kr) - \mathrm{i}HkI_m'(kr) \tag{6.31}$$

$$B_\theta^v = Gmk\frac{K_m(kr)}{kr} + Hmk\frac{I_m(kr)}{kr} \tag{6.32}$$

$$B_z^v = GkK_m(kr) + HkI_m(kr) \tag{6.33}$$

where G and H are constants.

The dispersion equation is found by using the fields given by Eqs. (6.19)-(6.21), and Eqs. (6.31)-(6.33), and using the boundary conditions at $r = a$ and $r = d$. Regarding k_{c-} as a function of ω, k and m, the dispersion equation is then obtained as (Collins, Cramer & Donnelly 1984):

$$\Delta_0 = \frac{1}{k_{c-}}\left[\frac{J_m'(k_{c-}a)}{J_m(k_{c-}a)} + \frac{mu_2}{k_{c-}a(u_1 - k^2)}\right] - R = 0 \tag{6.34}$$

where

$$R = -\frac{1}{k}\left[\frac{I_m'(ka)K_m'(kd) - I_m'(kd)K_m'(ka)}{I_m(ka)K_m'(kd) - I_m'(kd)K_m(ka)}\right]. \tag{6.35}$$

Equation (6.34) yields ω as a function of k and m. The radial wavenumber k_{c-} is then determined by back substitution in Eq. (6.26) or Eq. (6.30). If $k_{c-}^2 \equiv -\kappa_-^2 < 0$, the wave fields in the plasma can be expressed in terms of modified Bessel functions. The first term on the right–hand side of Eq. (6.34) then becomes

$$-\frac{1}{\kappa_-}\left[\frac{I_m'(\kappa_-a)}{I_m(\kappa_-a)} + \frac{mu_2}{\kappa_-a(u_1 - k^2)}\right]. \tag{6.36}$$

In this case, because of the near exponential variation of the I_m Bessel function, we have a clear–cut surface wave.

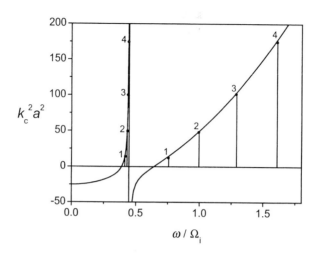

Figure 6.3: The first four pairs of frequencies of the discrete spectrum of waves in a plasma cylinder of radius a with conducting boundary conditions. The cold Hall plasma model is used. The values of $k_c a$ are the first four zeroes of the J_1 Bessel function. The set of four lower frequencies belong to the spectrum of ion cyclotron modes (ICW), while the set of four higher frequencies belong to the spectrum of compressional Alfvén modes (CAW). Here $\alpha = 0.5$, and $\Omega_i a / v_A = 10$.

Zero Vacuum Width

Consider first the case of a vacuum width equal to zero ($d = a$), so that R is equal to zero. For an azimuthally symmetric mode, $m = 0$, the set of radial mode wavenumbers $k_{c-} = k_{cn}$ ($n = 0, 1, 2 \ldots$) is determined by the zeroes of the J_1 Bessel function, and the magnetic fields are

$$B_r = A k_{c-} J_0(k_{c-} r), \quad B_\theta = -iA k J_1(k_{c-} r), \quad B_z = C J_1(k_{c-} r). \tag{6.37}$$

Each value of k_{cn} leads to two frequency solutions, as shown in Figure 6.3, where the first four pairs of frequencies corresponding to the first four zeroes of $J_1(k_c a)$ are indicated on a plot of $k_c^2 a^2$ against f.

The fast wave frequencies, lying above the upper cutoff frequency, increase monotonically with radial mode number n, and form an infinite discrete spectrum of frequencies. For this family of frequencies, C/A given by Eq. (6.23) is small for small f, so B_θ is small, and the wave magnetic field consists primarily of the compressive components B_r and B_z, so the mode is essentially a *compressional* mode, called a Compressional Alfvén Wave (CAW).

On the other hand, the ion cyclotron mode frequencies, lying between the lower cutoff and the generalized Alfvén resonance, form a dense infinite discrete spectrum with an upper limit at the generalized Alfvén resonance frequency. These are the "ion cyclotron waves" (ICW). The ratio C/A becomes very large for these waves, so the dominant wave magnetic field component is B_θ. The transverse plasma fluid velocity components are proportional to the magnetic field components at low frequency, so that the ion cyclotron mode is associated primarily with shearing azimuthal, or *torsional*, oscillations of the plasma around the cylinder axis, and the mode is referred to as the torsional mode. As α approaches zero, the ICW spectrum becomes infinitely dense, and is equivalent to the Alfvén wave with indeterminate value of k_c. For $m \neq 0$, although the solutions of Eq. (6.34) for k_c are now nonconstant functions of ω, for fixed k, the ion cyclotron (torsional) and fast (compressional) mode spectra still have the same qualitative behaviour.

Nonzero Vacuum Width

For a nonzero vacuum width, the lowest frequency compressional solution of Eq. (6.34) is a mode whose frequency goes to zero as k goes to zero, provided $m \neq 0$, in contrast to the zero vacuum width case in which a finite cutoff frequency exists for all compressional modes. When k is small, the lowest frequency $\pm|m|$ compressional modes have identical frequencies, with ω proportional to k, and they correspond to the low–frequency Alfvén surface waves whose properties are summarized in Chapter 4. For a large vacuum region Eq. (6.34) gives, for low frequency and $ka < |m|/2$,

$$k_{c-} = k \quad \text{and} \quad \omega = \sqrt{2}v_A k \tag{6.38}$$

and plasma wave fields (normalized to B_z at $r = a$)

$$B_r = i\frac{|m|}{ka}\left(\frac{r}{a}\right)^{|m|-1}, \quad B_\theta = -\frac{m}{ka}\left(\frac{r}{a}\right)^{|m|-1}, \quad B_z = \left(\frac{r}{a}\right)^{|m|}. \tag{6.39}$$

These waves are termed surface waves by analogy with the equivalent waves in planar geometry, which have fields that decay exponentially with distance from the surface. The dispersion relation (6.38) is the same as the planar surface wave result given by Eq. (4.18) for $|k_y| \gg k_z$. The torsional waves on the other hand are essentially body waves, with field maxima within the plasma and little field present in the vacuum region.

In Figure 6.4 (Collins, Cramer & Donnelly 1984) the numerical solution of the dispersion relation (Eq. 6.34) is shown for $m = 0$ and $m = \pm 1$, and for two nonzero vacuum widths, as a plot of the frequency, normalized to Ω_i, against α. The solid lines correspond to the compressional modes. The lowest frequency $m = \pm 1$ modes are surface waves at low values of k (with an almost linear dispersion relation), and as $k \to 0$ their dispersion relations become degenerate. The torsional mode is indicated by the dashed lines, which in fact form the envelope of a densely packed infinite spectrum of waves. The spectra for $m = 0$ and $m = \pm 1$ torsional waves are all very close to each other. Figures 6.4(a) and (b) are for a vacuum width of 1 cm and 0.01 cm respectively, and in each case the plasma radius $a = 7.3$ cm. In (b) the dotted lines terminating at $\alpha = 0$ indicate the dispersion relations when the vacuum width is zero (i.e. conducting boundary conditions). The surface wave nature of the lowest $m = \pm 1$ compressional modes disappears in the limit of zero vacuum width, when the waves become cutoff at $\alpha = 0$. The torsional mode dispersion relations are little affected by the width of the vacuum.

We can summarize the effects of finite ion cyclotron frequency on the waves in cylindrical geometry as follows: (henceforth we refer only to the lowest–frequency fast wave (CAW) modes)

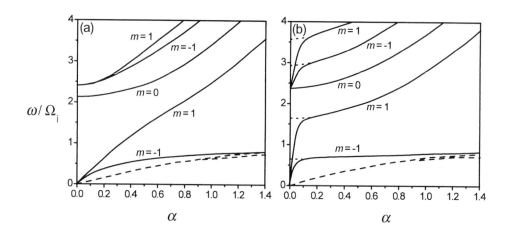

Figure 6.4: The normalized frequency as a function of $\alpha = v_A k/\Omega_i$ for the first two radial eigenmodes of the compressional wave when $m = \pm 1$ and for the first eigenmode when $m = 0$. (a) vacuum width is 1 cm, (b) vacuum width is 0.01 cm. The dashed lines represent the envelope of the torsional waves for $m = 0$ and $m = \pm 1$, and the dotted lines indicate the compressional wave dispersion relations for zero vacuum width. (Reproduced from Collins, Cramer & Donnelly 1984.)

i) The $m = +1$ wave has an appreciably higher frequency than the $m = -1$ wave once $\omega \geq 0.2\Omega_i$. This is shown explicitly for the case of the semi-infinite plasma with plane plasma-vacuum boundary for which the dispersion relation corresponding to Eq. (6.34) has an analytic solution given by Eq. (4.20), where positive k_y corresponds to $m = +1$ and negative k_y corresponds to $m = -1$.

ii) The $m = +1$ wave makes a smooth transition from surface to body wave, the transition occurring when $k_{c-}^2 \simeq (m/a)^2$. This is also derived by analogy with the behaviour at a plane interface. Consider the wave field behaviour of a cylindrical mode at large radius a. A local cartesian system may be set up with the y-axis corresponding to the local poloidal direction and wavenumber $k_y \simeq m/a$. The radius vector has x and y components in this local system, so that we can identify k_{c-}^2 with $k_x^2 + k_y^2$. The surface wave in plane geometry exists when $k_x^2 < 0$, the body wave when $k_x^2 > 0$. The transition is illustrated in Figure 6.5(a), where the $m = +1$ lowest compressional eigenmode magnetic field profiles are shown for three wavenumbers k. The $m = -1$ fields are shown in Figure 6.5(b), with the torsional wave fields shown in Figure 6.5(c) for comparison. Transition in the $m = +1$ case corresponds approximately to $\alpha \equiv v_A k/\Omega_i = 0.196$ for this particular case.

Even though it is evident from Figures 6.5(a) and (b) that B_r and B_θ do not apparently show surface wave behaviour in the plasma for $m = \pm 1$, the monotonic increase of $|B_z|$ towards the surface is an indicator of a surface wave. For $\alpha > 0.196$ there starts to develop a maximum of $|B_z|$ at a radius $< a$, and $|B_r|$ and $|B_\theta|$ start to decrease towards the surface, indicators of body wave behaviour. The surface wave nature of the modes at low k or α is of course much more obvious for $|m| \geq 2$, because then from Eq. (6.39) all the field components

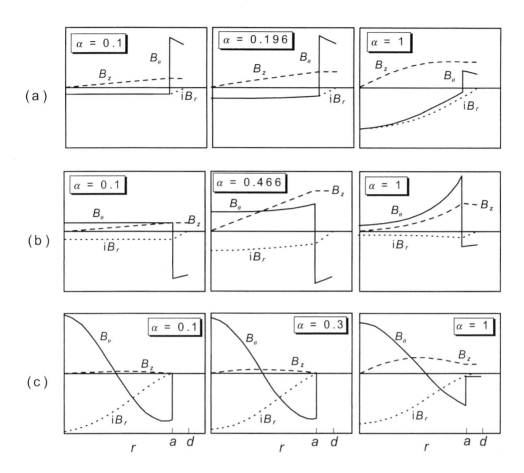

Figure 6.5: The first radial eigenmode field profiles are shown for a 1 cm vacuum width, for three values of α. (a) the $m = 1$ compressional wave, (b) the $m = -1$ compressional wave, (c) the $m = -1$ torsional wave. (Reproduced from Collins, Cramer & Donnelly 1984.)

increase monotonically towards the surface. In the plasma the $m = +1$ mode is purely right–hand circularly polarized (RHP) at low frequencies, with the degree of RHP decreasing at higher frequencies. The torsional ICW wave fields shown in Figure 6.5(c) are characteristic of body waves in the plasma.

iii) The $m = -1$ wave has a very individual behaviour. At low frequency its wavenumber k_{c-} is real and proportional to k, and the wave has the surface wave properties outlined above. As k increases, k_{c-} reaches a maximum and then decreases to zero, giving the fields illustrated in Figure 6.5(b) when $\alpha = 0.466$. Larger k gives imaginary k_{c-}; the wave fields are expressed in terms of modified Bessel functions and they become concentrated at the plasma surface, as illustrated in Figure 6.5(b) when $\alpha = 1$. The wave is purely left–hand circularly polarized at low frequencies but it becomes linearly polarized, with the B_θ field dominant, as ω approaches a limiting frequency. This limit occurs, with $k_{c-} \to i\infty$ and $\omega \to \omega_A$ (the Alfvén resonance frequency), at a limiting value of k at which point the wave has merged with the upper frequency limit of the torsional wave branch.

The existence of the limiting value of k for the negative m surface wave was established for the plane interface model by Cramer & Donnelly (1983): from Eq. (4.20), when $s = k/k_y$ has an upper bound value $s_1 = \alpha^{-1/2}$, the wavenumber $k_{c-} \to i\infty$ and the frequency has the Alfvén resonance value

$$f_1 = \alpha^{1/2}(1+\alpha)^{-1/2}. \tag{6.40}$$

In the cylindrical geometry the limiting value of k is then, on identifying k_y with m/a,

$$k_1 \simeq \left(\frac{\Omega_i \, |m|}{v_A a} \right)^{1/2}. \tag{6.41}$$

The effect of varying the vacuum region width on this limiting point is discussed by Collins, Cramer & Donnelly (1984). It is notable that the "surface wave" nature of this $m = -1$ mode is preserved even in the case of vanishing vacuum width: even though the wave is now cut off, that is, it has finite ω for $k \to 0$, the wavenumber k_{c-} is imaginary over most of the range of k, and the fields are concentrated at the plasma-wall interface. In this case the limiting wavenumber $k_1 \to \infty$ and the limiting frequency $\omega_1 \to \Omega_i$. The behaviour predicted for the $m = \pm 1$ surface waves has been verified in experiments in a cylindrical plasma device by Amagishi, Saeki & Donnelly (1989).

6.3.3   Nonideal Effects

We now include thermal and electron inertia effects in a low–β plasma, and consider the resulting modifications of the spectra of the waves in a cylindrical uniform density plasma discussed above. In this case the Ohm's law is given by Eq. (1.31), and the dielectric tensor components are as written in Eqs. (2.73) and (2.74). The short–wavelength QEW mode in the plasma must now be included in the analysis, in addition to the compressional mode in the plasma and the TE mode in the vacuum. Thus three boundary conditions are required at $r = a$, the continuity of E_θ, B_z and B_θ (or equivalently B_r, B_z and B_θ). For a finite value of u_3, the component E_z is nonzero in the plasma and couples to E_z of the transverse magnetic (TM) mode in the vacuum. For a phase velocity much less than the speed of light, however, the magnetic field components of the TM mode are negligible, and so are ignored.

It is instructive to calculate the approximate effect of the QEW on the dispersion relation by a perturbation theory, as in the planar geometry surface wave case discussed in Section 4.7.1, and assume the frequency (for a fixed value of k) changes by a small real and/or imaginary part. A similar technique can be used to calculate the perturbation of the frequency for a resistive plasma, for which the short–wavelength resistive mode must be taken into account (Collins, Cramer & Donnelly 1984). If the QEW has $k_{c+}^2 < 0$ for the perturbed frequency, as for the low–temperature IAW case of Figure 3.5(b), the short–wavelength mode is evanescent in the plasma, no energy is lost from the surface and the frequency is perturbed by a real amount. If the QEW has $k_{c+}^2 > 0$, as for the higher–temperature KAW case of Figure 3.5(a), then the KAW will propagate energy away from the surface into the plasma. In the case of a planar surface separating two semi–infinite regions, damping of the surface wave occurs because of this loss of radiated energy, that is, the frequency acquires a small (imaginary) part, and the surface wave has become a leaky wave. However, in the cylindrical collision-less plasma case with conducting walls, no energy can be lost from the system because of reflection from the walls, so that the frequency is shifted by a real amount for both the IAW and KAW cases. The same is also true for a cylindrical collisionless plasma surrounded by a vacuum with no conducting wall, because the wave fields in the vacuum are purely evanescent in the radial direction and no energy can be lost.

If $|k_{c+}r| \gg 1$ the magnetic field components of the QEW may be written in cylindrical geometry as

$$B_r = iCm\frac{J_m(k_{c+}r)}{k_{c+}r}\left[1 + \mathcal{O}\left(\frac{fk}{mk_{c+}}\right)\right] \tag{6.42a}$$

$$B_\theta = -CJ'_m(k_{c+}r)\left[1 + \mathcal{O}\left(\frac{mfk^3}{k_{c+}^3}\right)\right] \tag{6.42b}$$

$$B_z = CJ_m(k_{c+}r)\mathcal{O}\left(\frac{fk}{k_{c+}}\right) \tag{6.42c}$$

so when $f \ll 1$, the dispersion equation derived from continuity of $\boldsymbol{B}$ at $r = a$ is written as the following determinant equation:

$$\begin{vmatrix} kaJ'_m(k_{c-}a) & kaAK'_m(ka) & (m/k_{c+}a)J_m(k_{c+}a) \\ mJ_m(k_{c-}a) & mBK_m(ka) & J'_m(k_{c+}a) \\ k_{c-}aJ_m(k_{c-}a) & -kaBK_m(ka) & 0, \end{vmatrix} = 0 \tag{6.43}$$

where A and B are constants that can be derived from the vacuum fields, using the condition that $B_r(d) = 0$. Keeping the wavenumber $k = k_0$ fixed, we write the complex eigenfrequency as $\omega = \omega_0 + \delta\omega$, where ω_0 is the Hall model wave eigenfrequency, and $\delta\omega$ is the perturbation to the frequency due to the introduction of the QEW mode. We assume $|\delta\omega| \ll \omega_0$. Separating out the term proportional to k_{c+}, we can express Eq. (6.43) in the form

$$k_{c+}aJ'_m(k_{c+}a)\Delta_0(\omega) - mJ_m(k_{c+}a)\Delta_1(\omega) = 0. \tag{6.44}$$

As the Hall model dispersion equation is $\Delta_0(\omega_0) = 0$, as given by Eq. (6.34), it follows that

$$\Delta_0(\omega) \simeq \delta\omega \frac{\partial\Delta_0}{\partial\omega}\bigg|_{\omega_0}$$

and Eq. (6.44) gives

$$\delta\omega = \left[\frac{m}{k_{c+}a}\frac{J_m(k_{c+}a)}{J'_m(k_{c+}a)}\frac{\Delta_1(\omega)}{(\partial\Delta_0/\partial\omega)}\right]_{\omega_0}. \tag{6.45}$$

Using the low–frequency relation

$$\omega_0^2 = v_A^2(k_{c-}^2 + k_0^2) = 2v_A^2 k_0^2 \tag{6.46}$$

for the ideal MHD surface wave for a large vacuum region, $d \gg a$, as follows from Eq. (6.38), and

$$k_{c+}^2 k_{c-}^2 = \frac{u_3}{u_1}(u_1 - k^2)^2 \tag{6.47}$$

from Eq. (6.22) for $f \ll 1$, we have

$$k_{c+}(\omega_0) = (u_3/2)^{1/2} = k_0/\mu^{1/2} \tag{6.48}$$

where μ is defined in Eq. (4.86).

For the IAW, $\mu < 0$ and k_{c+} is imaginary, that is, the IAW is an evanescent wave, while for the KAW, $\mu > 0$ and k_{c+} is real, that is, the KAW is a propagating wave as mentioned above. The complex frequency perturbation is then

$$\delta\omega/\omega_0 = \mu^{1/2} \left[\frac{m^2}{(k_0 a)^4} \frac{(1+k_{c-}/k_0)}{(1+k_{c-}^2/k_0^2)} \left\{ \frac{d}{dz}\left(\frac{J_m'(z)}{zJ_m(z)}\right) \right\} \right]_{z=k_{c-}a}^{-1} \frac{J_m(k_{c+}a)}{J_m'(k_{c+}a)}. \tag{6.49}$$

Defining M as

$$M = J_m(k_{c+}a)/J_m'(k_{c+}a) \tag{6.50}$$

we find that for the IAW, $M \to -i$ and $\mu^{1/2} = i|\mu|^{1/2}$, so there is a real shift to the frequency, indicating that there is no radiative loss of energy and hence no wave damping. For the KAW, M and $\mu^{1/2}$ are both real, so that again the frequency is shifted by a real amount. This result is in contrast to the damping of the surface wave due to radiation of the KAW mode in a planar geometry found in Section 4.7.1: the reason is that for a cylindrical column plasma, the KAW radiates towards the centre of the column but cannot escape from the column, and so contributes to an undamped standing wave structure in the radial direction. Note that if k_{c+} is such that $J_m'(k_{c+}a) \simeq 0$ then the above theory is invalid, since the QEW would then be a resonant mode of the system as can be established from the dispersion equation (6.43), and the change of frequency would no longer be a perturbation to the Hall model wave frequency.

6.4   Nonuniform Plasmas

We now return to the case of the cylindrical plasma with a radially dependent density $\rho_0(r)$. For a wave mode with a given ω, m and k, the generalized Alfvén resonance condition may be met at some point r_0 in the density profile, where resonant absorption will occur. For example, if the surface separating the plasma from the vacuum region considered in the previous section is not sharp, but has a width b small compared with the axial wavelength, such that $\varepsilon = kb \ll 1$, the same perturbation technique as employed in Section 4.3.2 to calculate surface wave damping can be used to calculate the resonant damping of the mode. The result for the damping rate for the cold Hall plasma model is (Donnelly & Cramer 1984)

$$\gamma = \frac{\pi k(m/a + k^2 R\omega_0/\Omega_i)^2}{(du_1/dr)_a(\partial\Delta_0/\partial\omega)_{\omega_0}} \tag{6.51}$$

where R is defined in Eq. (6.35). The damping rate shows the same qualitative effects of finite ion cyclotron frequency as in the corresponding planar surface result given by Eq. (4.22). From Eq. (6.51) it is found that the presence of a conducting wall increases the Alfvén resonance damping rate over that found in the planar surface case.

However, if the width of the transition region in the stratified plasma where a resonance is encountered is not small, that is, when ε is not small, a numerical solution is usually necessary. An alternative approach is to model the nonuniformity by a series of step profiles in density and match boundary conditions across the interfaces (Paoloni 1978, Cross & Murphy 1986), although in this case Alfvén resonance damping will be absent, because we have seen, for example in Eq. (4.19), that the resonance damping is proportional to the width of the density jump.

If we omit the effects of electron inertia and electron pressure, $E_z = 0$ and the four wave equations (6.15)-(6.18) reduce to a pair of coupled first–order differential equations:

$$(u_1 - k^2)\frac{1}{r}\frac{d}{dr}(rE_\theta) = \frac{m}{r}u_2 E_\theta + \left(u_1 - \frac{m^2}{r^2} - k^2\right)p_z \tag{6.52}$$

$$(u_1 - k^2)\frac{dp_z}{dr} = -[(u_1 - k^2)^2 - u_2^2]E_\theta - \frac{m}{r}u_2 p_z. \tag{6.53}$$

With a local or WKB approximation, we can write the solution of Eqs. (6.52) and (6.53) in the form

$$p_z \propto J_m\left(\int_0^r k_c(r')dr'\right) \tag{6.54}$$

which reduces to the solution given by Eq. (6.37) in the case of a uniform plasma. Here $k_c = k_{c-}$ for the Hall-MHD mode is given formally by Eq. (6.26), but is now a function of r due to the dependence of u_1 on the plasma density.

Typical density and temperature distributions which drop to zero at the plasma edge $r = r_p$ are of the form

$$\rho_0(r) = \rho_0(0)(1 - r^2/r_p^2)^{\alpha_d} \tag{6.55}$$

where α_d is a constant. Figure 6.6 shows typical plots of k_c^2 versus r for the case of parabolic distributions ($\alpha_d = 1$). Both a low–temperature case ($\beta(r_0) < m_e/m_i$, where r_0 is the Alfvén resonance position) and a high–temperature case are shown. The cutoff-resonance-cutoff triplet discussed in the MHD approximation of Section 3.2.1 is assumed to be present in the plasma. The compressional wave, the ion cyclotron wave, the IAW and the KAW are propagating waves in the regions of positive k_c^2 shown.

For a single ion species plasma with ion charge number Z and mass number A, the Alfvén resonance position r_0 may be found from the condition

$$R = \frac{\rho_0(r_0)}{\rho_0(0)} = \frac{5.2 \times 10^{-4} k^2(1 - f^2)}{n'_{e0}f^2} \tag{6.56}$$

where k is in m^{-1}, $n'_{e0} = Z n_{e0}/A$, and the central electron density n_{e0} is in units of 10^{20} m^{-3}. The two cutoff positions r_1 and r_2 are given by the conditions

$$\frac{\rho_0(r_1)}{\rho_0(0)} = \frac{R}{(1 - f)}, \quad \text{and} \quad \frac{\rho_0(r_2)}{\rho_0(0)} = \frac{R}{(1 + f)}. \tag{6.57}$$

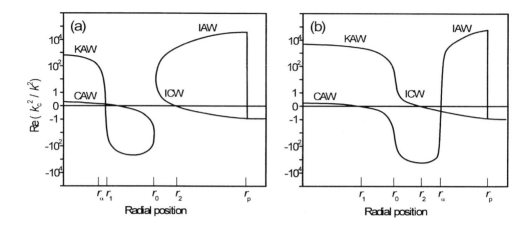

Figure 6.6: Real(k_c^2) for a cylindrical plasma with parabolic density and temperature profiles, and with $f = 0.2$ and $k = 3.5$ m^{-1}. (a) low temperature, (b) high temperature. Note that the ordinate scale changes from linear to logarithmic at 1. (Reproduced from Donnelly, Clancy & Cramer 1986.)

There is also a threshold frequency f_{th} at which $r_0 = 0$, that is, the Alfvén resonance just appears on the axis of the plasma column, defined by:

$$f_{\text{th}}^2 = \frac{5.2 \times 10^{-4} k^2}{n'_{e0} + 5.2 \times 10^{-4} k^2}. \tag{6.58}$$

For a wave mode with fixed k and a density distribution given by Eq. (6.55), r_0 moves from the centre to the edge of the plasma as f increases from f_{th} to a value $f_m = 1$. At the position r_α, the condition $\beta = m_e/m_i$ holds.

In the frequency range $f_{\text{th}} < f < f_m$, energy flows from an external antenna into the Alfvén resonance region via the Hall-MHD mode. The actual absorption process is determined by the short–wavelength modes introduced by nonideal effects. However, the absorption can be described in the context of ideal MHD (with $\omega \ll \Omega_i$). If the evolution in time of a nonaxisymmetric wave excited by an antenna is considered, the wave energy steadily increases and becomes concentrated in a layer of ever–decreasing width at the resonant surface (Tataronis 1975; Appert *et al.* 1980). Using an artificial damping term to remove the singularity that occurs at the Alfvén resonant position, equivalent to replacing ω by $\omega + i\nu$, with a pseudo–collisional term ν, Eqs. (6.52) and (6.53) can be solved numerically for a steady–state solution (Appert *et al.* 1980). If ν is decreased, the steady–state wave fields become more concentrated around r_0. However, provided that $\nu \ll \omega$, the power absorbed is independent of ν and it agrees with the power flow into the resonant layer predicted by the time–dependent calculation.

A qualitative understanding of many aspects of Alfvén wave heating can be obtained with the aid of Figure 6.6 (Donnelly, Clancy and Cramer 1985, 1986). The compressional Alfvén wave (CAW) has a radial component of propagation in the region $0 \leq r < r_1$, and, under the

WKB approximation, the radial eigenmodes of this wave satisfy conditions of the form

$$\int_0^{r_1} k_c(r')\mathrm{d}r' = n'\pi, \quad n' = 1, 2, \ldots \tag{6.59}$$

Provided that $f < f_m$, these eigenmodes always experience Alfvén resonance damping. Some of the CAW energy tunnels through the evanescent region $r_1 < r < r_0$ and is absorbed at the Alfvén resonance, such as via mode conversion into the short–wavelength IAW or KAW modes, thereby damping the wave (Karney, Perkins & Sun 1979). When the eigenmode conditions are not satisfied, the amplitude of the wave mode excited by the antenna is, in general, considerably smaller than the eigenmode amplitude, so the energy which is absorbed at r_0 (when $f_{th} < f < f_m$) is also smaller. The latter case has been called "mush heating" (Stix 1980).

The WKB treatment indicates that the ion cyclotron wave (ICW) propagates in the region $r_0 < r < r_2$. The ICW in a nonuniform plasma, which corresponds to the torsional Alfvén wave at low frequency, has also been called the *global Alfvén wave* (Appert *et al.* 1982), the *discrete Alfvén wave* (de Chambrier *et al.* 1983), or the *stable kink mode* (Ross, Chen & Mahajan 1982) when the effects of a background axial current are included, as discussed in the following section. If the Alfvén resonance is present, the absorption of energy at r_0 is always sufficiently large to damp out completely any ICW eigenmodes. Thus the ICW eigenmodes occur at frequencies just below f_{th}, provided the condition

$$f_{th}kr_p > \alpha_d^{1/2} \tag{6.60}$$

is satisfied. When global damping processes, such as resistivity, are included, the ICW eigenmodes exist only for frequencies approaching Ω_i. This is because at low frequency, r_2 approaches zero, so that very high values of k_c are needed and the waves are damped very efficiently.

Let us now consider the numerical solution of the wave equations with a density profile given by Eq. (6.55). For given boundary conditions, the dissipationless equations (6.52) and (6.53) yield eigensolutions if the eigenfrequency is such that the Alfvén resonance is not encountered in the plasma. These are eigenmodes of the CAW. If the resonance is encountered, then a damped CAW eigensolution with a complex eigenfrequency can in principle be found by searching in the complex frequency plane, giving a (negative) imaginary part to ω because of the resonance damping.

A more realistic method of determining the frequency and damping of the eigenmodes is to determine the resistance of an appropriate antenna as a function of frequency, and determine the quality factor Q of any antenna resonance peaks. A small collisional term, for example due to ion-neutral collisions, can be introduced to limit the size of the peaks. This corresponds more closely to the experimental situation of wave excitation by an external antenna of real frequency. A cavity eigenmode with real frequency $\hat{\omega}$ and damping rate γ has $Q \equiv \hat{\omega}/2\gamma$. Provided that Q is sufficiently large that the eigenmode gives the dominant contribution to the antenna resistance, then the eigenmode frequency equals the frequency where the resistance is a maximum, and the resistance of the antenna is proportional to γ^{-1}. For a mode that is affected by the Alfvén resonance, the damping is much higher than that due to the small collisional effect. The same technique can be used for waves affected by electron pressure and inertia, and described by Eqs. (6.15)-(6.18). In this case, for a mode affected by the Alfvén resonance, the peak in the antenna resistance is limited by the mode conversion into a KAW or IAW mode shown in Figure 6.6. In addition to the CAW eigenmodes, a separate family of spatial eigenmodes of the KAW or IAW in the bounded plasma may occur.

If the waves are excited by an antenna at a fixed radius r_a between the plasma and the wall, the antenna can be represented by a linear combination of divergence–free sheet currents of the form (Donnelly & Cramer 1984)

$$J(r,t) = I_{mk}\delta(r - r_a)\frac{k\hat{\theta} - (m/r_a)\hat{z}}{(k^2 + (m/r_a)^2)^{1/2}}\exp\left[i(m\theta + kz - \omega t)\right] \qquad (6.61)$$

where I_{mk} is a constant, depending on the spatial mode numbers m and k contributing to the antenna current. In each of the plasma-antenna and antenna-wall vacuum regions the wave fields can be expressed as combinations of the modified Bessel functions and their derivatives. Equations (3.65)-(3.66), or Eqs. (6.15)-(6.18), are solved as an initial value problem, and on applying the boundary conditions across the plasma-vacuum interface and the antenna, the antenna impedance is calculated. The real part of the impedance, that is, the antenna resistance, plotted against frequency shows peaks corresponding approximately to eigenmodes of the system for small collisional damping.

Typical results for the antenna resistance R, from Donnelly & Cramer (1984) for a parabolic density profile given by Eq. (6.55) with $\alpha_d = 1$, are shown in Figure 6.7, which compares results from the Hall-MHD model (Eqs. 3.65-3.66) and the two–fluid Hall model (Eqs. 6.15-6.18), for modes with azimuthal mode numbers $m = \pm 1$ and axial mode number (number of wavelengths around the torus of the Tokamak) $n = 1$. The MHD model predicts a large increase in R at the threshold frequency ω_{th} when the Alfvén resonance first enters the plasma, a low Q resonance for a damped eigenmode with fields corresponding to a surface wave, and a large decrease in R when the Alfvén resonance position reaches the plasma edge, $r_0 = r_p$. In comparison, the two–fluid model predicts a similar increase in R at ω_{th}, a higher Q eigenmode resonance, and no abrupt decrease in R when $r_0 = r_p$. The differences arise because of the different structure of the wave fields near r_0: the MHD model predicts a very large current flow over a very narrow region around r_0, and the Alfvén resonance absorption is only affected by the presence of the plasma edge when r_0 becomes very close to r_p.

Considering now the case of a hot plasma, we have seen, for example in Section 4.8 when describing the theory of surface waves on sharp surfaces, that the description of waves in a nonuniform plasma with kinetic theory effects such as collisionless damping is a formidable problem. However, rather than numerically solving the full kinetic Vlasov equations, a fourth–order set of differential equations incorporating warm and hot plasma effects can be employed (Donnelly, Clancy & Cramer 1986; Clancy & Donnelly 1986). This set of equations replaces the low–temperature set of Eqs. (6.15)-(6.18). The dielectric tensor becomes a differential operator, leading to higher–order differential equations, reflecting the additional wave modes that arise. In addition to finite electron inertia, pressure and collisional effects, the new set of equations incorporates finite ion Larmor radius effects, electron Landau damping and cyclotron damping.

An iterative expansion, in the finite ion Larmor radius (FLR) parameter $\lambda_i = k_x^2 V_i^2/\Omega_i^2$ (assumed much less than unity), is made on the wave equation (6.14) with the kinetic theory expressions for the components of the dielectric tensor (see Appendix). The procedure

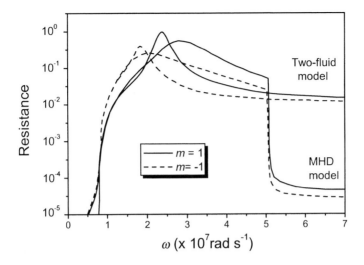

Figure 6.7: A comparison of the MHD and the two–fluid Hall model predictions of the antenna resistance (in ohms) as a function of the antenna frequency, for the $m = \pm 1$ and $n = 1$ modes. Here $\Omega_i = 6.7 \times 10^7$ rad s^{-1}. (Reproduced from Donnelly & Cramer 1984.)

consists of the replacement of k_x by $-i\partial/\partial r$, where care must be taken to ensure that the replacement is unambiguous, so as to satisfy energy conservation. The resulting equations are similar to Eqs. (6.15)-(6.18) with coefficients modified to include FLR effects. A term Λ arises, which incorporates the FLR effects and is a function of the plasma dispersion function defined in Eq. (A.13). For KAW modes

$$\Lambda k^2 \simeq 0.75 (T_i/T_e)(1 - f^2) \tag{6.62}$$

but for MHD and IAW modes $\Lambda k^2 \simeq 0$. The wave equations, valid for $|\Lambda k^2| < 0.5$, are:

$$
\begin{aligned}
u_1(1 + \Lambda k^2)\frac{\mathrm{d}E_\theta}{\mathrm{d}r} &= -\frac{1}{r}\left\{u_1(1 + \Lambda k^2) - mu_2)\right\} E_\theta + \frac{m}{r}k(1 + \Lambda u_1)p_\theta \\
&\quad + \left\{u_1(1 + \Lambda k^2) - (\frac{m}{r})^2\right\} p_z
\end{aligned} \tag{6.63}
$$

$$u_1(1 + \Lambda k^2)\frac{\mathrm{d}E_z}{\mathrm{d}r} = ku_2 E_\theta - (u_1 - k^2)p_\theta - \frac{m}{r}p_z \tag{6.64}$$

$$\frac{\mathrm{d}p_\theta}{\mathrm{d}r} = \frac{m}{r}k E_\theta + \left\{u_3 - (\frac{m}{r})^2\right\} E_z - \frac{1}{r}p_\theta \tag{6.65}$$

$$
\begin{aligned}
u_1(1 + \Lambda k^2)\frac{\mathrm{d}p_z}{\mathrm{d}r} &= \left\{u_2^2 - u_1(u_1 - k^2)(1 + \Lambda k^2)\right\} E_\theta + ku_2(1 + \Lambda u_1)p_\theta \\
&\quad - \frac{m}{r}ku_1(1 + \Lambda u_1)(1 + \Lambda k^2)E_z - \frac{m}{r}u_2 p_z.
\end{aligned} \tag{6.66}
$$

Numerical integration of these equations then reveals the mode conversion of the fast MHD wave launched by an antenna into either the KAW or IAW at the Alfvén resonance position, as discussed above for the fluid models, but with collisionless damping included (Clancy & Donnelly 1986; Donnelly, Clancy & Cramer 1986).

6.5   Effects of Current

Tokamak plasmas necessarily have a nonzero distributed equilibrium current along the toroidal axis to produce the confining magnetic field, so it is important to consider the effects of such a current on the propagation of waves in such bounded plasmas. Here we assume a straight cylindrical plasma, and in the next section we treat the unique features that a toroidal geometry introduces. The current modifies the spectrum of CAW eigenmodes, including surface–type modes, discussed in the previous sections. The current can also induce the existence of a dense spectrum of body wave–type modes, the discrete Alfvén waves, at frequencies just below the lower edge of the Alfvén continuum; these modes are discussed in the next section. The results are also relevant to the modes of cylindrical current–carrying flux tubes in the solar atmosphere (see Chapter 7).

We consider a cold, magnetized, current–carrying cylindrical plasma column of radius r_p, surrounded by a vacuum region, and an infinitely conducting metal wall at $r = r_w$. The equilibrium magnetic field $\boldsymbol{B}_0$ is assumed to be helically twisted around the cylinder axis, so that it has a poloidal component $B_{0\theta}(r)$ and an axial component $B_{0z}(r)$, but has no radial component. The current density also has poloidal and axial components:

$$J_{0\theta} = -\frac{1}{\mu_0}\frac{\partial B_{0z}}{\partial r}, \quad J_{0z} = \frac{1}{\mu_0 r}\frac{\partial}{\partial r}(r B_{0\theta}).$$
(6.67)

The plasma is pressureless, so the current must be force–free, that is, the current must be parallel to the magnetic field:

$$\boldsymbol{J}_0 \times \boldsymbol{B}_0 = 0$$
(6.68)

which leads to the following condition that the components of $\boldsymbol{B}_0$ must satisfy:

$$\frac{\partial}{\partial r}(B_{0\theta}^2 + B_{0z}^2) + \frac{2B_{0\theta}^2}{r} = 0.$$
(6.69)

In discussions of toroidal plasmas, but using straight cylindrical coordinates with periodicity in the z–direction as a first approximation, it is useful to employ the safety factor q, defined as

$$q = r B_{0z}/R B_{0\theta}$$
(6.70)

where R is the major radius of the torus. Strictly, for a straight cylindrical plasma, $R \to \infty$, but in this approximation R measures the periodicity scale along z.

If the pitch length of the helical magnetic field is independent of r, then q is also independent of r. The field in this case has zero shear, because all field lines rotate through the same angle in traversing a given distance along the axis (Cross 1988). An example of a shearless field that satisfies Eq. (6.69) is

$$B_{0\theta} = \frac{B_c J r}{r_p(1 + J^2 r^2/r_p^2)}, \quad B_{0z} = \frac{B_c}{1 + J^2 r^2/r_p^2} \tag{6.71}$$

where B_c is the field at the centre of the cylinder and J is an arbitrary constant.

An example of a field that has shear is that produced by a uniform axial current density $J_{0z} = 2B_c J/\mu_0 r_p$. To satisfy Eq. (6.69) there must also be a poloidal current, and the magnetic field components are

$$B_{0\theta} = B_c J r/r_p, \quad B_{0z} = B_c(1 - 2J^2 r^2/r_p^2)^{1/2}. \tag{6.72}$$

For small J the fields given by Eqs. (6.71) and (6.72) are the same to order J^2.

A third force–free magnetic field, with shear, that is used to describe diffuse linear pinches with highly twisted fields is the Bessel function model:

$$B_{0\theta} = B_c J_1(\kappa r), \quad B_{0z} = B_c J_0(\kappa r) \tag{6.73}$$

where κ is a constant.

The linearized wave equations for a wave with given mode numbers m and k, in a cold plasma with multiple ion species, and with arbitrary radial density $\rho_0(r)$ and current density profiles given by Eq. (6.67), have been derived by Sy (1985) using local field line coordinates (see also Cross (1988)). The frequency range encompasses the ion cyclotron frequencies but is well below the electron cyclotron frequency. The resulting equations are:

$$\frac{A}{r}\frac{d}{dr}(rQ) = C_1 Q - C_2 P$$

$$A\frac{dP}{dr} = C_3 Q - C_1 P \tag{6.74}$$

where

$$P = \mathbf{B} \cdot \mathbf{B}_0, \quad Q = E_\perp/i\omega B_0 \tag{6.75}$$

with $\mathbf{B}$ the wave magnetic field, and $E_\perp$ the wave electric field tangential to the magnetic surface $r = $ constant and normal to $\mathbf{B}_0$. The coefficients in Eq. (6.74) are

$$A = \frac{B_0^2 \omega^2 S}{c^2} - F^2 \tag{6.76a}$$

$$C_1 = \frac{2B_{0\theta}^2 \omega^2 S}{rc^2} + \frac{G\omega^2 B_0 D}{c^2} - \frac{2mF B_{0\theta}}{r^2} \tag{6.76b}$$

$$C_2 = \frac{\omega^2 S}{c^2} - \left(k^2 + \frac{m^2}{r^2}\right) \tag{6.76c}$$

$$C_3 = A\left\{A + \left(\frac{2B_{0\theta}^2}{rB_0}\right)^2 + r\frac{d}{dr}\left(\frac{B_{0\theta}}{r}\right)^2\right\}$$

$$\quad - \left(\frac{B_0^2 \omega^2 D}{c^2} + \frac{2B_{0\theta} B_{0z} F}{rB_0}\right)^2 \tag{6.76d}$$

where

$$F = \frac{mB_{0\theta}}{r} + kB_{0z}, \quad G = kB_{0\theta} - \frac{mB_{0z}}{r} \tag{6.77}$$

and S and D are the components of the cold plasma dielectric tensor defined in Eqs. (2.80) and (2.81). Defining a local wavenumber $\boldsymbol{k} = (0, m/r, k)$ in a given magnetic surface, we can write $F = \boldsymbol{k} \cdot \boldsymbol{B}_0$ and $G = -|\boldsymbol{k} \times \boldsymbol{B}_0|$.

Equations (6.74) yield the two second–order differential equations

$$\frac{\mathrm{d}}{\mathrm{d}r} \left(\frac{A}{rC_2} \frac{\mathrm{d}}{\mathrm{d}r} (rQ) \right) + \left\{ \frac{(C_2 C_3 - C_1^2)}{AC_2} + \frac{C_1}{rC_2} - \frac{\mathrm{d}}{\mathrm{d}r} \left(\frac{C_1}{C_3} \right) \right\} Q = 0 \tag{6.78}$$

$$\frac{\mathrm{d}}{\mathrm{d}r} \left(\frac{A}{rC_3} \frac{\mathrm{d}P}{\mathrm{d}r} \right) + \left\{ r \frac{(C_2 C_3 - C_1^2)}{AC_3} - \frac{\mathrm{d}}{\mathrm{d}r} \left(\frac{rC_1}{C_3} \right) \right\} P = 0. \tag{6.79}$$

A singularity of Eqs. (6.78) and (6.79) occurs where $A = 0$ (the generalized Alfvén or hybrid resonance), leading to resonance damping of CAW modes and the existence of the Alfvén continuum of singular modes. An apparent singularity occurs where $C_3 = 0$, but this has been shown not to lead to an extra continuous spectrum (Appert, Gruber & Vaclavik 1974; Swanson 1975; Goedbloed 1998).

In the limit of frequency much lower than any ion cyclotron frequency, when D tends to zero, the coefficients become

$$A = \frac{\mu_0 \rho_0 \omega^2}{v_A^2} - F^2 \tag{6.80a}$$

$$C_1 = \frac{2\omega^2 B_{0\theta}^2}{r v_A^2} - \frac{2mFB_{0\theta}}{r^2} \tag{6.80b}$$

$$C_2 = \frac{\omega^2}{v_A^2} - \left(k^2 + \frac{m^2}{r^2} \right) \tag{6.80c}$$

$$C_3 = A \left\{ A + \left(\frac{2B_{0\theta}^2}{rB_0} \right)^2 + r \frac{\mathrm{d}}{\mathrm{d}r} \left(\frac{B_{0\theta}}{r} \right)^2 \right\} - \left(\frac{2B_{0\theta} B_{0z} F}{rB_0} \right)^2 \tag{6.80d}$$

where $v_A = B_0/(\mu_0\rho_0)^{1/2}$ is the local Alfvén speed. Equations (6.74) with Eq. (6.80) are equivalent to the low–frequency equations originally derived by Hain & Lüst (1958) and subsequently employed for MHD stability and wave heating calculations (e.g. Appert *et al.* 1982).

In the low–frequency limit, Eq. (6.78) reduces to (Goedbloed & Hagebeuk 1972; Appert *et al.* 1982)

$$\frac{\mathrm{d}}{\mathrm{d}r} \left(\frac{A}{rC_2} \frac{\mathrm{d}}{\mathrm{d}r} (rQ) \right) + \left\{ A - r \frac{\mathrm{d}}{\mathrm{d}r} \left(\frac{B_{0\theta}}{r} \right)^2 - \frac{4k^2 B_{0\theta}^2}{r^2 C_2} + r \frac{\mathrm{d}}{\mathrm{d}r} \left(\frac{2kB_{0\theta}G}{r^2 C_2} \right) \right\} Q = 0. \tag{6.81}$$

We may write the Alfvén resonance condition $A = 0$ in the low–frequency case as

$$\omega^2 = \omega_A^2(r) = v_A^2 k_{||}^2 \tag{6.82}$$

thus defining a local Alfvén frequency ω_A, with a local wavevector parallel to the local equilibrium magnetic field defined as

$$k_\parallel = F/B_0 = \boldsymbol{k} \cdot \boldsymbol{B}_0/B_0 = \left(kB_{0z} + \frac{mB_{0\theta}}{r} \right) /B_0 . \tag{6.83}$$

Thus, if $B_{0\theta} \neq 0$, the Alfvén resonance condition depends on the poloidal mode number m as well as the axial wavenumber k.

The eigenmodes of a current–carrying plasma column with a vacuum layer and a metal wall can now be obtained, provided the boundary conditions at the plasma-vacuum interface are adjusted to take account of the twisted magnetic field there. The boundary conditions are that P and Q are continuous across the plasma edge. Allowing for the fact that the electric field parallel to $\boldsymbol{B}_0$ is zero in the plasma because of the neglect of resistivity, thermal effects and electron inertia, the wave fields in the vacuum may be expressed in terms of the TE and TM modes (Cramer & Donnelly 1984). The result for the dispersion equation is

$$\frac{Q}{P}(r = r_{\mathrm{p}}) = \frac{kR}{k_\parallel^2 B_0^2} \tag{6.84}$$

where R is defined in Eq. (6.35). Here B_0 is evaluated at $r = r_{\mathrm{p}}$.

Low Current

It is instructive to consider the solution of Eq. (6.84) in the low current case (Cramer & Donnelly 1984; Sy 1985; Cross 1988), when the parameter $J \ll 1$ and the three field models given by Eqs. (6.71)-(6.73) reduce to the uniform axial current model with, correct to order J^2,

$$B_{0\theta} = B_{\mathrm{c}} Jr/r_{\mathrm{p}}, \quad B_{0z} = B_{\mathrm{c}}(1 - J^2 r^2/r_{\mathrm{p}}^2). \tag{6.85}$$

An analytic solution of the wave equations (6.74) is possible if small terms of order J^2 are neglected. It is interesting to note here that an exact solution of the wave equations is possible for an incompressible plasma with a constant axial current (Bennett, Roberts & Narain 1999); see Section 7.7 for a further discussion of this case in an application to twisted solar flux tubes.

The plasma density is assumed uniform. We then have that A, C_1 and C_3 are constants, and

$$k_\parallel = k + mJ/r_{\mathrm{p}}. \tag{6.86}$$

From Eq. (6.79) we find then that P satisfies the Bessel equation (6.29), with a solution $P \propto J_m(k_c r)$, and Q is proportional to the same combination of Bessel functions giving B_r in Eq. (6.19), and

$$k_c^2 = \frac{C_2 C_3 - C_1^2}{A^2} . \tag{6.87}$$

For $J = 0$, Eq. (6.87) reduces to the result in Eq. (6.30) for the Hall model in the absence of current, while for low frequency in the presence of current, we have

$$A = B_{\mathrm{c}}^2 \left(\frac{\omega^2}{v_A^2} - k_\parallel^2 \right) \tag{6.88}$$

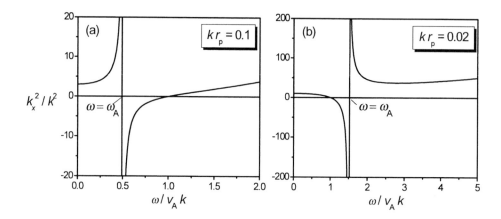

Figure 6.8: Square of the perpendicular wavenumber k_c^2, divided by k^2, as a function of the normalized phase velocity $\omega/v_A k_z$, for a cylindrical plasma with a small axial current. Here $m = -1$ and $J = 0.05$. The Alfvén resonance $\omega = \omega_A$ is indicated. (a) $kr_p = 0.1$ ($k > -mJ/2r_p$), (b) $kr_p = 0.02$ ($k < -mJ/2r_p$). (Reproduced from Cramer & Donnelly 1984.)

where v_A is based on the field B_c, and Eq. (6.87) becomes

$$k_c^2 = \left(\frac{\omega^2}{v_A^2} - k^2 \right) \left(1 + \frac{4 B_c^2 J^2}{A r_p^2} \right). \tag{6.89}$$

Figure 6.8 shows the low–frequency result given by Eq. (6.89) for k_c^2/k^2 plotted against the normalized phase velocity $\omega/v_A k$, for $m = -1$, $J = 0.05$ and for two values of kr_p. We see that the presence of the current has introduced a singularity of k_c^2 at the Alfvén resonance where $A = 0$, that is where

$$\omega = \omega_A = v_A k_{\parallel}. \tag{6.90}$$

In this respect the current plays a similar role to a finite ion cyclotron frequency. However, the character of the variation of k_c with frequency differs according to the size and direction of the current. If $k > -mJ/2r_p$, the variation is similar to that for the case of zero current and finite cyclotron frequency, as shown in Figure 6.8(a) for $m = -1$, with $k_c^2 \to +\infty$ as $\omega \to \omega_A$ from below. If $k < -mJ/2r_p$ the reverse applies (Figure 6.8(b)), and $k_c^2 \to +\infty$ as $\omega \to \omega_A$ from above. The dispersion relation is the same for wavenumbers (m, k) as for $(-m, -k)$.

For a large vacuum region, $r_w \gg r_p$, the lowest–frequency eigenmode solution of Eq. (6.84), that is, the lowest radial eigenmode of the CAW spectrum, is a surface wave with dispersion relation (for $\omega \ll \Omega_i$)

$$\omega^2 = 2 v_A^2 k_{\parallel}^2 \left(1 - \frac{m}{|m|} \frac{J}{k_{\parallel} r_p} \right). \tag{6.91}$$

This mode's dispersion relation is shown in Figure 6.9 for $m = -1$ (the solid curve), using the above analytic model. Numerical solution with either the sheared (Eq. 6.72) or shearless

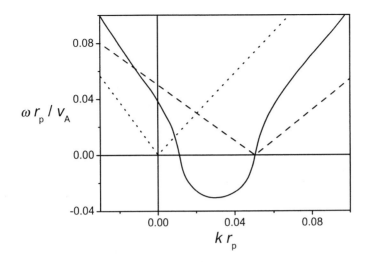

Figure 6.9: Normalized frequency $\omega r_{\mathrm{p}}/v_{\mathrm{A}}$ of the surface wave on a cylindrical current–carrying plasma column as a function of $k r_{\mathrm{p}}$. Below the origin the frequency is replaced by the negative of the growth rate of the unstable mode. The solid curve is the surface wave with current parameter $J = 0.05$, the dotted curve is the surface wave with $J = 0$, and the dashed curve indicates the Alfvén resonance condition. (Reproduced from Cramer & Donnelly 1984.)

(Eq. 6.71) magnetic fields also yields this dispersion relation very closely. For comparison, the dispersion relation for zero current is shown by the dotted curve. The dashed curve indicates the Alfvén resonance locus defined by Eq. (6.90). This curve is broadened out into an Alfvén continuum by higher–order effects of the current, because then A becomes a function of r. Where the surface wave curve crosses the Alfvén continuum, the wave will be resonance damped, and so will not be a pure eigenmode. An instability, that is $\omega^2 < 0$, occurs in the range

$$k = (m/|m| - m)J/r_{\mathrm{p}} \quad \text{to} \quad k = -mJ/r_{\mathrm{p}} \quad (\text{i.e. } k_{\|} = 0). \tag{6.92}$$

This is the current–driven external kink instability (Chance *et al.* 1977), which we see is the unstable branch of the Alfvén surface wave in a current–carrying plasma.

6.6   Discrete Alfvén Waves

An equilibrium current has another important consequence, besides the modification of the CAW spectrum that we have seen in the previous section. It also induces the existence of a family of discrete (i.e. noncontinuum) modes, which may be identified in the limit of short perpendicular wavelength with the shear or torsional Alfvén wave (Goedbloed 1975; Appert *et al.* 1982; Ross, Chen & Mahajan 1982). These modes are called the global Alfvén eigenmodes (GAE) or the discrete Alfvén waves (DAW); we shall mostly use the latter term.

We have seen that at low frequency in a current–free plasma, ideal MHD theory predicts that the shear Alfvén wave propagates at the local Alfvén speed in a direction parallel to the equilibrium magnetic field (see Section 3.3.1). However, there are several effects which lead to the coupling of the Alfvén wave with the CAW mode, and thus with other Alfvén waves on adjacent field lines. This can result in a global mode, the DAW mode, which propagates with a phase speed that is independent of position in the nonuniform plasma, but which has wave fields characteristic of the shear Alfvén wave rather than the CAW mode. One such effect is that of finite ion cyclotron frequency, which is responsible for the ICW modes discussed in Section 6.3.2. Thus the ICW modes are sometimes referred to as a particular type of DAW mode. Another effect giving rise to discrete modes is resistivity, as is discussed in Section 6.2.1.

In addition, the DAW modes can be present in a current–carrying plasma at low frequencies, $\omega \ll \Omega_i$, as is shown in this section. The DAW modes, as well as the ICW modes, are always found to have frequencies below the threshold frequency defined in the previous section, above which the Alfvén resonance is present in the plasma. At higher frequencies in a current–carrying plasma, the DAW mode spectrum is a combined effect of the current and the finite ion cyclotron frequency. Figure 6.10 shows schematically the dispersion relations of the DAW and ICW spectra for a cylindrical plasma column with a shearless magnetic field of the form given in Eq. (6.71) and a small axial current, for frequencies approaching the ion cyclotron frequency. The spectra occur just below the lower edge of the Alfvén continuum, which is indicated by the shaded region. Just as the surface CAW mode is related to the external kink instability, it is found that the DAWs or GAEs belong to the same family as another type of instability, the current–driven internal kink instability (Goedbloed & Hagebeuk 1972); the growth rates of these unstable modes are plotted below the real frequency axis in Figure 6.10.

If we inspect the plot of k_c^2 for the low–current model in Figure 6.8, it might be deduced that a dense spectrum of DAW modes should occur in the range of positive k_c^2 close to the resonance frequency (below the resonance frequency for $k > -mJ/2r_p$ and above the resonance frequency for $k < -mJ/2r_p$), in analogy to the dense ICW spectrum that occurs just below the resonance frequency for zero current but finite ion cyclotron frequency (see Figure 6.3). However, in contrast to the CAW spectrum, the existence of the DAW modes is very sensitive to the precise details of the density and magnetic field profiles, in particular to the terms of second order in J, and the low–current model results are misleading, in that the correct theory shows that there may be only a finite number of DAW modes below the edge of the Alfvén continuum. We show this in the following discussion, where we consider DAW modes for a nonuniform plasma in both planar and cylindrical geometries.

6.6.1   Slab Plasma

It is instructive first to consider a planar slab plasma, that is, a plasma bounded by a vacuum at the two planes $x = \pm R$. The analysis follows that of Davis & Donnelly (1986). The plasma is cold and has a single positive ion species, and extends infinitely and uniformly in the y and z–directions. The magnetic field lies only in the y-z plane but can otherwise vary in magnitude and direction with x. The equilibrium plasma density $\rho_0(x)$ and current density $J_z(x)$ are assumed to be peaked with zero slope at $x = 0$ and to fall monotonically and

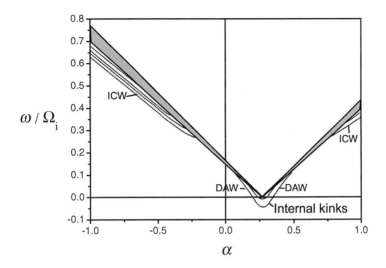

Figure 6.10: The DAW and ICW dispersion relations (dotted curves) for a cylindrical plasma column with a shearless magnetic field and small axial current. The frequency normalized to the ion cyclotron frequency is plotted against $\alpha = v_A k/\Omega_i$, and the real frequency is continued to the growth rate below the origin. The Alfvén continuum is indicated by the shaded region. The unstable internal kink modes are indicated. (Reproduced from Cramer & Donnelly 1984.)

symmetrically to zero at $x = \pm R$. The particular profiles used here, which are characteristic of laboratory plasmas, are

$$\rho_0(x) = \rho_0(0)\left[1 - \left(\frac{x}{R}\right)^2\right] \tag{6.93}$$

$$J_z(x) = J_z(0)\left[1 - \left(\frac{x}{R}\right)^2\right]^2. \tag{6.94}$$

The force–free condition given by Eq. (6.68) implies that the total field strength B_0 is independent of x, and the field components are given by

$$B_{0y}(x) = \mu_0 \int_0^x J_z(x')\,\mathrm{d}x', \quad B_{0z}(x) = (B_0^2 - B_{0y}^2)^{1/2} \tag{6.95}$$

where the boundary condition that $B_{0y} = 0$ at $x = 0$ is chosen. This implies that B_{0y} is antisymmetric in x and B_{0z} is symmetric. We also define

$$J^* \equiv \mu_0 J_z(0) = \frac{\mathrm{d}B_{0y}}{\mathrm{d}x}(0). \tag{6.96}$$

The linearized wave equations in the planar geometry may be written in a form analogous to those in the cylindrical geometry given by Eqs. (6.74):

$$
\begin{aligned}
A\frac{dQ}{dx} &= C_1 Q + C_2 P \\
A\frac{dP}{dx} &= C_3 Q - C_1 P
\end{aligned}
\tag{6.97}
$$

where $Q = E_\perp$ and $P = i\omega B_\parallel$, and $E_\perp$ and $B_\parallel$ are components of the wave electric and magnetic fields in a local coordinate system defined by the unit vectors in the y-z plane, perpendicular and parallel to the local equilibrium magnetic field:

$$
\begin{aligned}
\hat{e}_\perp &= (B_{0z}/B_0)\hat{y} - (B_{0y}/B_0)\hat{z} \\
\hat{e}_\parallel &= (B_{0y}/B_0)\hat{y} + (B_{0z}/B_0)\hat{z}.
\end{aligned}
\tag{6.98}
$$

The coefficients in Eqs. (6.97) are considerably simpler in the planar case than in the cylindrical case given by Eq. (6.80):

$$
\begin{aligned}
A &= u_1 - k_\parallel^2 \\
C_1 &= k_\perp u_2 \\
C_2 &= A - k_\perp^2 \\
C_3 &= u_2^2 - A^2
\end{aligned}
\tag{6.99}
$$

where u_1 and u_2 are defined in Eq. (2.73), and $k_\parallel$ and $k_\perp$ are the locally parallel and perpendicular components of the wavevector $k = (0, k_y, k_z)$,

$$
k_\parallel = k \cdot B_0/B_0, \quad k_\perp = |k \times B_0|/B_0.
\tag{6.100}
$$

We note that Ω_i and $f = \omega/\Omega_i$ are independent of x because B_0 is constant. Also noting that $Q = v_x B_0$ in the limit $\omega \ll \Omega_i$, we find that Eqs. (6.97) reduce to the cold plasma versions of Eqs. (3.52) and (3.53) in that limit.

It is convenient to convert Eqs. (6.97) into a single second–order equation analogous to Eq. (6.78) or Eq. (6.79). Defining

$$
\psi = (A/C_2)^{1/2} Q
\tag{6.101}
$$

we obtain the Schrödinger–like equation

$$
\frac{d^2\psi}{dx^2} - V(x)\psi = 0
\tag{6.102}
$$

where the "potential" function is given by

$$
\begin{aligned}
V(x) = &-k_x^2 - \frac{1}{4A^2}\left(\frac{dA}{dx}\right)^2 \\
&+ \frac{1}{A}\left[\frac{1}{2}\frac{d^2A}{dx^2} + \frac{dC_1}{dx} - \frac{1}{2C_2}\frac{dC_2}{dx}\frac{dA}{dx} - \frac{C_1}{C_2}\frac{dC_2}{dx}\right] \\
&+ \frac{3}{4C_2^2}\left(\frac{dC_2}{dx}\right)^2 - \frac{1}{C_2}\frac{d^2C_2}{dx^2}
\end{aligned}
\tag{6.103}
$$

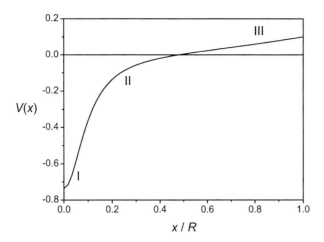

Figure 6.11: The "potential well" for DAW modes in a plasma slab with no current. Here $g = 1$, $f = 0.3$ and $\delta = 0.01$. The regions of approximation I, II and III in Eq. (6.106) are shown. The potential is symmetric for negative x.

where

$$-k_x^2(x) = k_\perp^2 - A + u_2^2/A. \tag{6.104}$$

When ρ is constant and there is no current, solutions of the form $\psi = \exp(\pm i k_x x)$ exist, with constant k_x given by Eq. (3.7). The DAW modes are not strongly dependent on the boundary conditions, so it is simply assumed that ψ vanishes at $x = \pm R$.

Equation (6.102) must generally be solved numerically as an eigenvalue problem. The DAW modes have frequencies below the lower edge of the continuum, and are characterized by the condition $A < 0$ everywhere in the plasma. For a given normalized frequency f, the eigenvalues of k_z corresponding to the DAW modes can be sought, or alternatively, the eigenvalues of

$$\delta = (k_z/k_{zc})^2 - 1 > 0 \tag{6.105}$$

where $k_{zc}(f)$ is the value of k_z at the lower continuum edge. We define $x_c(f)$ to be the position where the generalized Alfvén resonance ($A = 0$) at the lower continuum edge appears in the plasma. The quasi–potential $V(x)$ is then a function of δ.

Many properties of the eigensolutions and corresponding eigenvalues δ can however be deduced by finding the solutions corresponding to approximate forms of $V(x)$ that apply in limited regions of space. As an example, we consider the case of a current–free plasma and $k_y = 0$. The quasi–potential is then given by $V(x) = -k_x^2$, with k_x the local wavenumber defined in Eq. (6.104), and is a function of x purely through the dependence of the plasma density, assumed to be of the form given by Eq. (6.93), on x. We have $x_c = 0$ and $k_{zc}^2 = g^2 f^2/(1 - f^2)$. The DAW modes in this case are the ICW modes. The quasi–potential is then

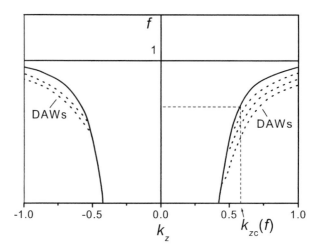

Figure 6.12: A typical Alfvén continuum (within the region bounded by the solid curves and the line $f = 1$) and DAW mode spectrum (dotted curves) for a diffuse slab plasma. (Reproduced from Davis & Donnelly 1986.)

approximated by (Davis & Donnelly 1986):

$$
\begin{array}{lll}
\text{I.} & x^2 \ll \delta R^2, & V(x) \simeq g^2 f^2 [\delta + 2f^2 - f^2/\delta + (1 + f^2/\delta^2)] \\
\text{II.} & \delta R^2 \ll x^2 \ll f R^2, & V(x) \simeq -R^2 g^2 f^4/x^2 \\
\text{III.} & f^2 R^2 \ll x^2 \ll R^2, & V(x) \simeq g^2 f^2 x^2/R^2
\end{array}
\tag{6.106}
$$

as shown in Figure 6.11. Here $g = \Omega_i/v_A$, where v_A is evaluated at $x = 0$.

The quasipotential $V(x)$ is then a "potential well", which deepens in regions I and II as the lower continuum edge is approached ($\delta \to 0$). Approximate solutions can then be written down for the three regions, and matched across the region boundaries. The result (Davis & Donnelly 1986) is that there are infinitely many DAW modes of each symmetry in x accumulating at the lower continuum edge, just as for the uniform density case discussed in Section 6.3.2, provided that $Rgf^2 > \frac{1}{2}$. However, just a single mode, of even symmetry, exists when $Rgf^2 < \frac{1}{2}$ (Leonovich, Mazur & Senatorov 1983). Including a current in the plasma, Eq. (6.102), with the profiles given in Eqs. (6.93) and (6.94), must be solved numerically. A typical resulting Alfvén continuum (bounded by the solid curves) and DAW spectrum (indicated by the dotted curves) for a slab plasma with a parabolic density and current profile is shown in Figure 6.12. The upper bound of the Alfvén continuum ($f = 1$) corresponds to Alfvén resonance at the plasma edge ($x = \pm R$), while the lower bound corresponds to resonance at $x = 0$ and a value of $k_z = k_{zc}$, which is a function of f. Note that only a finite number of DAW modes exists; as the frequency decreases the number decreases as each DAW mode disappears into the Alfvén continuum. We next consider the DAW modes for a current–carrying plasma in more detail in the cylindrical plasma case.

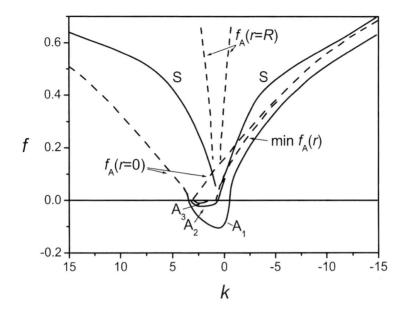

Figure 6.13: The frequency spectrum $f = \omega/\Omega_i$ of a current–carrying cylindrical plasma with a parabolic density profile, as a function of the axial wavenumber k. Below the origin the frequency is replaced by the growth rate of unstable modes. The Alfvén continuum is delineated by the dashed curves and the resonance–damped surface mode is indicated by S. The DAW modes, which become unstable, are indicated by A_1, A_2 and A_3. (Reproduced from Appert, Vaclavik & Villard 1984.)

6.6.2   Cylindrical Plasma

For a cylindrical plasma, just as for the slab plasma, the Alfvén resonance cannot be present in the plasma for a DAW mode with given (m, k), because otherwise the mode will be heavily resonance damped. It is incorrect to use the small–current analytic model result given by Eq. (6.87) for the DAW modes; the modes occur if the shearless field defined in Eq. (6.71) is used, but they disappear for the sheared magnetic field defined in Eq. (6.72), even for very small J (Cramer & Donnelly 1984). The question of the existence of DAW modes in a general magnetic field and plasma density profile for a cylindrical plasma, at low frequency, has been explored by Appert *et al.* (1982), and we now briefly discuss that analysis.

The lower edge of the Alfvén continuum corresponds to the Alfvén resonance occurring on the axis of the plasma column, $r = 0$, for a density profile peaked on the axis and for the low–current magnetic field models. This point corresponds to a minimum of the Alfvén frequency given by Eq. (6.82), so that

$$\frac{d\omega_A^2}{dr} = 0, \quad \frac{d^2\omega_A^2}{dr^2} > 0 \quad \text{at } r = 0. \tag{6.107}$$

The existence of DAW eigenmodes with eigenvalues ω^2 below the lower edge of the Alfvén continuum is related to the behaviour of the solutions around $w^2 = \text{Min}(w_A^2(r))$. The edge of the continuum is an accumulation point of a discrete spectrum if the solution is oscillatory at $r = 0$. Expanding the function A around $r = 0$, we have

$$A \propto w^2 - w_A^2 \simeq -\frac{1}{2}(w_A^2)'' r^2 \tag{6.108}$$

and Eq. (6.81) reduces to

$$r^2 Q'' + 5r Q' + g(0)Q = 0 \tag{6.109}$$

where

$$g(r) = 3 - m^2 - \frac{2C}{(w_A^2)''} \tag{6.110}$$

$$C = \frac{G^2}{\mu_0 \rho_0 B_0^2} \left\{ \frac{4B_{0\theta}^4}{r^2 B_0^2 (1-H)^2} + r\frac{d}{dr}\left[\frac{B_{0\theta}^2(1+H)}{r^2(1-H)} \right] \right\} \tag{6.111}$$

and

$$H = \frac{F B_{0z}}{k B_0^2}. \tag{6.112}$$

With a solution of the form $Q \sim r^\alpha$, α has an imaginary part, that is, the solution is oscillatory, if the following condition holds:

$$g(0) - 4 > 0. \tag{6.113}$$

Similar conditions can be derived for the cases where w_A^2 has a minimum at a point $r \neq 0$, and where $(w_A^2)'' = 0$ (Appert *et al.* 1982). For a given azimuthal mode number m, and for given magnetic field and density profiles, these conditions imply a range of allowable axial wavenumbers k for the DAW spectrum to exist.

A circular cross–section Tokamak with a large aspect ratio R/a, where R is the major radius and a is the minor radius, can be approximated as a straight cylindrical plasma column with the correspondence $k = n/R$ with the toroidal wavenumber n. Thus there is a range of possible n values for the DAW modes to exist. The following profiles of equilibrium current and density are typical for ohmically heated Tokamaks (Appert *et al.* 1982):

$$J_{0\theta} = 0, \quad J_{0z} = J(0)\left[1 - \left(\frac{r}{a}\right)^2\right]^4 \tag{6.114}$$

$$\rho_0 = \rho_0(0)\left\{ 0.95\left[1 - \left(\frac{r}{a}\right)^2\right] + 0.05 \right\}. \tag{6.115}$$

As in the slab geometry, the wave fields of the DAW modes are concentrated within the plasma, with little influence of the boundary conditions at the vacuum region and at the walls, so these waves are body waves, in contrast to the surface–type CAW lowest eigenmodes. The spectrum of the modes in a cylindrical plasma for $m = -1$ is shown in Figure 6.13

(Appert, Vaclavik & Villard 1984). The Alfvén continuum is characterized by the values of the frequency f_A in the centre ($r = 0$) and at the plasma edge ($r = R$). The minimum value of f_A is also shown. The surface CAW mode (S) exists inside the continuum and so is resonance–damped. The DAW modes (below the lower edge of the continuum) connect to unstable modes; the mode of highest growth rate (A_1) is interpreted as the unstable external kink mode, while the low growth rate modes (A_2 and A_3) are the unstable internal kink modes (Goedbloed & Hagebeuk 1972).

6.7   Toroidal Alfvén Eigenmodes

With a cylinder modelling of a Tokamak of major radius R_0, the parallel wavenumber $k_{||}$ defined in Eq. (6.83) can be written as

$$k_{||} = \frac{1}{R_0}\left(n + \frac{m}{q}\right) \tag{6.116}$$

where n is the toroidal mode number, m is the poloidal mode number and

$$q = \frac{r}{R_0}\frac{B_{0\phi}}{B_{0\theta}} \tag{6.117}$$

is the safety factor, defined in terms of the toroidal ($B_{0\phi}$) and poloidal ($B_{0\theta}$) components of the equilibrium magnetic field. The low–frequency Alfvén resonance condition can then be written as

$$\omega^2 = \omega_A^2(r) = \frac{v_A^2(r)}{R_0^2}\left[n + \frac{m}{q(r)}\right]^2 \tag{6.118}$$

which defines the position where a mode of given frequency and mode numbers m and n experiences Alfvén resonance.

The effect of the toroidicity of the Tokamak is that the DAWs arising from current effects in a straight cylinder ("cylindrical" DAWs), described in the previous section, are now coupled together for different poloidal mode numbers m, because of the bending of the cylinder. The modes corresponding to the discrete magnetoacoustic wave spectrum, such as surface waves, are also coupled together due to the toroidal effects. There are now usually several resonance surfaces in the plasma, each absorbing a fraction of the total power of a magnetoacoustic wave of given frequency launched by an external antenna. The DAWs that would have been undamped eigenmodes in a straight cylindrical plasma may now lie inside the continua, introducing a resonance damping, which is much higher than other damping mechanisms for DAWs, such as the Landau and TTMP damping predicted for a hot cylindrical plasma (Villard *et al.* 1995). This makes power deposition in the centre of the discharge difficult, particularly for small aspect ratio machines, because a body wave with high fields in the centre cannot be excited. The cylindrical DAWs can be destabilized by resonant interaction with high–energy α particles, but the toroidicity tends to stabilize them.

Gaps appear in the Alfvén continuous spectrum at the crossing points of the modes of different poloidal mode number m, due to the coupling effects of the toroidicity and ellipticity

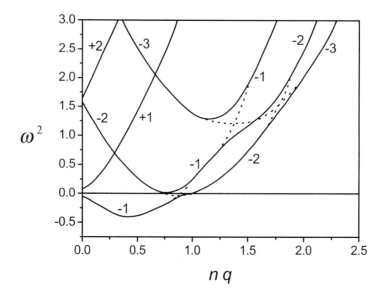

Figure 6.14: Alfvén wave spectrum of an elliptical toroidal plasma. The solid curves show the spectrum after mode coupling due to toroidal and elliptical effects, and the dotted curves show the frequencies before mode coupling. The labels next to each curve give the value of the poloidal mode number m for the unperturbed modes. (Reproduced from D'Ippolito & Goedbloed 1980.)

of a Tokamak. The gap formation is similar to the formation of forbidden energy bands for an electron in a periodic potential in a crystal in solid–state theory. The magnetic field gives the plasma structure, just as the lattice imposes structure on the crystal. The toroidicity couples the two modes together, with a gap in the spectrum appearing at the crossing points in the dispersion relation defined by Eq. (6.125) (D'Ippolito & Goedbloed 1980), as shown in the dispersion relation diagram of Figure 6.14. A similar splitting of the discrete spectrum occurs in a straight cylindrical plasma with a noncircular (in particular elliptic) cross–section (Dewar *et al.* 1974).

In the low–current model at low frequency, because q is independent of r, there is no localized Alfvén resonance point in the plasma. However, for general profiles, the range of Alfvén resonance positions defines the Alfvén continuum of resonant frequencies. In a true toroidal geometry, the resonance condition defined by Eq. (6.118) predicts reasonably well the position of the Alfvén resonances when the continuum frequencies of different m's are well separated from each other, but fails when they are close to each other, as seen in Figure 6.14. Toroidal coupling, due to a nonuniform toroidal magnetic field over a magnetic surface, removes the degeneracy of shear Alfvén modes of different m's, and gaps break up the Alfvén continuum at the physical crossing points of the interacting modes. For example, the coupling of a cylindrical continuum mode with m, with a continuum mode with $m + 1$, creates gaps

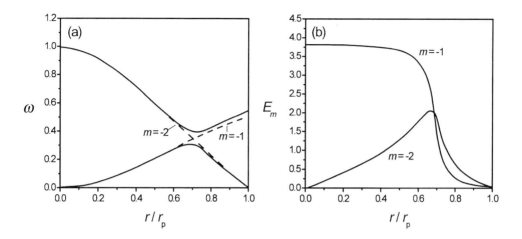

Figure 6.15: (a) Shear Alfvén continuous spectrum with gap (solid curves), for a parabolic safety factor profile and a constant density profile. The cylindrical spectra are shown by the dashed curves. (b) Radial profiles of the dominant poloidal harmonics of the poloidal electric field for the $n = -1$ TAE mode, for the same equilibrium as in (a). (Reproduced from Fu & Van Dam 1989.)

where $k_{\|m} = -k_{\|m+1}$, that is, at rational q magnetic surfaces

$$q = (|m| + 1/2)/|n|. \tag{6.119}$$

The continuum shear waves may be described by two coupled second–order eigenmode equations for the poloidal electric fields of the two dominant poloidal mode numbers m and $m + 1$ (Fu & Van Dam 1989). One equation is given by

$$
\left[\frac{d}{dr} r^3 \left(\frac{\omega^2}{v_A^2} - k_{\|m}^2 \right) \frac{d}{dr} - (m^2 - 1)r \left(\frac{\omega^2}{v_A^2} - k_{\|m}^2 \right) \right.
$$
$$
\left. + \frac{d}{dr} \left(\frac{\omega^2}{v_A^2} \right) r^2 \right] E_m + \left(\epsilon \frac{d}{dr} \frac{\omega^2}{v_A^2} \frac{r^4}{r_p} \frac{d}{dr} \right) E_{m+1} = 0 \tag{6.120}
$$

and the second equation is of the same form as Eq. (6.120), except for the substitutions $m \to m + 1$ and $m + 1 \to m$. Here $\epsilon = 3r_p/2R_0$. (The inverse aspect ratio r_p/R_0 is assumed small.)

In cylindrical geometry ($\epsilon = 0$), the two poloidal modes are decoupled, and there are two cylindrical shear Alfvén continua, given by $\omega_1^2 = v_A^2 k_{\|m}^2$ and $\omega_2^2 = v_A^2 k_{\|m+1}^2$. However, in the Tokamak geometry, the poloidal mode numbers are no longer good "quantum" numbers, because of the coupling of the modes. The singularity of the resulting fourth–order differential equations occurs where the determinant of the coefficients of the second–order derivative terms in Eqs. (6.120) equals zero, which yields the following two continuum frequency branches (Fu & Van Dam 1989):

$$\omega_{\pm}^2 = v_A^2 \frac{(k_{\|m}^2 + k_{\|m+1}^2) \pm \sqrt{(k_{\|m}^2 - k_{\|m+1}^2)^2 + 4\epsilon^2 x^2 k_{\|m}^2 k_{\|m+1}^2}}{2(1 - \epsilon^2 x^2)} \tag{6.121}$$

where $x = r/r_p$. Thus at the crossing point of the two cylindrical continua a gap appears with a width

$$\Delta\omega = \omega_+ - \omega_- \simeq 2\epsilon x |v_A k_{\|m}| \tag{6.122}$$

evaluated at the crossing point, as shown by the Alfvén continua in Figure 6.15(a), in the form of a plot of the radially dependent continuum frequency against r. In the absence of toroidal coupling, the cylindrical continua derived from Eq. (6.118) are shown by the dashed curves. The gap size is proportional to x^2 when the poloidal mode numbers differ by more than one (Cheng & Chance 1986).

We consider next the effect of toroidicity on the discrete magnetoacoustic wave spectrum. If we assume the low–current straight cylinder model of Section 6.5, for which we have

$$q = \frac{r_p}{R_0 J} \tag{6.123}$$

the lowest–frequency surface waves have the dispersion relation, from Eq. (6.91),

$$\omega^2 = \frac{2v_A^2}{R_0^2} \left((n + q/m)^2 - \frac{m}{|m|} \frac{(n + q/m)}{q} \right). \tag{6.124}$$

Two modes with the same toroidal mode number n, but with poloidal mode numbers m_1 and m_2, would have the same frequency when

$$q = \frac{m_2 - m_1 + 1}{2}. \tag{6.125}$$

This mode crossing leads to gaps in the magnetoacoustic wave spectrum similar to the gaps in the Alfvén continuum shown in Figure 6.14.

Another important effect of both toroidicity and ellipticity is that a different family of DAWs or GAEs can exist in the toroidicity and ellipticity induced gaps in the Alfvén continuum; these DAWs are called TAEs (Toroidal Alfvén Eigenmodes) and EAEs (Elliptical Alfvén Eigenmodes), or "gap modes". The TAE was first reported by Cheng, Chen & Chance (1985) and Cheng & Chance (1986), the EAE by Betti & Freidberg (1991). Such global eigenmodes can be found by numerical solution of the two coupled differential equations of the form of Eq. (6.120). The number of TAEs is proportional to n, and the spectrum fills up the continuum gap as n becomes large. The modes involve several poloidal harmonics, so that the parallel phase speed can be either greater or smaller than v_A at different radial locations. The field E_θ for the mode is peaked at the location of the gap, near the crossover of the cylindrical continua. Both electron and ion Landau damping can be strong for these modes, but the TAEs may be destabilized by fast particles in Tokamak plasmas (Fu & Van Dam 1989). Resonant absorption of the TAE occurs where the frequency matches a frequency of the Alfvén continuum. For the case shown in Figure 6.15(a), a TAE mode exists inside the continuum gap, at $\omega = 0.93|v_A k_{\|m}|_{q=1.5}$. The radial profiles of the two dominant poloidal harmonics for this TAE mode are shown in Figure 6.15(b); the mode is peaked at the location of the gap.

6.8   Current Drive

The continuous operation of a Tokamak fusion reactor requires a continuous toroidal current to provide the confining poloidal magnetic field. Such operation is preferred to a pulsed operation, where the current is induced by an external time–varying magnetic field. Various means of producing a continuous current have been suggested, such as beams of neutral particles that convert by charge exchange collisions into fast ions, and various types of waves (Fisch 1987). The use of Alfvén waves to drive a current was first suggested by Wort (1971). In that scheme the current is produced by a linear wave-particle (Landau) interaction that drives electrons with low velocity $v_{\parallel}$ parallel to the magnetic field. However, such electrons are trapped in magnetic wells in Tokamaks, and so are prevented from carrying the current.

Renewed hope for efficient current drive using low–frequency waves appeared with the work of Chan, Miller & Ohkawa (1990a,b) on current drive through magnetic helicity injection, which depends on collisional and nonlinear fluid effects. The magnetic helicity density is defined as

$$H = \langle \boldsymbol{A} \cdot \boldsymbol{B} \rangle \tag{6.126}$$

where $\boldsymbol{A}$ and $\boldsymbol{B}$ are respectively the wave magnetic vector potential and the wave magnetic field, and the brackets denote an average over space and over a period of the wave. The magnetic helicity is independent of the choice of gauge for the magnetic vector potential. For a circularly polarized wave propagating in the direction of the equilibrium magnetic field, with real frequency ω and complex wavenumber $k = k_0 + ik_{\mathrm{I}}$, the helicity density in the wave is

$$H = k_0 A^2 \exp(-2k_{\mathrm{I}} z). \tag{6.127}$$

The creation of plasma current can be attributed to the transfer of the wave helicity present in a circularly polarized wave to the plasma. The current arises from a collision–induced phase shift between the fluid velocity and the wave magnetic field, producing a real component of $\boldsymbol{v} \times \boldsymbol{B}$ in the direction of the static magnetic field. For a plasma with resistivity η and viscosity μ, the mean steady current is given by (Taylor 1989),

$$2\eta \langle j_z \rangle = (\mu - \eta) k_0^2 H. \tag{6.128}$$

Thus for a purely resistive plasma, the helicity dissipated in the current drive is equal in magnitude but opposite in sign to the helicity input; two units of input helicity are dissipated by the fluctuating currents. The viscosity, on the other hand, does not dissipate the helicity. The transport of wave helicity requires that the wave structure is localized transverse to the direction of propagation; a plane wave possesses a finite helicity density but does not transport helicity across surfaces (Mett & Tataronis 1989).

Different types of localized low–frequency waves can be used for current drive: discrete Alfvén waves (DAW), kinetic Alfvén waves (KAW), inertial Alfvén waves (IAW) and fast magnetoacoustic waves. Enhanced current drive may be achieved if the wave is close to the Alfvén resonance condition, because of the larger wave field amplitudes, and the waves can also simultaneously cause internal transport barriers in the plasma that can suppress plasma turbulence. It is interesting that a link has been made between helicity injection in current drive and the "α–dynamo" effect, which is believed to play an important role in the generation of cosmic magnetic fields (Litwin & Prager 1998).

6.9 Localized Alfvén Waves

Another aspect of experimental studies of Alfvén waves is the direct launching of localized shear Alfvén waves, rather than the global modes used in resonance heating and current drive. As we have discussed in Chapters 3 and 4, short perpendicular wavelength modes related to Alfvén shear waves are normally generated in laboratory experiments via mode conversion at the Alfvén resonance from propagating magnetoacoustic waves or surface waves. However, shear Alfvén waves have also been excited directly by a small antenna localized inside a cylindrical nonuniform plasma waveguide (Cross 1983; Borg 1994, 1998; Morales & Maggs 1997; Gekelman *et al.* 1997; Gekelman 1999). The Alfvén wave can be strongly guided in laboratory plasmas in regions of high density gradient and along strongly curved magnetic fields. In these experiments, the shear Alfvén wave was observed to be localized initially, in agreement with ideal MHD theory, but was found to disperse radially as it propagated away from the antenna. Such experiments are also relevant to space physics observations, as is discussed in Chapter 7.

For a collisionless plasma, a description of this phenomenon in terms of the KAW and IAW modes is given in Section 2.8.5, where it is shown that a continuous wave disturbance excited by a point source propagates on a resonance cone at a small angle to the field and as a wave train at the local Alfvén speed along the field. Direct excitation of the IAW can have some novel applications in laboratory plasmas, such as efficient helium removal from the edge regions of fusion plasmas, and as a primary ionization source in Stellarators (Borg 1998). Steep density gradients typical of a laboratory plasma periphery have little effect on the guidance for the perpendicular wavenumbers necessary to form a transversely localized disturbance, so direct antenna coupling can be analysed in terms of a locally uniform plasma. It is found that direct antenna excitation of the Alfvén wave is more efficient than that by resonant mode conversion of the fast magnetoacoustic wave.

In a collisional plasma, resistive effects can account for the diffusion of the shear wave fields away from the initial magnetic field lines (Sy 1978, 1984), and we now analyse this situation for a straight cylindrical plasma column of radius r_0 with a uniform axial magnetic field. An equation describing axisymmetric ($m = 0$) shear or torsional Alfvén waves of frequency ω in a resistive, pressureless, fully ionized plasma with a radially dependent density, can be derived from Eqs. (1.12) and (6.5):

$$\left(1 - \frac{i\eta k_A^2(r)}{\mu_0\omega}\right)\frac{\partial^2 B_\theta}{\partial z^2} + k_A^2(r)B_\theta = \frac{i\eta k_A^2(r)}{\mu_0\omega}\frac{\partial}{\partial r}\left[\frac{1}{r}\frac{\partial}{\partial r}(rB_\theta)\right] \tag{6.129}$$

where

$$k_A^2(r) = \frac{\mu_0\omega^2\rho_0(r)}{B_0^2} = \frac{\omega^2}{v_A^2(r)} \tag{6.130}$$

with $v_A(r)$ the local Alfvén velocity. Just as in the cartesian geometry case of a shear wave propagating solely along the magnetic field described in Section 3.3.1, the torsional wave decouples from the magnetoacoustic wave only if $m = 0$. If normal modes with fixed axial wavenumber k_z in an axially uniform cylindrical waveguide are sought, the second derivative in z in Eq. (6.129) can be replaced by $-k_z^2 B_\theta$, and the resulting radial equation may be solved, with the appropriate boundary conditions, as an eigenvalue problem for the radial modes, as in Sections 6.2 and 6.4, resulting in a spectrum of eigenfrequencies. However, if we are interested in the axial propagation of a wave of given frequency from an axially localized source, the two–dimensional differential equation (6.129) must be solved.

In the absence of resistivity, a Fourier analysis of Eq. (6.129) into axial components $\exp(ik_z z)$ leads to the dispersion relation

$$\omega^2 = k_z^2 v_A^2(r) \tag{6.131}$$

showing that for a given wavenumber k_z, a continuous spectrum of shear wave frequencies is allowed in a nonuniform plasma. Note that for any allowed frequency the wave field B_θ can have any radial dependence at the source. Alternatively, for a fixed frequency ω, which is more appropriate to localized antenna studies, the solution of Eq. (6.129) with $\eta = 0$ is

$$B_\theta(r, z) = A(r) \exp(\pm i k_A(r) z) \tag{6.132}$$

with $A(r)$ an arbitrary function determined by initial conditions (Pneuman 1965; Pridmore-Brown 1966). This result indicates that the radial profile of a constant phase wave front becomes progressively distorted as the disturbance moves away from an antenna, as was discussed in Section 3.3.1 (see Figure 3.12). This wave front distortion was observed experimentally by Cross & Lehane (1968).

Let us now reintroduce (small) resistivity, and define a small parameter

$$\varepsilon = \frac{\eta}{\mu_0 \omega r_0^2} = \frac{1}{R_m} \tag{6.133}$$

where R_m is the magnetic Reynolds number. Now the right–hand side term of Eq. (6.129) can be regarded as a term giving "resistive radial coupling" to the shear wave. If we neglect it, the solution is approximately

$$B_\theta \simeq A(r) \exp\left(i k_A z - \frac{\varepsilon k_A^3 r_0^2}{2} z \right) \tag{6.134}$$

showing the usual resistive damping of the shear Alfvén wave. The right–hand side of Eq. (6.129) is of diffusive form, and Eq. (6.129) may be solved using the method of multiple scales for a radially and axially localized source of axisymmetric shear Alfvén waves with the initial condition

$$A(r, t = 0) = A_0 \delta(z) \delta(r - a) \tag{6.135}$$

where a is the radius of the localized antenna. The resulting solution is (Sy 1984)

$$B_\theta(r, z) = \frac{\text{const.}}{\varepsilon k_A z} \exp\left\{ i k_A z - \frac{\varepsilon k_A^3 r_0^2 z}{2} - \frac{\varepsilon k_A (k_A')^2 r_0^2 z^3}{6} \right\}$$

$$\times \ \exp\left\{ -\frac{(r - a)^2}{2 \varepsilon k_A r_0^2 z} \right\} e^{-\alpha} I_1(\alpha), \tag{6.136}$$

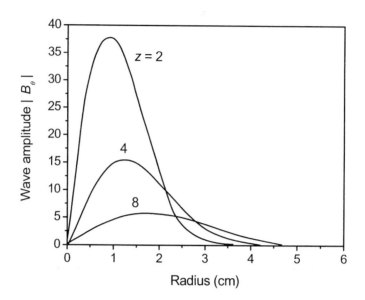

Figure 6.16: Radial profiles of the wave amplitudes at various normalized axial distances from the antenna at 0.5 MHz. The antenna radius is 0.75 cm and the hydrogen plasma has radius 7.5 cm, number density 2×10^{21} m^{-3}, $B_0 = 0.7$ T and $\eta = 3 \times 10^{-4}$ ohm m. (Reproduced from Sy 1984.)

where $\alpha = ra/\varepsilon k_A r_0^2 z$ and I_1 is a first–order modified Bessel function. The solution given by Eq. (6.136) indicates a resistive radial diffusion as the wave propagates away from the antenna. For a parabolic radial density profile, the radial wave field profiles obtained from Eq. (6.136) at various axial distances from the antenna are shown in Figure 6.16. The wave field profile appears to relax to a J_1 Bessel function profile. Ion-neutral collisions (see Section 2.5.1) also modify the frequency dispersion and damping of directly excited shear Alfvén waves. Features such as anomalous dispersion were observed in experiments on directly excited waves in a cylindrical partially ionized plasma reported by Amagishi & Tanaka (1993).

7 Space and Solar Plasmas

7.1 Introduction

Alfvén and magnetoacoustic waves are commonly observed in the natural environment of plasmas in space, such as the Earth's magnetosphere and the interplanetary plasma, and in the solar atmosphere. Such observations may be via direct measurement of electric and magnetic fields in the waves by artificial satellites, or via optical evidence, in the case of waves in the Sun's atmosphere. The presence of waves in some regions, even though not directly observed, is invoked to explain some phenomena, such as the heating of the solar corona. In this chapter we briefly survey some of the occurrences of the waves in space and solar plasmas, starting with the Earth's magnetospheric and ionospheric plasma and proceeding to the interplanetary and cometary plasmas, including waves in dusty plasmas. Finally we explore the properties of the waves in the corona, chromosphere and photosphere of the Sun.

7.2 The Magnetosphere

Oscillations in the geomagnetic field were observed in the 19th century, when ultra–low–frequency (ULF) fluctuations, with periods of seconds to minutes, were observed on the ground (see, e.g., Kivelson 1995b). Dungey (1954) was the first to discuss these variations in terms of resonant oscillations in the geomagnetic field and low–frequency waves in the magnetized plasma surrounding the Earth. The waves may be excited in a continuous fashion, via instabilities at the boundary between the magnetosphere and the solar wind. They may also be produced impulsively when sudden changes occur in the convection flow of the magnetosphere, causing surges in current and pressure, and the waves propagate to relax the stress in the system.

7.2.1 Micropulsations

Hydromagnetic waves are observed to occur in the Earth's magnetosphere in the frequency range of 1 mHz up to the local proton cyclotron frequency ≥ 1 Hz. The waves are detected as small oscillations of the earth's magnetic field, and so are called geomagnetic micropulsations. They are classified by period and continuous or irregular characteristics, as Pc–1 up to Pc–5, and Pi–1 to Pi–2. The Pc pulsations are continuous pulsations with well–defined spectral peaks, and frequencies ranging from 0.2 Hz (Pc–1) to 7 mHz (Pc–5), while the Pi pulsations are irregular pulsations with power at many different frequencies, and frequencies

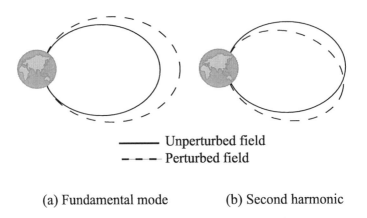

———— Unperturbed field
– – – – Perturbed field

(a) Fundamental mode   (b) Second harmonic

Figure 7.1: Schematic diagram of the two lowest–frequency modes of standing oscillations on a magnetic field line in the Earth's magnetosphere. (Adapted from Kivelson 1995b)

of 0.025 Hz (Pi–1) to 25 mHz (Pi–2) (Kivelson 1995b). Corresponding waves have also been observed in the magnetospheres of other planets. The pulsations in the Earth's magnetosphere are detected on the ground with magnetometers, and in addition have been observed directly by the AMPTE, Dynamic Explorer and Viking satellites. The source of the oscillating or pulsating magnetic fields observed at the surface is thought to be MHD waves standing along magnetic field lines in the outer atmosphere and reflected at the ionosphere at the two ends, that is, similar to bounded modes in laboratory plasmas as discussed in Chapter 6. The waves may be pictured as displacements of the magnetic field lines, as illustrated for the first two modes in Figure 7.1.

The waves are thought to be excited through the interaction between the solar wind and the magnetosphere. For example, a shear in the flow of plasma across the magnetosphere can produce ULF surface waves through the Kelvin-Helmholtz instability, as described in Section 5.2.1. The surface waves compress the magnetosphere, and the perturbations generate compressional magnetoacoustic waves that decay or propagate across magnetic field lines. Mirror instabilities due to anisotropic temperatures (Section 5.2.2) may also play a role, because in much of the tenuous plasma of the magnetosphere $T_\perp > T_\parallel$ is satisfied. On the other hand, the higher–frequency wave classes (Pc–1, some Pc–2, and Pi–1) are thought to arise from a local resonant interaction with ions present in beams, such that the Doppler shifted frequency equals the ion cyclotron frequency. The magnetospheric plasma is in general nonuniform, so that energy may be coupled from the magnetoacoustic waves into the shear waves via "field line resonances", that is, at the local Alfvén resonance points (Hasegawa & Chen 1974).

In the approximate dipole background magnetic field of the Earth, with radial variation of the plasma density, the Alfvén and magnetoacoustic wave spectrum may be analysed, just as for the cylindrical and toroidal geometries discussed in Chapter 6. The field and the density can be assumed to be azimuthally symmetric about the magnetic axis, and the waves vary with the longitudinal or toroidal angle ϕ as $\exp(im\phi)$. Just as for the Cartesian geometry case with $k_y = 0$ (Section 3.3.1) and cylindrical geometry case with $m = 0$ (Section 6.9), if $m = 0$ in the dipole geometry the shear wave is decoupled from the magnetoacoustic wave, has magnetic field and velocity components polarized in the toroidal direction, and propagates along field lines with the local Alfvén speed. In general, however, for $m \neq 0$ the shear wave and the magnetoacoustic waves are coupled, and the magnetoacoustic wave, for example excited as a propagating wave or a surface wave by a Kelvin-Helmholtz instability, can lose energy into the shear wave via Alfvén resonance absorption.

To model the main effects of such coupling, it is sufficient to use the Cartesian slab coordinate system of Sections 3.3 and 3.4. As was assumed by Southwood & Hughes (1983), we consider a plasma slab, with uniform magnetic field $B_0 \hat{z}$ but plasma density $\rho_0(x)$ a function of x, with the x–direction corresponding to the radial direction in the Earth's magnetosphere. The plasma extends to $z = \pm l$, so the magnetic lines are finite in length, with the end points corresponding to the footpoints of the field lines at the ionosphere. At the ionosphere there is collisional damping, so for wave fields that are driven at a given real frequency, the wavenumber k_z acquires an imaginary part; $k_z = k* = k_R - i\kappa$, with $k_R = n\pi/2l$ where n is an integer.

As mentioned above, as the shear Alfvén waves propagate along magnetic field lines towards the surface of the Earth, they encounter the ionosphere, the electrically conducting, but collisional, region above about 100 km altitude. The ionosphere acts as a boundary, where some of the wave energy is reflected back into the magnetosphere, some energy is dissipated due to collisional damping of the Pedersen currents (see Section 2.5.2), and some energy survives, highly modified, to be detected as magnetic perturbations on the ground. There is a strong collisional Hall effect in the ionosphere, because the ion collision time is of the same order as the cyclotron period, particularly in the E region of the ionosphere. The signal below the ionosphere is thought to be due to the ionospheric Hall currents, that filter out the signals that vary horizontally on short scale–lengths (Southwood & Hughes 1983).

Alfvén waves have been found to play an important role in the coupling of energy between the magnetosphere and the ionosphere of the Earth. The dynamics of the coupling between planetary ionospheres and magnetospheric plasmas may in fact be viewed as the consequences of the reflection of Alfvén waves at a sharp boundary between the two regions (Buchert 1998). Another example of Alfvén waves in a planetary magnetosphere is provided by the planet Jupiter. Finite amplitude standing Alfvén waves were detected by the spacecraft Voyager 1 in the Io plasma torus, as predicted by a nonlinear Alfvén wave model (Gurnett & Goertz 1981).

7.2.2 Kinetic and Inertial Alfvén Waves

We have seen in Chapters 3 and 4 that a surface wave, such as that excited by the Kelvin-Helmholtz instability at the magnetosphere/solar wind boundary, can encounter the Alfvén resonance, and thence mode convert to the KAW or IAW when the scale–length of the resonance region approaches the ion gyroradius or the electron inertia length respectively. In terms of magnetospheric physics, this mode coupling will occur when the field line resonance is excited. The field–aligned electric field present in the short perpendicular wavelength KAW and IAW modes can accelerate charged particles, and so can lead to the precipitation of free electrons via heating or acceleration. There are possible connections between MHD field line resonances and the observations of discrete auroral arcs, and the KAW and IAW modes are now widely believed to be a fundamental ingredient in the physics of the arcs (Seyler, Wahlund & Holback 1995). Acceleration by Landau damping of the modes is not sufficient to account for the observed electron energy of tens of KeV, because the parallel phase velocity is less than the electron thermal speed. However, possible mechanisms for acceleration of electrons to several times their thermal speed are: trapping by wave and associated nonlinear acceleration, bounce resonance acceleration (Hasegawa 1976), and formation of solitary KAWs (Goertz 1984), in which electric fields appear similar to those in double layers.

Evidence for solitary nonlinear KAW and IAW waves (SKAW) in the Earth's magnetosphere has been provided by the Swedish–German Freja satellite, launched in 1992, whose high–resolution data showed that auroral low–frequency electromagnetic turbulence is dominated by spikes in the electric and magnetic fields (Louarn *et al.* 1994). The spikes have the appearance of solitary structures, with either humps or dips in the plasma density (of up to several tens of per cent), and have been interpreted as KAW and IAW solitons. Many earlier satellite observations of such electric field signatures were made, and linked to auroral arcs, but the observations were low resolution and the events were referred to as "electrostatic shocks" (Seyler, Wahlund & Holback 1995).

The transverse scale of the spikes is of the order of the electron inertial length δ_e (tens of metres to a few hundred metres), which is also the transverse scale of discrete auroral arcs, suggesting that the SKAW are related to localized plasma energization. The Poynting flux of the SKAW has been estimated to be large enough to heat the plasma effectively if it were dissipated (Louarn *et al.* 1994). The ratio between the electric and magnetic field amplitudes of the spikes is close to the local Alfvén velocity ($\simeq 10^7 \mathrm{ms}^{-1}$), as is required for an SKAW. Time–dependent solutions of the two–fluid model by Seyler, Wahlund & Holback (1995) show best agreement in waveform shape with the data during the transient nonlinear steepening stage, and not with the steady–state SKAW. The simulations also indicate that larger–amplitude initial conditions for the IAW result in wave breaking, and not in a steady–state solitary wave. Spikes observed with weaker magnetic fields have been postulated to be due to solitary waves on the ion-acoustic branch of the waves (see Eq. 2.233) (Wu & Wang 1996).

The altitude range explored by Freja (600 - 1700 km) forms a transition region between the innermost magnetosphere and the upper auroral ionosphere, where the plasma parameters vary rapidly, and where the transition condition $\beta = m_e/m_i$ separating the KAW from the IAW could be encountered. The transition condition is equivalent to $V_i \simeq v_A$, or $z_0^e \simeq 1$ (see Eq. 2.156), so electron Landau damping will be strong, and such damping could play a role in the auroral plasma energization processes (Wu *et al.* 1996).

The problem of the origin of the SKAWs in the magnetosphere has been addressed by Stasiewicz *et al.* (1997) and Seyler *et al.* (1998). We have seen in Section 2.8.5 that the dispersive nature of linear kinetic and inertial Alfvén waves produces a resonance cone, at which most of the wave energy propagates when emitted from a distant localized wave source, such as an electron beam in the magnetosphere. The angle of the cone to the magnetic field is given by Eq. (2.266) or Eq. (2.267).

7.3 Solar and Stellar Winds

7.3.1 Turbulent Waves in the Solar Wind

The solar wind is a plasma with a mean flow that varies with distance from the sun, and also with a nonuniform magnetic field. Representative solar wind parameters are: $n_e \simeq 10$ cm^{-3}, $T_e \simeq T_p \simeq 10^5$ K, $B_0 \simeq 10^{-8}$ T. The identification of Alfvén waves in the solar wind by means of spacecraft measurements was first achieved in the late 1960s (Belcher & Davis 1971). The predominant component of the Alfvén waves observed in the solar wind is of solar origin, while there is a secondary population of waves arising from local regions such as cometary and planetary exospheres. The physics of the secondary population of waves tends to be better understood than the primary population (Wu 1995).

The assumption of incompressibility is often used in the description of the waves in the solar wind. The feature that the perturbation velocity and magnetic field for incompressible Alfvén waves are parallel and proportional to each other (see Section 2.2), is used in the identification of Alfvén waves from the observational data. Other useful "transport" quantities, that help to identify modes from measurements of waves in space plasmas (Gary 1986), are the compressibility, and the magnetic helicity, the cross–correlation of the oscillating magnetic vector potential with the oscillating magnetic field that is also of use in the description of current drive in laboratory plasmas (see Eq. (6.126) in Section 6.8).

Strong correlations between the magnetic and velocity perturbation fields are observed, with the peak perturbation magnetic field being comparable in magnitude to the ambient solar magnetic field. The waves were found to propagate outwards in the solar wind frame, with the largest amplitude fluctuations being observed in the leading edge of high–velocity streams. The waves have a broad wavenumber spectrum, with a magnetic power spectrum of the form $k^{-\alpha}$ where the spectral index α lies between 1 and 2 (Coleman 1967). It is thus appropriate to call the observed waves Alfvénic turbulence.

The waves of solar origin are thought to originate from regions close to the Sun. In the source regions both Alfvén and magnetoacoustic waves would be generated, but because the obliquely propagating magnetoacoustic waves are strongly linearly damped by collisionless transit time magnetic damping (Section 2.7.3), only the Alfvén waves would persist. The parallel propagating waves are still subject to damping due to viscosity, thermal conductivity and resistivity; in the solar wind, where the plasma β varies from very small close to the Sun, up to $\beta \simeq 1$ at the Earth's orbit, for parallel propagation, the damping of the Alfvén wave and the fast magnetoacoustic wave is equal up to $\beta \simeq 1.2$, and then it is much higher for the fast wave (Whang 1997). If the waves are generated by a resonant beam instability, the unstable waves would have parallel wavenumbers close to that determined by the resonant condition, from Eqs. (5.20) and (5.22),

$$k \simeq |(\omega \pm \Omega_i)/V_0| \tag{7.1}$$

where V_0 is the beam velocity of the streaming ions. This is, however, contradictory to the broad wavenumber spectrum observed. The broad spectrum could be a result of propagation and evolution effects on the waves, including inhomogeneities of the medium and non-linearities.

Nonresonant instabilities (Section 5.2.1) have been invoked for the generation of the waves, but all suffer inadequacies. Stream-shear driven instabilities, such as the Kelvin-Helmholtz instability, cannot lead to the observed propagation away from the Sun. The firehose instability due to a temperature anisotropy (Section 5.2.2) cannot strongly excite waves because the observed anisotropy at 1 A.U. is rather weak, $T_{\parallel}/T_{\perp} \simeq 1.5$.

More recent observations confirm that Alfvén waves in interplanetary space are predominantly travelling away from the Sun in the solar wind frame; however, there is also a minor component of waves that travel in the direction of the Sun. The origin of the inward propagating waves is not clear, but the nonlinear interaction of these waves with the outward propagating waves may be responsible for the observed Alfvénic turbulent spectrum. The inward waves may be produced by the parametric decay of large–amplitude waves described in Section 5.5.2; however, it was shown there that the decay instability cannot occur in the relatively high–β solar wind plasma. Indeed, one would expect that much of the energy of the outward propagating waves should have been converted to outward propagating sound waves and inward propagating Alfvén waves. Another possible source of waves is via pick–up ion instabilities of neutral hydrogen and helium atoms of interstellar origin (Lee & Ip 1987) (see Section 7.5).

7.3.2   Wind Acceleration

In addition to the study of the turbulent Alfvén wave spectrum in the solar wind, the propagation of Alfvén waves in the solar corona and stellar coronae has been analysed in connection with the acceleration of the solar and stellar winds. For example, the initial value problem in a spherically symmetric isothermal and gravitationally stratified stellar atmosphere with a time–dependent linear MHD numerical model has been studied by An *et al.* (1990). Their model, which we describe here, provides a good example of the use of the Alfvén wave equations in a spherically symmetric system with gravity.

Consider a star with a spherically symmetric isothermal atmosphere. In the stellar wind region, the equilibrium magnetic field can be approximated to be directed radially outwards from the star; however, the steady outflow of matter is neglected in this model. The (monopole) magnetic field is current–free. Using spherical polar coordinates with origin at the centre of the star, it is given by

$$B_{0r} \propto \frac{1}{r}. \tag{7.2}$$

The equilibrium pressure in the atmosphere satisfies the following variant of the force balance equation (2.1):

$$\nabla p_0 = -\rho_0 \frac{GM}{r^2}\hat{r} \tag{7.3}$$

where the gravitational force due to the star has been included, but there is no magnetic force. Here G is the gravitational constant and M is the stellar mass.

Using the isothermal gas law with the constant temperature T, and converting to the radius normalized to the stellar radius $r_{\rm s}$, Eq. (7.3) can be expressed as

$$\frac{\partial p_0}{\partial r} = -\frac{\alpha p_0}{r^2} \tag{7.4}$$

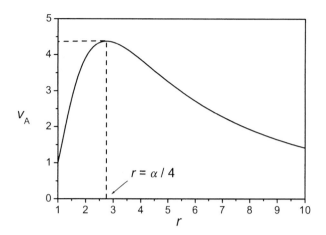

Figure 7.2: The Alfvén speed profile in a spherically symmetric plasma around a star, against the radius r, normalized to the star's radius r_s, as obtained from Eq. (7.7). The Alfvén speed is normalized to the value at r_s. The parameter $\alpha = 11$.

where

$$\alpha = \frac{GM}{RTr_s} \tag{7.5}$$

and R is the gas constant. The equilibrium pressure and gas density have the following normalized radial profile:

$$p_0(r), \rho_0(r) = e^{-\alpha(1-1/r)}. \tag{7.6}$$

The resulting local Alfvén speed has the normalized form

$$v_A(r) = \frac{e^{(\alpha/2)(1-1/r)}}{r^2} \tag{7.7}$$

with the normalization $v_A(r) = 1$ at the stellar surface $r = 1$. The Alfvén speed is shown as a function of r in Figure 7.2, for $\alpha = 11$, relevant to the Sun. It is important to note that there is a maximum value of v_A, occurring at $r = r_m = \alpha/4$.

Assuming a perturbation of the velocity only in the poloidal (θ) direction, that is, considering only the shear Alfvén wave, we have the following wave equations:

$$\rho_0 \frac{\partial v_{1\theta}}{\partial t} = \frac{\mu_0}{r} \frac{\partial (r B_{1\theta})}{\partial r} \tag{7.8}$$

$$\frac{\partial B_{1\theta}}{\partial t} = \frac{1}{r} \frac{\partial}{\partial r} (r v_{1\theta} B_{0r}) \tag{7.9}$$

which may be rewritten as normalized equations

$$\frac{\partial V}{\partial t} = v_A^2 \frac{\partial b}{\partial r} \tag{7.10}$$

$$\frac{\partial b}{\partial t} = \frac{\partial V}{\partial r} \tag{7.11}$$

in terms of the variables $V = rB_{0r}v_{1\theta}$ and $b = rB_{1\theta}$. For a wave of fixed frequency, Eqs. (7.10) and (7.11) reduce to

$$\frac{d^2 V}{dr^2} + \frac{\omega^2}{v_A^2(r)} V = 0. \tag{7.12}$$

For short wavelengths a local WKB approximation can be used to solve Eqs. (7.10)-(7.12). However, the equations must be solved numerically for wavelengths comparable to the scale–length of the spatial variation of the Alfvén speed, and the existence of the maximum of the Alfvén speed means that the wave can be strongly reflected at the point $r = r_m$. Thus the numerical solution of Eqs. (7.10) and (7.11), for steady and pulsed waves, yields wave trapping between the stellar surface and the point of maximum v_A, and a large gradient in the wave magnetic field amplitude near $r = r_m$.

The acceleration of the stellar wind by the Alfvén waves can then be found by the following reasoning. The force exerted on the plasma by the wave can be derived from the nonlinear momentum equation: the acceleration in the radial direction is

$$a_r = -\frac{1}{\rho_0}\frac{\partial}{\partial r}\left(\frac{B_{1\theta}^2}{2}\right) + \frac{v_{1\theta}^2}{r} - \frac{1}{\rho_0}\frac{B_{1\theta}^2}{r}. \tag{7.13}$$

The first term on the right–hand side of Eq. (7.13) is a result of magnetic pressure. The second and third terms are due to centripetal force and magnetic tension respectively, and cancel each other in the small–wavelength WKB approximation. In the trapping region there are (geometrical) resonance frequencies, and the large wave magnetic field gradient yields a local enhancement of the Alfvén wave pressure force given by Eq. (7.13), which is a maximum at the resonance frequencies. This may lead to the driving of massive winds along the open magnetic field lines in the atmospheres of cool stars. Earlier work on stellar winds employed the WKB approximation for the waves, and so was unable to encompass reflection problems.

The effect of a steady outflow of plasma from the star was also considered by An et al. (1990), and it was found that the trapping occurs in the region where the outflow has a negligible effect on the Alfvén speed. In the transition region of a stellar atmosphere there is a steep temperature gradient that will significantly affect the trapping of waves launched below the transition region. If the star has a corona, the temperature variation there must also be taken into account.

7.4   Dusty Space Plasmas

The physics of dusty plasmas has been studied intensively because of its importance for a number of applications in space and laboratory plasmas (Goertz 1989; Mendis & Rosenberg 1994). The general topic of waves in dusty space plasmas has been reviewed by Verheest (2000). A dusty plasma is characterized as a low–temperature ionized gas whose constituents are electrons, ions and micron–sized dust particulates. The latter are usually negatively charged due to the attachment of the background plasma electrons on the surface of the dust grains via collisions. The presence of dust particles (grains) changes the plasma parameters and affects the collective processes in such plasma systems. In particular, the charged dust grains can effectively collect electrons and ions from the background plasma. Thus in the state of equilibrium the electron and ion densities are determined by the neutrality condition, which is given by

$$-en_e + en_i - Z_d e n_d = 0 \qquad (7.14)$$

where $n_{e,i,d}$ is the concentration of plasma electrons (with the charge $-e$), ions (for simplicity, we consider singly charged ions), and dust particles, respectively. Here we have assumed a single characteristic charge on the dust particle $-Z_d e$. This charge can vary significantly depending on the plasma parameters and the properties of the grain, such as its size; in reality there will also always be a range of grain sizes, and so also a range of dust charges. In the approximation that the grains are charged purely by plasma particles colliding with the grain surface, the dust charge number Z_d can be estimated from an equation expressing the balance of ion and electron currents onto the grain (Cramer, Verheest & Vladimirov 1999), and is a function of the grain radius for a spherical grain. Because of the much higher thermal speed of the electrons, the grain will usually be negatively charged.

The charged dust grains will be coupled to the plasma via the electric and magnetic fields, and so will affect the spectrum of very low–frequency Alfvén and magnetoacoustic waves through their inertia and cyclotron motion (Pilipp *et al.* 1987). At higher frequencies, well above the dust cyclotron frequencies, when the dust particles can be assumed to be stationary, the influence of the dust on the charge imbalance between ions and electrons may still radically alter the dispersion relation of Alfvén and magnetoacoustic waves, simply because of the soaking up of negative charge by the grains. This leads to an imbalance in the ion and electron Hall currents, even at frequencies well below the ion cyclotron frequency.

With a negligibly small charge on the dust grains, the waves have the usual shear and compressional Alfvén wave properties, while for a nonzero charge on the grains the waves are better described as circularly polarized waves (for parallel propagation). The left–hand polarized wave has whistler or helicon wave properties extending to low frequencies (Mendis & Rosenberg 1994), in analogy to the case of electromagnetic waves in plasmas in solids with unequal electron and hole numbers (Baynham & Boardman 1971). The right–hand polarized wave has a cutoff frequency at infinite wavelength.

The above properties of the waves follow from simply treating the dust grains as an additional, albeit extremely heavy, secondary negative ion species, the effects of which were discussed in Section 2.6. In the charged dust case, the ratio $|g|$ of dust cyclotron frequency Ω_d to ion cyclotron frequency is extremely small; for a single charge on the grain, $|g|$ is typically $\simeq 10^{-9}$. We define the proportion of negative charge residing on the grains as

$$\delta = \frac{Z_d n_d}{n_i} = b|g| \qquad (7.15)$$

where b is defined as the ratio of charged dust mass density to plasma ion mass density, generalizing the definition in Section 2.6.

The parameter δ can be typically of the order of 10^{-4} in interstellar dust clouds or cometary atmospheres. The wavenumber for parallel propagation in the dusty plasma may be derived from the bi–ion result given by Eq. (2.134) in the case of frequency much lower than the primary ion cyclotron frequency, that is $f \ll 1$, and for $\delta \ll 1$, and is given by

$$k_z^2 = \frac{\omega^2}{v_A^2}\left(\frac{\omega \pm \Omega_m}{\omega \pm \Omega_d}\right) \tag{7.16}$$

showing, for the right–hand polarized mode, the cutoff where $k_z = 0$ at $\omega = \Omega_m$ (Eq. 2.142), and a dust cyclotron resonance where $k_z \to \infty$ at $\omega = \Omega_d$ (see Figure 2.7(b) for the dispersion relations with a negative secondary ion). Here the Alfvén speed v_A is defined in terms of the plasma ion mass. The right–hand polarized wave cutoff frequency (Eq. 2.142) in this case is given by

$$\Omega_m = \frac{\Omega_d + \Omega_i \delta}{1 - \delta} \simeq \Omega_d + \Omega_i \delta. \tag{7.17}$$

The cutoff frequency can therefore be much higher than the dust cyclotron frequency, because we may have $\Omega_i \delta \gg \Omega_d$, showing that dispersive effects are introduced in this frequency regime not by the motion of the dust grains, but by the soaking up of the electrons by the dust.

Another consequence of the very large dust (or negative ion) mass is that, for the frequency range $\Omega_d \ll \omega \ll \Omega_m \ll \Omega_i$, the left–hand polarized wave has a whistler–type dispersion relation (Mendis & Rosenberg 1994; Verheest & Meuris 1995; Vladimirov & Cramer 1996), obtained from Eq. (7.16);

$$\omega = \frac{k_z^2 v_A^2}{\Omega_i \delta}. \tag{7.18}$$

This feature of the "dust–whistler" is shown in the dispersion relation plotted in Figure 7.3. Comparing this whistler mode with the usual whistler mode in the right–hand polarized mode well above the ion cyclotron frequency, we note that the heavy negatively charged dust grains play the role of the ion species, and the much lighter positively charged plasma ions play the role of the electrons.

For waves propagating obliquely to the magnetic field in a dusty plasma, the treatment of waves in multi–ion plasmas in Section 3.2.3 may be used. The resonance condition defined in Eq. (3.26) becomes, for $\omega \ll \Omega_i$,

$$k_z^2 = \frac{\omega^2}{v_A^2}\left(1 + \frac{b}{1 - \omega^2/\Omega_d^2}\right). \tag{7.19}$$

There are again two resulting resonance frequencies: in addition to the Alfvén resonance frequency, we may define a *dust-ion hybrid resonance frequency*, in analogy with the ion-ion hybrid resonance frequency given by Eq. (3.30):

$$\omega_H \simeq \Omega_d (1 + b)^{1/2} \simeq (\Omega_d \Omega_m)^{1/2} \tag{7.20}$$

for small k_z. For large k_z, the hybrid resonance frequency is close to Ω_d. These features are shown by the resonance curves for the bi–ion plasma in Figures 3.7(b) and 3.8. A magneto-acoustic wave propagating into a dusty plasma may encounter a resonance point where the

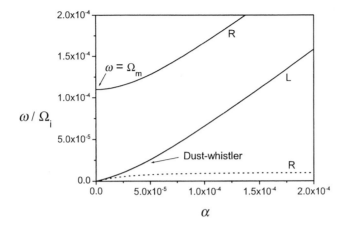

Figure 7.3: The normalized frequency $f = \omega/\Omega_i$ of the parallel propagating left–hand (L) and right–hand (R) circularly polarized waves in a plasma with negatively charged dust grains of a single mass and charge, plotted against the normalized wavenumber $\alpha = v_A k_z/\Omega_i$. The density ratio of dust to ions $b = 10$, $g = -10^{-5}$, and $\delta = 10^{-4}$. The cutoff at $\omega = \Omega_m$ is indicated. The dust mass is taken exaggeratedly high in order to show the dust cyclotron (lower R) mode.

local $k_x^2 \to \infty$. This resonance may be interpreted as an Alfvén resonance or a dust-ion hybrid resonance according to its position on the dispersion diagram (Figure 3.7(b)). The singularity in k_x will be removed by nonideal effects, as discussed in Chapter 3, and mode conversion into a short–wavelength mode may occur.

There are a number of features that distinguish charged grains from simple heavy ions. One is the fact that there is usually present a spectrum of dust grain sizes. If there is a continuous spectrum of dust grain sizes, the dielectric tensor elements u_1 and u_2 do not have the simple forms given by Eqs. (2.132) and (2.133), but involve integrals over the dust size with singularities at $\omega = \Omega_d$, where Ω_d is a function of the dust size (Bliokh, Sinitsin & Yaroshenko 1995; Tripathi & Sharma 1996; Wardle & Ng 1999; Verheest 2000). For a single dust grain size, u_1 and u_2 are real functions (provided $f \neq 1$ and $f/|g| \neq 1$), whereas for a size spectrum, the wave frequency may become equal to a dust cyclotron frequency within the integrals, resulting in an imaginary part to u_1 and u_2, and leading to dust cyclotron damping of the wave.

Another unique feature of the dust grains is in the grain charging process itself. It has been shown (Tsytovich & Havnes 1993) that this process introduces a form of collisionless damping with a characteristic dust charging frequency, in addition to Landau damping, of waves that possess a longitudinal electric field component. Such damping has been mostly discussed in connection with low–frequency acoustic waves in unmagnetized plasmas, but such an electric field is also present in the short–wavelength mode, such as the KAW or IAW mode, that is excited at the resonance point in Alfvén (or dust-ion hybrid) resonance absorption, as discussed in Section 2.8. Dust charging, in common with other dissipative processes such as resistivity, also induces the existence of a short perpendicular wavelength highly damped mode, into which a magnetoacoustic wave will mode convert as it encounters a resonance (Cramer & Vladimirov 1998). In a sense, this short–wavelength mode is the shear Alfvén wave, modified by the dust charging dissipative process.

Surface waves will exist on sharp interfaces between dusty magnetized plasmas of differing properties, for example, that between a dusty plasma and a vacuum (Cramer & Vladimirov 1996b; Cramer, Yeung & Vladimirov 1998). The dispersive effects on the surface waves of the dust grains, treated as negative ions of a single mass and charge, can be deduced from the treatment of negative secondary ions in Section 4.4 for the case of very small $|g|$. Thus for positive k_y there is a surface wave associated with the cutoff of the right–hand polarized mode, and a second, low–frequency surface wave associated with the dust cyclotron mode, while for negative k_y there is a surface wave associated with the left–hand polarized mode (see Figures 4.4-4.7). For a nonzero width surface, damping may occur due to Alfvén or dust-ion hybrid resonance damping (see Section 4.4). If there is a spectrum of dust sizes, the surface wave may be damped by dust cyclotron damping, as discussed above. Another damping effect is that, even in the case of a sharp surface and thus with no resonance damping, the short–wavelength mode induced by the grain charging (discussed above) will couple to the usual surface mode because of the boundary conditions at the surface (see Section 4.7), with a resulting modification of the dispersion relation and a global damping due to the dust charging. Provided the dust charging frequency is small compared with the wave frequency, these effects will not change the results of this section significantly.

Possible effects of charged dust on Alfvén waves in the interplanetary plasma are discussed in the following section. Further discussion of the effects of charged dust on parallel and obliquely propagating waves, and resonance absorption, can be found in Section 8.2, in connection with the plasmas that are collisionally coupled to a neutral gas in dusty interstellar clouds.

7.5   Cometary Plasmas

7.5.1   Ion Ring–Beam Instability

Comets continuously eject neutral gas, leading to the formation of the cometary atmosphere, which has a typical radius of several million kilometres. The concept of the loading of the solar wind by ionized molecules, such as water, from cometary atmospheres was introduced by Biermann, Brosowski & Schmidt (1967). The mass loading of the solar wind is accompanied by instabilities, giving rise to MHD turbulence. Such strong turbulence around comets was discovered by satellite observations, in an extended region surrounding the comets (of size greater than 10^6 km for comet Giacobini-Zinner) (Tsurutani & Smith 1986). The turbulence is apparently correlated with the presence of energetic heavy ions resulting from the ionization by the solar wind particles of cometary neutral molecules in the "pick–up" process, and the wave spectrum exhibits a peak near the water group ion cyclotron frequency, in the spacecraft (or cometary) frame (Tsurutani & Smith 1986; Thorne & Tsurutani 1987). The waves making up the observed turbulence are in the magnetoacoustic mode (Tsurutani *et al.* 1987).

The plasma surrounding the comet is made up of solar wind ions (predominantly protons), electrons and energetic heavier ions, such as water, of cometary origin. The neutral molecules from the comet all have roughly the same velocity relative to the magnetic field of the solar wind (their outflow velocity from the cometary nucleus being about 1 km s^{-1} (Galeev 1991)). The cometary pickup ions, immediately following ionization, have a gyration velocity of

$$V_{c\perp} = V_{SW} \sin \alpha \qquad\qquad\qquad (7.21)$$

about the magnetic field, and a drift velocity

$$V_{c\parallel} = V_{SW} \cos \alpha \qquad\qquad\qquad (7.22)$$

along the magnetic field in the frame of the solar wind, where V_{SW} is the solar wind velocity and α is the local angle between the solar wind velocity and the magnetic field (Winske *et al.* 1985; Thorne & Tsurutani 1987). The ions thus form a combined gyrating ring and a field–aligned beam distribution, which has been shown to excite a number of instabilities involving Alfvén and magnetoacoustic waves (Wu & Davidson 1972; Winske *et al.* 1985; Sharma & Patel 1986) .

The MHD waves predicted to be excited have the largest growth rate for parallel propagation, so only that case is considered here. There are two distinct types of instability predicted for the ring beam. The nonresonant firehose instability is a fluid instability, as discussed in Section 5.2.2, arising from an imbalance in the effective parallel and perpendicular pressures. The resonant instability is due to a resonant interaction between the cometary ion and the electromagnetic wave that is left–hand polarized in the ion frame of reference, such that the Doppler–shifted wave frequency equals the ion cyclotron frequency. The nonresonant instability requires either a large beam velocity or a high concentration of beam ions, conditions that are unlikely to occur except immediately cometward of the cometary bow shock ($\simeq 10^5$ km from comet Giacobini-Zinner) (Tsurutani *et al.* 1987). Because waves are observed over a more extended region, the resonant instability is apparently the dominant process. The resonant instability can be for left–hand or right–hand polarized waves in the solar wind frame of reference, depending on the angle α. Nongyrotropic ion distribution functions, such as partially filled rings, are commonly observed, and also give rise to instabilities (Brinca, Borda de Auga & Winske 1993). The waves excited in the instabilities react back on the velocity distribution of cometary ions, and quasilinear theory has been used to predict the diffusion in velocity space of the cometary ions interacting resonantly with the waves (Galeev 1991). A shell distribution in velocity space is formed rapidly from the initial ring distribution, within a few cometary ion gyroperiods.

It is instructive to derive here the ion ring–beam instability, as another example of the hydrodynamic microinstabilities considered in Section 5.2.2.

7.5.2   The Dispersion Equation

Consider an initial frame of reference fixed to the solar wind, with the solar wind plasma consisting of solar wind ions (protons) and electrons, with each species having zero equilibrium fluid velocity. Positive cometary ions (of the water group, with charge number Z_c and ion mass 16.8 times the proton mass) and electrons, produced by the pickup process, have a fluid

velocity along the background magnetic field and relative to the solar wind plasma given by
Eq. (7.22). The newly born cometary ions and electrons also have a gyration velocity given by
Eq. (7.21) perpendicular to the magnetic field, with a resulting ring–beam velocity distribution
function that is azimuthally symmetric about the magnetic field, and that is of the form

$$f(\boldsymbol{v}) = \frac{1}{2\pi V_{c\perp}}\delta(v_\perp - V_{c\perp})\delta(v_\| - V_{c\|}). \tag{7.23}$$

We can also allow for the presence of a population of charged dust grains in the cometary
atmosphere. Every dust grain is taken to have the same mass and (negative) charge. The
dust grains are assumed to be stationary in the comet's frame of reference, and so also have a
streaming velocity V_{SW} through the solar wind plasma. The grains are assumed to acquire a
negative charge by collisional attachment of electrons and ions from the charged particle populations produced in the pickup process and from the solar wind electron and ion populations.

All species are assumed cold for the purposes of calculating the dispersion relation, and
overall charge neutrality is expressed by

$$n_i + Z_c n_{ci} = n_e + n_{ce} + Z_d n_d \tag{7.24}$$

where n_s is the equilibrium number density of the species: s = e, i, ce, ci and d for the solar
wind electrons, solar wind protons, newborn cometary electrons, newborn cometary ions and
dust grains respectively, each with charge q_s. The charge numbers on the dust grains and the
cometary ions are given by $Z_d = |q_d|/e$ and $Z_c = q_{ci}/e$ respectively, and the dust grains could
acquire negative charge from both the solar wind and cometary electron populations. We now
proceed to neglect the effects of the dust, but reintroduce it in Section 7.5.4. We consequently
have the neutrality conditions $n_i = n_e$ and $Z_c n_{ci} = n_{ce}$.

The dispersion equation for electromagnetic waves of frequency ω and wavenumber k
along the magnetic field, in a plasma with a number of stationary cold background charged
species, and with zero–order ring–beam velocity distributions of cometary ions and electrons
given by Eq. (7.23), has been calculated by Winske *et al.* (1985). It is straightforward to gen-
eralize the cold collisionless plasma theory of Section 2.4 to allow for a velocity distribution
of the form given by Eq. (7.23). If we neglect the contribution of the charged dust, the resulting generalization of the parallel dispersion equation (2.66) may be written, in the solar wind
frame, as

$$c^2 k^2 = \omega^2 - \sum_{s=e,i}\frac{\omega_{ps}^2\omega}{\omega \pm \Omega_s} - \sum_{s=ce,ci}\frac{\omega_{ps}^2(\omega - kV_{c\|})}{\omega - kV_{c\|} \pm \Omega_s} - \sum_{s=ce,ci}\frac{\omega_{ps}^2 k^2 V_{c\perp}^2}{2(\omega - kV_{c\|} \pm \Omega_s)^2} \tag{7.25}$$

where ω_{ps} is the plasma frequency, and $\Omega_s = q_s B/m_s$ is the (signed) cyclotron frequency, of
species s. The second term on the right–hand side of Eq. (7.25) is the contribution of the solar
wind particles, the third term is the contribution of the newborn cometary species streaming
along the magnetic field through the solar wind, and the last term is the contribution of the
gyrational motion of the newborn cometary species. The upper (lower) sign refers to waves
with right(left)–hand circular polarization.

If the mass densities in the solar wind ions and the cometary ions are ρ_i and ρ_{ci} respectively, and we define $\rho_0 = \rho_i + \rho_{ci}$, we can introduce the Alfvén velocity for the combined

plasma (excluding the dust): $v_A = B_0/(\mu_0\rho_0)^{1/2}$. We also define the fractional solar wind mass density $\sigma = \rho_i/\rho_0$.

It is now convenient to change to the cometary frame of reference, in which the cometary plasma has zero velocity along the magnetic field, as used by Verheest & Meuris (1998). This is also approximately the observing spacecraft reference frame. Thus the wave frequency is transformed according to $\omega \to \omega + kV_{c\parallel}$. Then, using Eq. (7.34) and assuming the phase velocity is much less than c, we obtain from Eq. (7.25):

$$
\frac{v_A^2 k^2}{\sigma\Omega_i^2} + \frac{\omega + kV_{c\parallel}}{\omega + kV_{c\parallel} \pm \Omega_i} \mp \left(\frac{n_e}{n_i}\frac{\omega + kV_{c\parallel}}{\Omega_i} + \frac{n_{ce}}{n_i}\frac{\omega}{\Omega_i} \right)
$$
$$
+ \frac{Z_c^2 m_i n_{ci}}{m_{ci} n_i} \left(\frac{\omega}{\omega \pm \Omega_{ci}} + \frac{k^2 V_{c\perp}^2}{2(\omega \pm \Omega_{ci})^2} \right) = 0.
\tag{7.26}
$$

It is convenient to cast the dispersion equation (7.26) into dimensionless form by defining a number of dimensionless parameters measuring the relative sizes of various terms, and dimensionless variables, as done by Winske *et al.* (1985). Thus we define:

- Normalized frequency and wavenumber: $z = \omega/\Omega_i$ and $q = kv_A/\Omega_i$,

- Normalized solar wind velocities: $U = -V_{c\parallel}/v_A$ and $V = V_{c\perp}/v_A$ (the negative sign is used in the definition of U to conform to the notation of Verheest & Meuris (1998)),

- Ratio of cometary ion to proton charge density $a = Z_c n_{ci}/n_i$,

- Ratio of cometary ion to proton cyclotron frequencies: $e = \Omega_{ci}/\Omega_i = Z_c m_i/m_{ci}$, with $a = e/(1/\sigma - 1)$.

Eq. (7.26) then becomes

$$
\frac{q^2}{\sigma} - \frac{(z - qU)^2}{1 \pm (z - qU)} - a\frac{z^2}{e \pm z} + \frac{ae}{2}\frac{q^2 V^2}{(e \pm z)^2} = 0.
\tag{7.27}
$$

Considering first the dispersion relation for stable waves in the case of zero solar wind velocity, that is, with $U = V = 0$ in Eq. (7.27), we can solve for the normalized frequency z in terms of the normalized wavenumber q, and obtain a bi–ion dispersion relation of the form shown in Figure 2.7(a) (Thorne & Tsurutani 1987). The left–hand polarized L–mode, or Alfvén-cyclotron mode, experiences a resonance at the proton cyclotron frequency ($z = 1$) and at the water cyclotron frequency ($z = 0.056$), and a cutoff, from Eq. (2.140), at $z = (e + a)/(1 + a)$, that is at

$$
\omega_{co} = \Omega_b\frac{1 + \rho_{ci}/\rho_i}{1 + a} > \Omega_{ci}.
\tag{7.28}
$$

The small stop–band for L–mode propagation for $\Omega_{ci} < \omega < \omega_{co}$ has been shown to lead to significant modification of the conditions for wave-particle resonance by Thorne & Tsurutani (1987). The right–hand polarized R–mode, or magnetoacoustic mode, has no resonance.

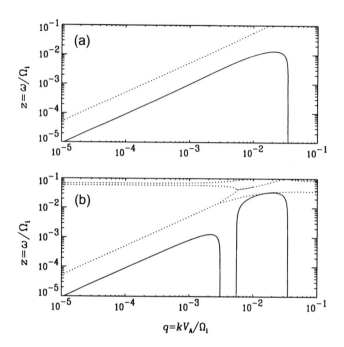

Figure 7.4: Normalized frequencies (dotted curves) and growth rates (solid curves) for right–hand circularly polarized waves in the solar wind plasma with nonzero solar wind velocity, plotted against the normalized wavenumber. $U = V = 6$. (a) right–hand polarized mode, (b) left–hand polarized mode. (Reproduced from Cramer, Verheest & Vladimirov 1999.)

7.5.3   Unstable Waves

Resonant instabilities can occur if the denominators in the second, third and fourth terms of Eq. (7.27) vanish. For positive frequency, there is no resonantly unstable mode for the R–mode, because the wave cannot resonate with the (positively charged) cometary ion cyclotron motion. However, there is a nonresonant unstable firehose mode, extending to low positive frequencies, shown in Figure 7.4(a) by the growth rate (the solid curve) for the R–mode (e.g. Winske *et al.* 1985). For the left–hand polarized mode, a resonance occurs at $z = e$, that is $\omega = \Omega_{ci}$. This is a cyclotron resonance in the cometary ions, and it gives rise to the unstable mode shown in Figure 7.4(b) by the growth rate branch on the right side of the figure. This mode has maximum growth rate at $z \simeq q \simeq e$, and has a higher maximum growth rate than the unstable mode shown in the left of the figure, which is a nonresonantly unstable firehose mode. Even though the firehose mode has a smaller growth rate than the maximum growth rate of the resonant mode, it is dominant at low wavenumbers.

A resonance in the R–mode can occur for negative frequency, corresponding to an anomalous Doppler shift to a positive frequency, right–hand polarized wave in the solar wind frame of reference. Such waves must propagate towards the Sun with a phase speed lower than the ion beam velocity, are blown back away from the Sun, and are detected with left–hand polarization in the cometary frame of reference (Thorne & Tsurutani 1987).

The low-wavenumber approximation to the nonresonant firehose dispersion relation can be obtained by assuming $z \ll e$ and $z - qU \ll 1$ in Eq. (7.27), which becomes

$$\frac{q^2}{\sigma} - (z - qU)^2 - \frac{az^2}{e} + \frac{a}{2e}q^2 V^2 = 0. \tag{7.29}$$

This can be written in the following form:

$$[z - \sigma qU]^2 = q^2 \left[1 - \sigma(1 - \sigma)U^2 + (1 - \sigma)V^2/2 \right]. \tag{7.30}$$

Eq. (7.30) is equivalent to the equations obtained by Verheest & Meuris (1998) using a fluid model where, instead of the term in V^2 introduced by the ring–beam model that drives the instability, there is a term proportional to the difference of the perpendicular and parallel gas pressures. The requirement for instability to occur can be written from Eq. (7.30) as (Verheest & Meuris 1998)

$$(1 - \sigma)(\sigma U^2 - V^2/2) > 1. \tag{7.31}$$

This criterion is satisfied for the numerical solutions shown in Figure 7.4.

The small z approximation for the firehose instability breaks down when the resonant region of q is approached, and an upper limit to q for the firehose mode appears, that is not accounted for by the fluid theory of Verheest & Meuris (1998). In other words, as the wavenumber increases and the frequency approaches the cometary ion cyclotron frequency, fluid theory breaks down, but the collisionless plasma theory shows that there is an upper limit to the wavenumber for the firehose mode. For the L–mode, this upper limit of q occurs before the onset of the resonant mode (Figure 7.4(b)). For the R–mode (Figure 7.4(a)), the firehose growth rate increases with q until $z \simeq e$, after which the size of the driving term in V^2 in Eq. (7.27) rapidly declines, as does the growth rate.

Another resonant instability occurs at high frequency, $\omega \gg \Omega_i$, due to the Doppler–shifted whistler dispersion relation of the mode (Winske *et al.* 1985). Setting $|z - qU| \gg 1$ in Eq. (7.27) in the absence of the resonant cometary terms yields the dispersion relation,

$$z' = -(z - qU) \simeq \frac{q^2}{\sigma} \tag{7.32}$$

where z' is the normalized frequency of a (right–hand polarized) whistler wave in the solar wind frame of reference. An instability of this mode now arises in the presence of the cometary ions if the denominators of the cometary ion terms in Eq. (7.27) are resonant, that is if

$$e - qU + \frac{q^2}{\sigma} = 0 \tag{7.33}$$

which yields $q \simeq \sigma U$ as the condition on the wavenumber for resonance.

It has been suggested (Sharma, Cargill & Papadopoulos 1988) that the low–frequency MHD waves observed upstream of comets, and thought to be excited by the ring–beam instability described here, undergo resonant absorption (see Chapter 3) in the structured plasma near the cometary bow shock, and thus give rise to the observed rapid heating of the solar wind protons.

7.5.4   The Effects of Dust

Dust grains may also be present in the plasma where the pickup ions are produced, provided some of this production occurs close to the comet in the direction of the Sun, or downstream of the comet in the dust tail. The dust in the cometary environment can be charged due to radiation, and due to the flow of charged particles (electrons and ions) onto the grains from the background plasma. As we have seen in Section 7.4, even if the proportion of charge on the dust grains compared to that carried by free electrons is quite small, it can have a large effect on hydromagnetic shear and compressional Alfvén waves propagating at frequencies well below the ion cyclotron frequency.

In this section we consider the effects of the presence of charged dust grains on the instabilities of electromagnetic waves due to ring–beam cometary ion velocity distributions. For simplicity we again consider only waves propagating parallel to the magnetic field. The effect of charged dust on the nonresonant firehose instabilities predicted to be excited by such cometary distributions was investigated using a fluid model with the effect of the ring–beam distribution represented by an anisotropic gas pressure (Verheest & Meuris 1998; Verheest & Shukla 1995). This approach has been generalized (Cramer, Verheest & Vladimirov 1999) to include the effects of dust on both the nonresonant and the resonant instabilities, by using the collisionless plasma theory described in the previous section, and is described here. We neglect the effect of dust charging on the stability of the waves.

Overall charge neutrality including the dust grains is expressed by Eq. (7.24). There should strictly be a term in Eq. (7.25) resulting from the streaming of the dust grains through the solar wind, but this term is negligible as a result of the following considerations. Generally, the wave frequency can be in two distinct ranges, because of the large difference in ion and dust grain masses: frequency well above the dust–grain cyclotron frequency, and frequency comparable to the dust–grain cyclotron frequency. Here we assume the former case, that is, a frequency range such that the dust grains can be considered stationary, but the electrons are completely magnetized, that is

$$|\Omega_d| \ll \omega \ll |\Omega_e|. \tag{7.34}$$

The frequency can however be comparable to the proton cyclotron frequency and the cometary ion cyclotron frequency. The wave cannot therefore be resonant with the dust grains, and so the size of the streaming dust contribution to Eq. (7.25) would be of order ω_{pd}^2, where ω_{pd} is the dust–grain plasma frequency, which is very small. We note also that if the wave frequency were comparable to Ω_d for a typical grain, it would be necessary to take into account the distribution of dust grain masses and charges. However in the frequency range of interest here, we simply need the proportion of negative charge not residing in the electrons, and thus the average dust–grain charge.

If the dust grains only acquire negative charge from the slow moving (relative to the dust grains) cometary electron population, we have $n_e = n_i$ (Verheest & Meuris 1998; Cramer, Verheest & Vladinirov 1999). In this case the dispersion equation (7.27) is modified to read

$$\frac{q^2}{\sigma} - \frac{(z - qU)^2}{1 \pm (z - qU)} - a\frac{z^2}{e \pm z} \pm \frac{ad}{1 + d}z + \frac{ae}{2}\frac{q^2 V^2}{(e \pm z)^2} = 0 \tag{7.35}$$

where d is the ratio of dust grain to cometary electron charge density, $d = Z_d n_d / n_{ce}$.

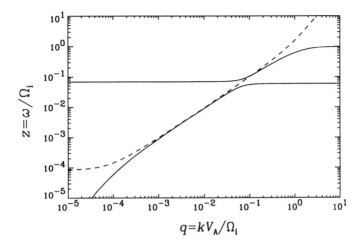

Figure 7.5: The frequency of waves in a plasma consisting of protons, electrons, positively charged cometary water group ions and negatively charged dust grains, normalized to the proton cyclotron frequency, plotted against the normalized wavenumber q, for propagation along the magnetic field. The solar wind velocity is zero. The solid curves correspond to the L–mode, and the dashed curve to the R–mode. (Reproduced from Cramer, Verheest & Vladimirov 1999.)

We consider first the dispersion relation for stable waves in the case of zero solar wind velocity, with $Z_c = 1$, $b = 10^{-2}$ and $d = 10^{-2}$. We define the ratio of the magnitude of charge on the dust grain fluid to the charge on the proton fluid,

$$\delta = \frac{Z_d n_d}{n_i} = \frac{bd}{1+d}. \tag{7.36}$$

In typical cometary atmospheres, we may have $\delta \simeq 10^{-4}$. The modification of the dispersion relation for the plasma with two positive ion species shown in Figure 2.7(a), due to the negatively charged dust, is shown in Figure 7.5 for the left–hand circularly polarized mode (solid curve) and the right–hand circularly polarized mode (dashed curve). The R–mode has no resonance on the scale of Figure 7.5 because we are assuming the dust grains are much more massive than protons and water ions, and condition (7.34) holds. However, it has a cutoff (for $\delta \ll a$) at

$$\omega_{do} \simeq \frac{\Omega_i e \delta}{e + a} \tag{7.37}$$

and a stop–band extending to very low frequencies. In the absence of the cometary ion beam, the R–mode cutoff is given by Eq. (7.17).

The consequences on the linear and nonlinear properties of waves in the dusty plasma, of the cutoff of the R–mode induced by the dust, and of the whistler–type dispersion relation of the L–mode, have been discussed in Section 7.4 and by Cramer & Vladimirov (1996a) and

Vladimirov & Cramer (1996). It is evident from Figure 7.5 that the dust will only affect the dispersion relation of modes at low wavenumber, up to $q \simeq \delta$, whereas the maximum growth rate of the resonant instabilities induced by the ring–beam distribution occurs at $q \simeq e$, that is at $\omega \simeq \Omega_{ci}$ (Winske *et al.* 1985). Nevertheless, it is important to determine the effects of the dust on the instabilities of the waves in the low wavenumber region, where the lower growth rate nonresonant firehose instabilities occur.

It has been shown using fluid theory (Verheest & Meuris 1998; Verheest & Shukla 1995) that the presence of dust affects the nonresonant firehose instabilities of both modes, that occur at low wavenumbers, where the effect of the dust is greatest. The cold plasma ring–beam theory verifies those results, and shows how the firehose instabilities relate to the resonant instabilities. Typical unstable solutions of the dispersion equation (7.35) are shown in Figure 7.6(a) (for right–hand polarized waves) and Figure 7.6(b) (for left–hand polarized waves). The value of d is chosen artificially high for illustrative purposes in these figures, but we discuss the more realistic cases later in this section. It is evident from Figures 7.4(b) and 7.6(b) that the presence of dust has no effect on the resonant left–hand polarized mode.

The low–wavenumber approximation to the nonresonant firehose dispersion relation (7.30) becomes

$$[z - \sigma(qU \pm \delta/2)]^2 = q^2 \left[1 - \sigma(1 - \sigma)U^2 + (1 - \sigma)V^2/2 \right] + \sigma^2 \delta \left[\delta/4 \pm qU \right]. \quad (7.38)$$

It was shown by Verheest & Meuris (1998), and may be deduced from Eq. (7.30), that the presence of dust has two effects on the behaviour of the low–frequency firehose instability. For both the L–mode and the R–mode, an initially unstable solution (a negative coefficient of q^2 on the right–hand side of Eq. (7.38)) can be made stable if q is sufficiently small, as shown in Figure 7.6. This lower value of q for firehose instability is different for the left–hand and right–hand polarized modes, and is found to increase as the amount of dust increases. On the other hand, if the criterion given by Eq. (7.31) is not satisfied and there is no instability in the dust–free case, that is, the coefficient of q^2 in Eq. (7.38) is positive, for the L–mode (the negative sign in Eq. (7.38)) there is a range of q in the presence of dust where instability occurs. As the amount of dust increases, the range of q for instability changes. The R–mode is always stable in this case. The presence of dust increases the upper limit of q for the firehose instability of the L–mode, until the firehose branch merges with the resonant branch (Figures 7.4(b) and 7.6(b)).

It was also shown by Verheest & Meuris (1998), using observations of comet Halley by the Giotto and VEGA spacecraft, that at a distance D of approximately 10^5 km from the comet, the lower critical wavenumber at which the firehose mode is made stable is $k \simeq 10^{-9} \times Z_d n_d$, where n_d has units m^{-3}. Using $Z_d \simeq 1$ and $n_d \simeq 10^{-3}$ at $D \simeq 10^5$ km, this corresponds to $\delta \simeq 10^{-10}$ and a wavelength of $\simeq 6 \times 10^9$ km. However, for a lower activity comet such as Grigg-Skjellerup, the bow shock where the firehose instability may be operating is only $\simeq 10^4$ km (Huddleston *et al.* 1993), where the dust density may be $\simeq 1$ m^{-3} and the average dust charge may be greater than unity, leading to a critical wavelength of $< 10^6$ km.

We note that at low wavenumber, one effect of the dust in a stationary plasma is to induce a whistler–like dispersion relation of the L–mode (Eq. 7.18) at frequencies much lower than the ion cyclotron frequency. The usual high frequency electron–whistler mode is also found to be unstable due to an ion ring–beam distribution (Winske *et al.* 1985). However, there does

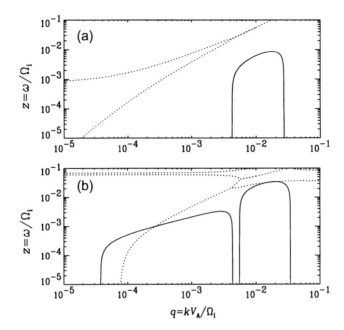

Figure 7.6: Normalized frequencies (dotted curves) and growth rates (solid curves) for right–hand circularly polarized waves in the solar wind plasma with nonzero solar wind velocity, plotted against the normalized wavenumber. $U = V = 6$. Charged dust is included, with $d = 0.1$. (a) Right–hand polarized mode, (b) left–hand polarized mode. (Reproduced from Cramer, Verheest & Vladimirov 1999.)

not appear to be a corresponding resonant instability in the dust-whistler mode, because the denominator in the last driving term of Eq. (7.35) cannot be made resonant for $q \ll e$. If there were a ring–beam distribution of dust grains, however, such an instability would arise in the whistler–like L–mode at a frequency much higher than the dust–grain cyclotron frequency.

Thus the dust has its greatest effect on the nonresonant firehose instability that is predicted by the theory of a dust–free plasma. The dust will cause an initially firehose–unstable wave to become stable at sufficiently small wavenumber, for both left and right–hand polarized waves. On the other hand, a left–hand polarized wave that is stable in the absence of dust will become firehose–unstable for a range of wavenumbers in the presence of dust. The right–hand polarized wave, if it is stable in the absence of dust, will always remain stable in the presence of dust. Even though the firehose instability is likely to occur only in a restricted region behind the cometary bow shock, the long–wavelength waves arising may play a role in accelerating ions to observed high energies, and the presence of dust may affect this process because it eliminates the long–wavelength modes. The presence of charged dust can also influence the absorption of the waves excited by the instabilities caused by the pickup ions, such as the resonant absorption process suggested by Sharma, Cargill & Papadopoulos (1988).

7.6   The Solar Corona

7.6.1   Heating of the Corona

The corona of the Sun is considerably hotter than the visible photosphere, and has a quite nonuniform temperature distribution. From X–ray images, the corona is hottest above regions where many sunspots are seen. It is therefore apparent that the magnetic fields must be playing an important role in the heating of the corona. The field above a bipolar magnetic region exists in the form of a loop, and a theory of coronal heating should also explain why magnetic loops seem to be the hottest regions in the corona. Heat cannot flow directly from the relatively cool visible solar surface to the hot corona, so the heating of the solar and stellar coronae must be a nonequilibrium process. The power needed to heat the corona is $\sim 10^7$ erg cm^{-2}s^{-1}, and so is small compared with the total power of $\sim 6 \times 10^{10}$ erg cm^{-2}s^{-1} radiated from the surface.

The first theories of coronal heating were based on the idea that acoustic waves are generated by convective turbulence on the surface, and propagate to the corona, where they undergo viscous dissipation. However, the acoustic waves refract or steepen, and dissipate too quickly in the lower corona. In addition, the energy flux of acoustic waves is insufficient to heat the corona. MHD waves were then invoked to accomplish the coronal heating (Osterbrock 1961); the waves could propagate along the field lines and carry the necessary energy. However, noting that linear Alfvén and magnetoacoustic waves in a uniform plasma would experience little resistive or collisionless damping under the hot coronal plasma conditions, a dissipation mechanism is not obvious. Regions of open field lines could be heated by noncompressive Alfvén waves, but not the closed loops, which are observed to be much hotter.

Several other dissipative mechanisms have been proposed, such as Alfvén resonant absorption, phase mixing, and anomalous resistivity caused by small–scale turbulence. Viscous damping of the waves could occur if short spatial scales arise, as may be provided by the phase mixing process. Nonlinear effects may not be important because of the low amplitude of the waves. Another possible mechanism for coronal magnetic loop heating is that as the loop footpoints move, current sheets may form as a result of the plasma trying to reach equilibrium configurations, and the magnetic energy so dissipated may heat the loop (Parker 1994). We discuss further in this section the resonant absorption and phase mixing processes, because they explicitly involve Alfvén and magnetoacoustic waves.

7.6.2   Resonant Absorption

Coronal active regions are a highly inhomogeneous environment, and are thought to consist of magnetic loop structures, with a typical height of 10^5 km. Alfvén surface waves should therefore thrive in these regions. An early model involving surface waves for the heating of the class of steady–state "coronal rain" magnetic loops radiating in the extreme ultraviolet (EUV), was that the chromospheric oscillations at a period of about 300 s would shake the footpoints of the loops, causing surface waves to propagate along the loop (Ionson 1978). These surface waves would be resonantly damped (see Section 4.5.2) in the sheath enveloping the loop. The absorption width was estimated to be about 1 km. The radius of a typical loop is 10^4 km, and the difference of Alfvén speeds inside and outside the loop should be sufficiently large to support Alfvén surface waves. The damping length of the surface wave along the loop axis direction exceeds the loop length, so that the absorption sheath should extend for the entire length of the loop. Because of the low collision frequency in the corona, the resonance absorption would proceed via mode conversion into the kinetic Alfvén wave which, having a longitudinal component of electric field and a short perpendicular wavelength, would be subsequently efficiently resistively and viscously damped. The heated sheath would induce a global convection within the loop, with hot plasma rising in a boundary layer about 100 km thick adjacent to the sheath and entering the loop's interior at the top of the loop via instabilities. Upon entering the loop's interior, the hot plasma would cool by radiating in the EUV, condense into cool clumps, and fall downward along the loop magnetic field lines.

The perpendicular structure of a KAW greatly increases viscous damping, because of the short length–scale, and so leads to irreversible heating of the ions. The rate at which energy is viscously dissipated per unit volume can be written as (Ionson 1978)

$$q_i = \rho \nu_r \left(\frac{\partial v_{\perp i}}{\partial x} \right)^2 \tag{7.39}$$

where $v_{\perp i}$ is the ion velocity perpendicular to the magnetic field and ν_r is the cross–field kinematic viscosity,

$$\nu_r = \frac{V_i^2 \nu_{ii}}{\Omega_i^2} \left(1 + \frac{\nu_{ii}^2}{\Omega_i^2} \right)^{-1} \tag{7.40}$$

where ν_{ii} is the ion-ion collision frequency. The heating rate is then

$$q_i = \lambda_i \nu_{ii} |B_\perp|^2 / 2\mu_0 \tag{7.41}$$

for $\nu_{ii} \ll \Omega_i$, with $B_\perp$ the wave magnetic field component perpendicular to the equilibrium magnetic field, and where $v_{\perp i} \simeq B_\perp / B_0$ has been used. The ion Larmor parameter λ_i is defined in Eq. (2.157).

The KAW has a field–aligned electric field $E_\parallel$, so the electrons are heated by resistive dissipation of the field–aligned electron currents. The heating rate is given by

$$\begin{aligned} q_e &= \frac{1}{2} Re(J_{\parallel e} E_\parallel) \\ &\simeq (m_e/m_i)^{1/2} \beta^{-1} q_i. \end{aligned} \tag{7.42}$$

Because $\beta \simeq (m_e/m_i)^{1/2}$ in the solar plasma, electrons and ions are heated at approximately the same rate there.

Since the early work of Ionson (1978) on coronal heating via Alfvén resonance absorption, extensive further analysis of the process has been reported. For example, the resistive MHD equations at low frequency have been numerically solved (Poedts, Goossens & Kerner 1989) to model the heating of solar coronal loops by Alfvén resonance absorption, using similar techniques to those discussed in Chapter 6 for laboratory plasmas. Thus the waves incident on the coronal loop are modelled by a periodic external driver analogous to an antenna, and the response of the coronal loop is calculated. As we have seen in Chapters 3 and 6, resistivity η causes the order of the wave field differential equations to double, i.e. an extra, resistive, mode is excited at the Alfvén resonance, which provides the mechanism for absorption. The resonant absorption is most efficient when the plasma is excited with a frequency near a normal mode frequency of the coronal loop, and the normal mode of lowest frequency is a surface wave. It is found by Poedts, Goossens & Kerner (1989) that the heating of the plasma is localized in a narrow layer of width $\propto \eta^{1/3}$, and the energy dissipation rate is almost independent of the resistivity, as is expected for the Alfvén resonance process.

7.6.3   Phase Mixing

It was pointed out in Section 3.3.1 that in a plasma with a nonuniformity transverse to the magnetic field, a shear Alfvén wave with an initially plane wavefront will develop a curved wavefront (surface of constant phase), because the wave propagates with the local Alfvén speed. In the transverse direction the phase of the wave fields may eventually vary spatially quite rapidly, and large gradients of the fields may develop. This phase mixing of the waves has been proposed as a coronal heating mechanism (Heyvaerts & Priest 1983), because the resulting small length scale variations may enable the normally ineffective collisional dissipative mechanisms such as resistivity and viscosity to become efficient means of damping the wave energy.

Consider an equilibrium magnetic field $B_0(x)\hat{z}$, with a corresponding local Alfvén speed $v_A(x)$. From Eqs. (3.59) and (3.60) with $k_y = 0$, the perturbation velocity component v_y satisfies the equation

$$\frac{\partial^2 v_y}{\partial t^2} = v_A^2(x)\frac{\partial^2 v_y}{\partial z^2} \tag{7.43}$$

with the local dispersion relation $\omega(x) = k_z v_A(x)$ for each magnetic surface oscillating with its own frequency. For the initial condition of a standing wave of fixed k_z, as time proceeds the phase of the wave on each field line changes, that is, we have phase mixing. Steep gradients are generated in the x-direction, because we may write the initial spatial derivative of v_y as

$$\frac{\partial v_y}{\partial x} = \frac{d\omega}{dx}v_y t \tag{7.44}$$

and after only a few wave periods, the gradients are steep enough for resistive and viscous dissipation to damp the wave. For propagating waves, phase mixing also occurs, but in space (see Figure 3.12), and gradients steepen in a few wavelengths.

7.7   Solar Flux Tubes

7.7.1   Modes in Flux Tubes

Magnetic flux tubes arise in two circumstances on the Sun: isolated flux tubes of the photosphere, and magnetic loops in the corona. The modes of perturbation of these solar magnetic flux tubes have been studied from two points of view. Firstly, the wave modes of tubes composed of straight axial magnetic field lines, that is, with no current along the axis of the tube, have been analysed with regard to their energy transport and dissipation properties (Edwin

& Roberts 1983). Waves excited at one end of a coronal flux tube, corresponding to the photospheric footpoint of the coronal loop, will propagate as magnetoacoustic body waves or surface waves along the tube, with their energy being damped by Alfvén resonance or cusp resonance absorption in the nonuniform plasma of the corona, and so contribute to the coronal heating, as discussed in Section 7.6.

Secondly, the equilibrium and stability properties of *twisted* magnetic flux tubes, that is, with an axial current and thus a poloidal, as well as an axial, component of magnetic field, have been studied with the aim of explaining the observed longevity of most coronal loops (Einaudi & van Hoven 1983), as well as providing a model for the sudden explosive release of energy in a solar flare (Hood & Priest 1979). The unstable modes are simply the propagating tube modes, made unstable by the presence of the axial current (Cramer & Donnelly 1985).

A flux tube may exhibit either sausage–like or kink–like modes (azimuthally symmetric or antisymmetric respectively), and outside the tube the disturbance may be either evanescent (surface wave) or propagating (leaky) (see Chapter 4). The theory of waves in cylindrical magnetic flux tubes has much in common with the theory of cylindrical laboratory plasmas discussed in Chapter 6. We first discuss the case of zero axial current in a uniform tube with a sharp boundary (Edwin & Roberts 1983). The plasma is described by the ideal MHD model with zero resistivity, so that a single type of mode (magnetoacoustic) exists (see Section 3.2.1). Surface waves in a planar geometry with the ideal MHD model have already been discussed in Section 4.5. The effects of gravity will be ignored.

A uniform cylindrical tube of radius a is assumed, with a uniform zero–order internal magnetic field $B_0\hat{z}$, and an external field $B_e\hat{z}$ for $r > a$. The zero–order particle pressures and densities inside and outside the tube are respectively p_0 and ρ_0, and p_e and ρ_e. The sound and Alfvén speeds are respectively c_0 and v_A, and c_e and v_{Ae}. For such uniform magnetic fields, the equation for balance of zero–order total pressure for a cylindrical tube takes the same form as for the planar interface case given by Eq. (3.48):

$$p_0 + B_0^2/2\mu_0 = p_e + B_e^2/2\mu_0. \tag{7.45}$$

The density ratio is given by the corresponding version of Eq. (4.33).

Following Edwin & Roberts (1983), a solution of the ideal MHD equations for the axial wave magnetic field B_{1z}, bounded on the axis of the tube, is

$$B_{1z} = R(r)\exp \mathrm{i}(kz + m\theta - \omega t) \tag{7.46}$$

where the radial solution may be written in terms of Bessel functions I_m and J_m of order m,

$$R(r < a) = A_0 \begin{cases} I_m(m_0 r) & \text{if } m_0^2 > 0, \\ J_m(n_0 r) & \text{if } n_0^2 = -m_0^2 > 0 \end{cases} \tag{7.47}$$

where A_0 is a constant. The sausage mode (cylindrically symmetric) is given by $m = 0$, while the kink modes are given by $m \neq 0$.

The radial wavenumber is given by

$$m_0^2 = \frac{(k^2 c_0^2 - \omega^2)(k^2 v_A^2 - \omega^2)}{(c_0^2 + v_A^2)(k^2 c_T^2 - \omega^2)} \tag{7.48}$$

where

$$c_T = \frac{c_0 v_A}{(c_0^2 + v_A^2)^{1/2}}. \tag{7.49}$$

These results are the cylindrical geometry analogues of the planar geometry wavenumber given by Eq. (3.2), with m_0^2 replacing $-k_x^2 - k_y^2$, and with c_T the "tube velocity", defined in Eq. (3.3).

In the region external to the flux tube, if there is no propagation of energy away from or towards the tube, the solution is in terms of the bounded K_m Bessel function,

$$R(r) = A_1 K_m(m_e r), \quad r > a \tag{7.50}$$

where A_1 is a constant and

$$m_e^2 = \frac{(k^2 c_e^2 - \omega^2)(k^2 v_{Ae}^2 - \omega^2)}{(c_e^2 + v_{Ae}^2)(k^2 c_{Te}^2 - \omega^2)} \tag{7.51}$$

with an external tube speed

$$c_{Te} = \frac{c_e v_{Ae}}{(c_e^2 + v_{Ae}^2)^{1/2}}. \tag{7.52}$$

Continuity of the radial velocity component and the total pressure across the cylinder boundary $r = a$ yields the dispersion relation discussed by Edwin & Roberts (1983). The case $m_0^2 > 0$ evidently yields a surface wave, because the field B_{1z} is concentrated at the tube boundary. It may be thought that the case $m_0^2 < 0$ always yields a body wave, because the wave fields are concentrated within the tube. However, it should be noted that the lowest–frequency mode for $m \neq 0$, even if $m_0^2 < 0$ and so with B_{1z} described by the J_m Bessel function, can be categorized as a surface wave because B_{1z} has a monotonic increase towards the surface for small k (see Eqs. (6.39) and the discussion following them). In the limit of a planar boundary ($ka \gg 1$ and m/a fixed to correspond to k_y) the wave is clearly a surface wave (see Chapter 4).

With the ordering $v_A > c_e > c_0 > v_{Ae}$, relevant to the solar photosphere, we have $c_T \simeq c_0$ and $\rho_e/\rho_0 \simeq \gamma v_A^2/2c_e^2$. In this case (and with $m = 0$ and $m = 1$) the wave modes are as shown in the dispersion relations of Figure 7.7 (Edwin & Roberts 1983):

- sausage ($m = 0$) and kink ($m = 1$) slow surface waves with speeds along the tube just above c_T but below c_0. There is also an infinite number of sausage and kink waves with speeds just above the surface waves, but below c_0.

- a fast kink surface wave at a speed

$$\simeq c_k = \left(\frac{\rho_0 v_A^2 + \rho_e v_{Ae}^2}{\rho_0 + \rho_e} \right)^{1/2}. \tag{7.53}$$

This result also follows from the planar dispersion relation (4.34) in the limit $k_y^2 \gg k_z^2$.

- a fast sausage surface wave with speed c_e at $ka < 1$ and speed $\simeq c_k$ at $ka > 1$.

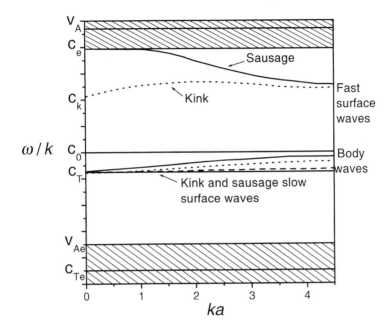

Figure 7.7: The phase speed of modes in a flux tube under solar photospheric conditions. The hatching denotes regions from which free modes (real ω and k) are excluded. (Only two of the infinitely many slow body waves are shown.) (Adapted from Edwin & Roberts 1983.)

A linear absorption mechanism for the kink waves is resonant absorption, at either the Alfvén resonance, at points in the density or magnetic field profile where the phase velocity parallel to the equilibrium magnetic field becomes equal to the local Alfvén speed, or the compressive (or cusp) resonance, where the phase velocity equals c_T, as discussed for the planar surface case in Section 4.5.2. The fast wave is found to damp only via the Alfvén resonance, while the slow wave damps via both resonances. The resonance damping vanishes for a sharp boundary, but will be present for a flux tube with diffuse profiles. The sausage mode is not resonance damped (see Section 3.4.1 for a discussion of the planar case with $k_y = 0$), essentially because the low–frequency magnetoacoustic wave cannot couple to the shear Alfvén wave if $m = 0$.

7.7.2   Twisted Flux Tubes

We now consider the dispersion relation of modes in a cylindrical flux tube with an axial current, that is, a twisted flux tube. The cases of a low–β plasma cylinder surrounded by a vacuum, or by a high–β plasma, have been discussed by Cramer & Donnelly (1983, 1984). As discussed in Section 6.5, in the limit of a small uniform axial current, the equilibrium magnetic fields can be written as, correct to order J^2 in the small constant J,

$$B_z \simeq B_{0z}(1 - J^2 r^2/a^2), \quad B_\theta \simeq J B_{0z} r/a. \tag{7.54}$$

If the plasma density is assumed uniform and terms of order J^2 in the wave equation (6.81) are neglected, solutions of the form of Eq. (7.47) in terms of Bessel functions are obtained, with $k_c^2 = -m_0^2$ of the form given by Eq. (6.89). It should be noted that this approximation leads to somewhat misleading results for the spectrum of discrete Alfvén wave (DAW) modes, with the character of body waves, as discussed in Section 6.6. However, the modes with phase speeds well separated from the Alfvén speed, such as surface waves, are well described by this model. As shown in Section 6.5, the major effect of the current is an asymmetry in the surface wave spectrum with respect to waves propagating in opposite directions along the magnetic field, provided $m \neq 0$. The dispersion relation for a plasma-vacuum interface is of the form given by Eq. (6.91), and is shown in Figure 6.9. The dispersion relation is symmetric in k about

$$k = \left(\frac{m}{2|m|} - m\right)\frac{J}{a}. \tag{7.55}$$

The low–frequency wave has a surface wave character, and the spectrum is connected to the unstable external kink mode where $\omega^2 < 0$. We may note that the surface wave has become dispersive, that is, ω/k_z is not constant, due to the plasma current.

If now a nonzero gas pressure in the plasma is allowed for, a possible exact equilibrium satisfying Eq. (2.1) is that of a flux tube with *uniform twist* (zero shear or constant safety factor q defined in Eq. (6.70)) and uniform axial field B_{0z}. The magnetic fields and gas pressure inside the tube are then

$$B_{0z} = B_0, \quad B_{0\theta} = \frac{B_0 J r}{a}, \quad p_0(r) = p_0(0) - \frac{B_0^2 J^2 r^2}{\mu_0 a^2} \tag{7.56}$$

with J a constant. The plasma density is assumed uniform. Outside the tube, the magnetic field is in the z–direction. An exact solution of the wave equations is possible with this equilibrium, with arbitrary size of J, if the incompressible limit for the plasma, $\gamma \to \infty$, is taken (Bennett, Roberts & Narain 1999). The solution is again of the form given by Eq. (7.47) in terms of Bessel functions, with

$$m_0^2 = k^2\left(1 - \frac{4J^2 v_A^4 k_\parallel^2}{a^2(\omega^2 - v_A^2 k_\parallel^2)^2}\right) \tag{7.57}$$

where

$$k_\parallel = k + mJ/a \tag{7.58}$$

and v_A is defined in terms of B_0.

The square of the radial wavenumber given by Eq. (7.57), and shown in Figure 7.8 plotted against the normalized phase velocity $\omega/v_A k$, for $m = 0$ and two values of the current parameter J, may be compared with the cold current–carrying plasma result (Eq. 6.89 and Figure

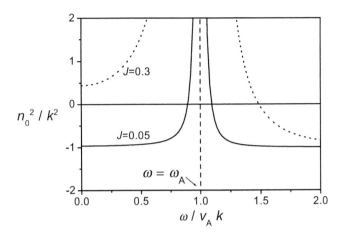

Figure 7.8: Square of the perpendicular wavenumber, $n_0^2 = -m_0^2$, divided by k^2, as a function of the normalized phase velocity $\omega/v_A k_z$, for a cylindrical incompressible plasma with an axial current. Here $m = 0$, $ka = 0.5$, and $J = 0.05$ (solid curves) and $J = 0.3$ (dotted curves). The Alfvén resonance $\omega = \omega_A$ is indicated by the dashed line.

6.8). Again a resonance occurs at $\omega^2 = v_A^2 k_{\parallel}^2$. The lowest–frequency mode is a surface wave ($m_0^2 > 0$); in contrast to the cold plasma case, when a surface wave occurs only for $m \neq 0$, in the incompressible case a surface wave occurs for $m = 0$ as well (see the discussion of the incompressible case for a planar surface in Section 4.5.1). It is interesting to note from Figure 7.8 that the dispersion relation is almost symmetric about the Alfvén resonance. This implies that a dense infinite spectrum of DAW body waves will occur at frequencies both below and above the Alfvén resonance frequency. This is in contrast to the cold plasma case, where the DAW spectrum occurs below the Alfvén resonance or the lower edge of the Alfvén continuum, and merges with the Alfvén continuum for a diffuse profile (see Section 6.6).

Another type of diffuse current–carrying flux tube may be modelled by a variant of the magnetic fields used in the cylindrical model of a Tokamak plasma, given by Eq. (6.71), but which allows for a nonzero plasma pressure:

$$B_\theta = \frac{B_m r/a}{1 + r^2/a^2}, \quad B_z = \frac{B_{0z}}{1 + r^2/a^2}, \quad p = p(0) + \frac{B_m^2 - B_{z0}^2}{2\mu_0(1 + r^2/a^2)^2}. \tag{7.59}$$

In this case the magnetic field vanishes as r tends to ∞, in contrast to that given by Eq. (7.56), so this corresponds to a magnetic flux tube with magnetic field balanced by the gas pressure of the surrounding field–free region. Modes on a diffuse flux tube with current are related to the modes of the flux tube with a sharp boundary, but can experience Alfvén or compressive resonance damping at points in the profile.

Finally, some analysis of nonlinear waves on cylindrical flux tubes has been done, as mentioned in Section 5.7. It was pointed out there that cylindrical geometry can introduce dispersion to the dispersion relation of linear waves in a flux tube, as can the effects of a current in the tube as pointed out above. This can possibly lead to solitary nonlinear wave structures with the nonlinearity balanced by the dispersion. For example, a weakly nonlinear sausage mode may propagate on a flux tube, governed by an equation related to the Benjamin-Ono equation (Edwin & Roberts 1983). Large twists and compressions at the footpoints of solar flux tubes can launch nonlinear Alfvén and magnetoacoustic waves, both surface modes and body waves, as shown by Sakai *et al.* (2000a).

7.7.3   The Interaction of Acoustic Waves with Flux Tubes

Acoustic oscillations have been observed in the regions surrounding sunspots (Braun, Duvall & LaBonte 1988). It was found that up to 70% of the inward travelling acoustic wave could be absorbed by the sunspot. A proposed mechanism for this absorption of wave energy is the Alfvén or compressive resonance absorption process (Hollweg 1988; Lou 1990; Sakurai, Goossens & Hollweg 1991b; Goossens & Poedts 1992; Goossens & Hollweg 1993; Stenuit, Poedts & Goossens 1993). A plane acoustic wave launched from outside the flux tube may be decomposed into a sum of cylindrical waves of different azimuthal mode (m) numbers, each of which will be absorbed, with a different damping rate, at the resonance (Sakurai, Goossens & Hollweg 1991b).

Any energy that remains after resonance absorption can produce a linear mode coupling of the acoustic wave with the natural modes of the flux tube. If the incident acoustic pulse has a scale–length in the z–direction of approximately a, then waves will be launched with $ka \simeq 1$ in both directions. The high–frequency $m = 0$ component of the acoustic pulse will couple to the $m = 0$ (sausage) fast surface wave, and the $m = \pm 1$ component will couple to the $m = \pm 1$ (kink) fast surface wave, with fields concentrated at the boundary. The excitation of nonlinear Alfvén and magnetoacoustic waves by an acoustic shock wave colliding with a flux tube has been investigated by Sakai *et al.* (2000b).

8 Astrophysical Plasmas

8.1 Introduction

Alfvén waves are believed to play major roles in some astrophysical processes in regions where magnetized plasmas are present, such as the environs of stars and interstellar clouds. We have already mentioned in Chapter 5 the crucial role of Alfvén waves in the shock acceleration of cosmic rays to high energies. Alfvén waves may even play a part in cosmology. Magnetic field fluctuations in the form of Alfvén waves in the primordial universe may influence the present cosmic microwave background anisotropy (Durrer, Kashnioshvili & Yates 1998). In this chapter we treat two problems in astrophysical plasmas where Alfvén waves arise, albeit under very different conditions: waves in partially ionized and dusty interstellar plasmas, and waves in the relativistic and very strongly magnetized pair plasmas of pulsar magnetospheres.

8.2 Interstellar Clouds

Magnetized partially ionized and dusty plasmas are present in stellar winds, interstellar molecular clouds and star–formation regions, and Alfvén and magnetoacoustic waves should play important parts in the transport and deposition of magnetic energy, and the transport of angular momentum in such regions. Hydromagnetic motions in interstellar clouds and star–formation regions are an important source of support against gravitational collapse to form protostars. It is important to know the speeds and damping rates of the hydromagnetic waves in interstellar clouds because the waves transport angular momentum during the cloud collapse. The ionization fraction of molecular clouds is typically only about 10^{-7}, and dust grains may contribute approximately 1% of the mass of the cloud.

The dynamics of the collapse of dense cloud cores of star–formation regions may be profoundly affected by the collisional processes occurring in them (Wardle & Ng 1999). A modification of the Alfvén wave propagation is an indication of a change in the dynamical behaviour of the gas. Alfvén waves can provide support along field lines against gravitational collapse, if they suffer minimal decay or loss into the surrounding medium, and cloud support requires the energy in Alfvén–like disturbances to remain comparable to the cloud's gravitational binding energy.

Magnetic fields can transport angular momentum in star–formation regions (e.g., Choudhuri (1998)). If there are magnetic field lines coming out of a rotating astrophysical object, it is possible to remove angular momentum from the object with the aid of magnetic stresses. Ferraro's law of isorotation for flux freezing states that a steady state is possible only if the angular velocity Ω_r is constant along the magnetic field lines, that is, the field lines rotate like rigid bodies. Magnetic braking can occur during star formation: any collapsing gas cloud in the Galaxy has some angular momentum arising from the galactic rotation, and is threaded by the general galactic magnetic field. If the angular momentum per unit mass, $\Omega_r r^2$, is conserved during the collapse, then Ω_r of the cloud increases. The field lines connecting the collapsing cloud with the surrounding plasma oppose this variation of Ω_r, and provide a braking effect to slow down the rapidly spinning collapsed cloud (Mestel 1968). The magnetic stresses, travelling at the Alfvén or magnetoacoustic speed, try to spin up the surrounding plasma to Ω_r.

In magnetized stellar winds, the magnetic field is in the form of spirals ("Parker spirals"). The distance up to which the magnetic energy remains larger than the kinetic energy of the flow is the Alfvén radius r_A (approximately ten times the stellar radius). The plasma rotates approximately like a solid body out to r_A, that is, the magnetic field is strong enough to force the plasma to rotate at the angular velocity $|\Omega_r|$ out to r_A, where the angular momentum per unit mass is $\Omega_r r_A^2$ (Weber & Davis 1967). There the angular momentum per unit mass is approximately 100 times that at the star's surface, and the wind is much more efficient at removing angular momentum than mass. Beyond r_A, the plasma rotates with an angular speed less than Ω_r, and the field lines become twisted. The associated stresses can eventually force the outflow to be collimated around the polar axes. In accretion disks around massive objects like black holes, the fields may cause a small amount of gas to be thrown from the disk into jets (Blandford & Payne 1982).

In a collapsing cloud the centrifugal force increases more rapidly than gravity. Hence, unless some angular momentum is removed, eventually the collapse would come to a halt. The field may slow down the rotation. However the field would become very strong if the flux is frozen in to the plasma, and the magnetic pressure also may halt the collapse. The magnetic flux needs to be removed. The outflow of ionized fluid through the neutral atoms is often called *ambipolar diffusion*. In this process, the magnetic pressure force acts on the ionized fluid alone. The ionized fluid carries the magnetic flux with it while spreading out of the central region, and so leads to a decrease of field in the star–forming region (Mestel & Spitzer 1956). The ambipolar diffusion also leads to damping of hydromagnetic waves, as we shall see in this section.

8.2.1 Alfvén and Magnetoacoustic Waves in Interstellar Clouds

We now consider the effect of neutral atoms and molecules, and charged dust, on the propagation of low–frequency hydromagnetic waves in magnetized interstellar clouds. There is some possible direct evidence of these hydromagnetic waves in such clouds, because they have been postulated to be the cause of the observed large widths of CO emission lines from molecular clouds (Arons & Max 1975), which would imply supersonic motions if the waves were purely acoustic, but allows for subsonic motions if the waves are hydromagnetic.

Interaction of the charged species with the neutral species in the interstellar cloud occurs via collisions. The damping of hydromagnetic waves in a partially ionized but dust–free self–gravitating plasma has been investigated by Balsara (1996). Dust can also be a major component of interstellar molecular clouds, and so it is of interest to know the effect of dust in the clouds on the speeds and damping rates of the waves. The effects of charged and neutral dust on small–amplitude magnetohydrodynamic waves in the clouds have been investigated for propagation parallel to the magnetic field (Pilipp *et al.* 1987; Wardle & Ng 1999) and for oblique propagation (Cramer & Vladimirov 1997). For interstellar molecular clouds the charge on the grains is negative, because electron attachment is the dominant process. An appreciable fraction of the negative charge in an overall neutral cloud may reside on the dust grains. For HII regions there can be several hundred electrons per grain, while for HI regions there are only a few per grain (Spitzer 1978). The theory of small–amplitude waves was applied to the problem of wave propagation in the dustless and dusty regions of stellar outflows such as for the star α Ori (Havnes, Hartquist & Pilipp 1989). It was argued that waves that propagate with no losses out to the zone of dust formation are dissipated rapidly in the dusty region. The effects of dust on the propagation of shock waves in interstellar clouds, including the rotation of the magnetic field in an obliquely propagating shock, have been considered by Pilipp & Hartquist (1994) and Wardle (1998).

As was pointed out in Section 7.4, even if the proportion of negative charge on the dust grains compared to that carried by free electrons is quite small (typically $\simeq 10^{-4}$ in interstellar clouds), it can have a large effect on hydromagnetic Alfvén and magnetoacoustic waves propagating at frequencies well below the ion cyclotron frequency. In particular, the right–hand circularly polarized mode experiences a cutoff due to the presence of the dust.

Here we consider waves propagating at an arbitrary angle with respect to the magnetic field. The wave frequencies are supposed to be much less than the ion cyclotron frequency. Waves that propagate obliquely to the magnetic field in a nonuniform plasma can encounter the Alfvén resonance, and the resonance absorption of waves in interstellar molecular clouds could play a role in their energy balance and in the magnetic braking of protostellar clouds.

8.2.2 Wave Equations with Collisions

A molecular cloud consists of neutral atomic and molecular species, ionized atomic and molecular species, electrons, neutral dust grains and negatively charged dust grains (Pilipp *et al.* 1987). The dust grains will in general have a range of sizes, but we assume dust grains of a single size, that is, of a single mass and size of charge (see Section 7.4). The cloud is assumed relatively cold here, so that the gas pressures of all the species may be neglected. As in the discussion of collisional effects in Section 2.5, fluid equations for the different species are used. For simplicity we neglect the motion of neutral dust grains that was included by Pilipp *et al.* (1987). The approach of Cramer & Vladimirov (1997) is closely followed here. Thus a four–fluid model of the plasma is used, that employs the linearized fluid momentum equations (1.5), written explicitly for the plasma ions (i) (singly charged), neutral molecules (n), charged dust grains (d) and electrons (e):

$$\frac{\partial v_i}{\partial t} = \frac{e}{m_i}(E + v_i \times B_0) - \nu_{in}(v_i - v_n) - \nu_{id}(v_i - v_d) \tag{8.1}$$

$$\frac{\partial v_n}{\partial t} = -\nu_{ni}(v_n - v_i) - \nu_{nd}(v_n - v_d) \tag{8.2}$$

$$\frac{\partial v_d}{\partial t} = \frac{-Z_d e}{m_d}(E + v_d \times B_0) - \nu_{dn}(v_d - v_n) - \nu_{di}(v_d - v_i) \tag{8.3}$$

$$0 = \frac{-e}{m_e}\left(\boldsymbol{E} + \boldsymbol{v}_e \times \boldsymbol{B}_0\right) - \nu_{en}\left(\boldsymbol{v}_e - \boldsymbol{v}_n\right) - \nu_{ed}\left(\boldsymbol{v}_e - \boldsymbol{v}_d\right) \tag{8.4}$$

where m_s is the species mass and $\boldsymbol{v}_s$ is the species velocity in the wave. The charged particles are coupled to the neutral particles via collisions, and the collision terms contain ν_{st}, the collision frequency of a particle of species s with the particles of species t.

We have neglected electron inertia, as well as momentum exchange between ions and electrons (because of the low degree of ionization in the cool clouds), but have included ion and neutral molecule inertia terms because we are mainly interested in the frequency regime above the dust cyclotron frequency, but well below the ion cyclotron frequency, where the ion and neutral molecule dynamics are important. To complete the system of equations, Maxwell's equations (1.1)-(1.4) ignoring the displacement current are used, with the conduction current density given by

$$\boldsymbol{J} = e\left(n_{i0}\boldsymbol{v}_i - n_{e0}\boldsymbol{v}_e - n_{d0}Z_d\boldsymbol{v}_d\right). \tag{8.5}$$

The background magnetic field $\boldsymbol{B}_0$ is taken to be in the z-direction, and the steady electron, ion and dust densities are n_{e0}, n_{i0} and n_{d0}. As in Section 7.4, the parameter $\delta = Z_d n_{d0}/n_{i0}$ measures the proportion of negative charge residing on the dust grains, with the remainder in the electrons, so that the total system is charge neutral according to Eq. (7.14). We assume for simplicity that the charge on the dust particles is not affected by the wave, that is, we neglect the dust charging effects discussed by, for example, Vladimirov (1994a,b).

We can define the transverse direction of wave field variation to be the x-axis, so without loss of generality the wave fields are assumed to vary as $\exp(ik_x x + ik_z z - i\omega t)$. Thus oblique propagation (i.e. nonzero k_x) is allowed for.

We employ the Alfvén speed based on the plasma ion density, $v_A = \sqrt{B_0^2/\mu_0\rho_i}$, where $\rho_i = m_i n_{i0}$, the ion cyclotron frequency $\Omega_i = B_0 e/m_i$ and the dust cyclotron frequency $\Omega_d = B_0 e/m_d$. Eliminating $\boldsymbol{E}$ and $\boldsymbol{v}_e$, we obtain, for $\omega \ll \Omega_i$,

$$-i\omega\boldsymbol{v}_i = -\Omega_{m0}\left(\boldsymbol{v}_i - \boldsymbol{v}_d\right) \times \hat{\boldsymbol{b}} + \frac{v_A^2}{(1-\delta)}(\nabla \times \boldsymbol{b}) \times \hat{\boldsymbol{b}} \tag{8.6}$$
$$-\nu_{in}\left(\boldsymbol{v}_i - \boldsymbol{v}_n\right) - \nu_{id}\left(\boldsymbol{v}_i - \boldsymbol{v}_d\right)$$

$$-i\omega\boldsymbol{v}_n = \nu_{ni}\left(\boldsymbol{v}_i - \boldsymbol{v}_n\right) + \nu_{nd}\left(\boldsymbol{v}_d - \boldsymbol{v}_n\right) \tag{8.7}$$

$$-i\omega\boldsymbol{v}_d = \frac{\Omega_d}{(1-\delta)}\left(\boldsymbol{v}_i - \boldsymbol{v}_d\right) \times \hat{\boldsymbol{b}} + \frac{\Omega_d}{\Omega_i}\frac{v_A^2}{(1-\delta)}(\nabla \times \boldsymbol{b}) \times \hat{\boldsymbol{b}} \tag{8.8}$$
$$-\nu_{dn}\left(\boldsymbol{v}_d - \boldsymbol{v}_n\right) + \nu_{di}\left(\boldsymbol{v}_i - \boldsymbol{v}_d\right)$$

$$-i\omega\boldsymbol{b} = \nabla \times \left[\frac{1}{(1-\delta)}\left(\boldsymbol{v}_i - \delta\boldsymbol{v}_d\right) \times \hat{\boldsymbol{b}}\right] \tag{8.9}$$

where b is the wave magnetic field, $\hat{\boldsymbol{b}} = \boldsymbol{B}_0/B_0$ and $\Omega_{m0} = \Omega_i\delta/(1-\delta)$ is related to the dust cutoff frequency Ω_m introduced in Eq. (7.17) by $\Omega_m = \Omega_d + \Omega_{m0}$. The neglect of gas pressure in all species implies that $v_{iz} = v_{dz} = v_{nz} = 0$.

8.2.3   The Dispersion Relation

We now write the fields in terms of mode amplitudes $b_\pm$:

$$b_\pm = b_x \pm i b_y, \qquad v_{i\pm} = v_{ix} \pm i v_{iy}, \quad \text{etc.} \tag{8.10}$$

These are the normal mode amplitudes for parallel propagation, where the $+(-)$ sign corresponds to the left (right) hand circularly polarized wave.

Using Eq. (8.10) in Eqs. (8.6)-(8.9) yields, after some algebra,

$$b_\pm = -\frac{k_z}{(1-\delta)\omega} C_\pm v_{i\pm} \tag{8.11}$$

and

$$(D_\pm - (k_z^2 + \frac{k_x^2}{2})) b_\pm = \frac{k_x^2}{2} b_\mp. \tag{8.12}$$

Here

$$C_\pm = \frac{\omega \pm (1-\delta)\Omega_d + i(\nu_{dn} + (1-\delta)\nu_{di}) + \nu'_{dd}}{\omega \pm \Omega_d + i(\nu_{dn} + \nu_{di}) + \nu'_{dd}} \tag{8.13}$$

$$D_\pm = \frac{(1-\delta)^2\omega^2}{v_A^2} \frac{\left[\varepsilon_1(\omega \pm \Omega_d + i(\nu_{dn} + \nu_{di}) + \nu'_{dd}) + \varepsilon_2(\pm\Omega_{m0} + i\nu_{id} - \nu'_{id})\right]}{(\omega \pm \Omega_d + i(\nu_{dn} + \nu_{di}) + \nu'_{dd})} \tag{8.14}$$

with

$$\omega_0 = \omega + i(\nu_{ni} + \nu_{nd}) \tag{8.15}$$

and

$$\varepsilon_1 = 1 + i\nu_{in}/\omega_0, \qquad \varepsilon_2 = 1 + i\nu_{dn}/\omega_0$$
$$\nu'_{dd} = \nu_{dn}\nu_{nd}/\omega_0, \qquad \nu'_{id} = \nu_{in}\nu_{nd}/\omega_0. \tag{8.16}$$

In our case of δ very small, $C_\pm \simeq 1$.

Equation (8.12) shows that for oblique propagation ($k_x \neq 0$), the amplitudes of opposite circular polarization are coupled together, that is, the modes are not purely circularly polarized. We then obtain from Eq. (8.12):

$$k_x^2 = \frac{(k_z^2 - D_+)(k_z^2 - D_-)}{(D_+ + D_-)/2 - k_z^2}. \tag{8.17}$$

This is the generalization of the result given by Eq. (3.25) to include collisional terms as well as charged dust. Note that there is still a single value of k_x^2, that is, a single magnetoacoustic mode; if electron-ion or electron-dust collisions were included, there would be an additional short–wavelength mode (see Section 3.2.4). The cutoffs of k_x (where $k_x = 0$) correspond to

$$k_z^2 - D_\pm = 0 \tag{8.18}$$

which describes the case of parallel propagation in interstellar clouds treated by Pilipp *et al.* (1987). For no collisions but with charged dust present, the parallel propagation dispersion relation is given by Eq. (7.16).

If k_z is fixed and there are no collisions, there is a resonance condition at which $k_x \to \infty$, given from Eq. (8.17) by the condition

$$k_z^2 = (D_+ + D_-)/2 \qquad (8.19)$$

which is equivalent to Eq. (7.19), and which defines the Alfvén resonance frequency and the dust-ion hybrid resonance frequency, as discussed in Section 7.4.

8.2.4 Conductivity Tensor

An instructive alternative approach to establishing the collisional wave equations and the dispersion relation is to employ a conductivity tensor that contains all the collisional effects, as in the discussion of Hall effects in a cold collisional plasma in Section 2.5.2. The fluid making up the interstellar cloud is weakly ionized, so the overall fluid velocity v is essentially the neutral gas velocity. The electric field in the frame of reference comoving with the fluid is

$$\boldsymbol{E}' = \boldsymbol{E} + \boldsymbol{v} \times \boldsymbol{B}_0 = \boldsymbol{\sigma}^{-1} \cdot \boldsymbol{J} \qquad (8.20)$$

where, in that frame, the conductivity tensor is defined by

$$\boldsymbol{J} = \boldsymbol{\sigma} \cdot \boldsymbol{E}'. \qquad (8.21)$$

This can be written as

$$\boldsymbol{J} = \sigma_{||} \boldsymbol{E}'_{||} + \sigma_{\mathrm{P}} \boldsymbol{E}'_{\perp} + \sigma_{\mathrm{H}} \boldsymbol{B}_0 \times \boldsymbol{E}'_{\perp}/B_0 \qquad (8.22)$$

thus defining the parallel, Pedersen and Hall conductivity components as in Eq. (2.115).

The conductivity components may be calculated from the momentum equations (8.1)-(8.4) written in the neutral gas frame. The collisions between each type of charged species and the neutral gas are included, with collision frequencies ν_{jn}, but to avoid unwieldy expressions, the collisions between all the charged species are neglected. The conductivity components are then the following generalizations of the results given by Eqs. (2.116)-(2.118):

$$\sigma_{\mathrm{P}} = \frac{e}{B_0} \sum_j n_{j0} Z_j \frac{\Omega_j (\nu_{jn} - i\omega)}{\Omega_j^2 + (\nu_{jn} - i\omega)^2} \qquad (8.23)$$

$$\sigma_{\mathrm{H}} = \frac{e}{B_0} \sum_j n_{j0} Z_j \frac{(\nu_{jn} - i\omega)^2}{\Omega_j^2 + (\nu_{jn} - i\omega)^2} \qquad (8.24)$$

$$\sigma_{||} = e^2 \sum_j \frac{n_{j0} Z_j^2}{m_j (\nu_{jn} - i\omega)} \qquad (8.25)$$

where the sums are over all the charged species (including the electrons), and Z_j is the (signed) charge number for the species j. For a single dust grain size, as in Sections 8.2.2 and 8.2.3, we have $j = i$, e and d. For a number of different grain sizes, there would be corresponding terms in the sums, and for a continuous dust size spectrum, these terms can be replaced by integrals over the grain size (see Bliokh, Sinitson & Yaroshenko (1995) and Tripathi & Sharma (1996) for the collisionless case, when cyclotron resonances can occur in the integrals).

If the wave frequency is well below all the collision frequencies and cyclotron frequencies, $\omega \ll \nu_{jn}, |\Omega_j|$ for all j, (the case considered by Wardle & Ng (1999)), the conductivity components become independent of ω, and can be written

$$\sigma_P = \frac{e}{B_0} \sum_j \frac{n_{j0} Z_j \beta_j}{1 + \beta_j^2} \tag{8.26}$$

$$\sigma_H = \frac{e}{B_0} \sum_j \frac{n_{j0} Z_j}{1 + \beta_j^2} \tag{8.27}$$

$$\sigma_\| = \frac{e}{B_0} \sum_j n_{j0} Z_j \beta_j \tag{8.28}$$

where β_j is the Hall parameter for the species j:

$$\beta_j = \Omega_j / \nu_{jn}. \tag{8.29}$$

Again, for a continuous dust grain size distribution, the dust contributions to the conductivities can be written as integrals as the grain size. Because the wave frequency is assumed much less than any cyclotron frequency, these integrals do not involve any cyclotron resonance contributions.

The conductivities given by Eqs. (8.26)-(8.28) have been calculated by Wardle & Ng (1999) for typical dense interstellar cloud conditions, with three grain models: single–size grains, an MRN (Mathis, Rumpl & Nordsiek 1977) power law size distribution with index -3.5, and an MRN model with an additional component of very small grains representing polycyclic aromatic hydrocarbons (PAH). It is found that the Hall contributions of the grains using the MRN model, with or without the PAH, are important for gas densities between 10^7 and 10^{11} cm^{-3}.

8.2.5   Damping and Dispersion of Alfvén Waves

Summing the momentum equations (8.1)-(8.4) gives the MHD equation (1.12) for the overall (cold) fluid:

$$\rho \frac{\partial v}{\partial t} = J \times B. \tag{8.30}$$

This equation, together with Eqs. (1.3), (1.4) and (8.20), leads to the following dispersion relation for parallel propagation (Wardle & Ng 1999):

$$\omega^2 \pm v_A^2 k^2 \exp(\pm i\theta)\omega/\omega_c - v_A^2 k^2 = 0 \tag{8.31}$$

where

$$\omega_c = \frac{B_0^2 \sigma_\perp}{\rho} \tag{8.32}$$

and

$$\theta = \cos^{-1}(|\sigma_{\mathrm{H}}|/\sigma_{\perp}) \tag{8.33}$$

with the total conductivity perpendicular to the magnetic field defined as

$$\sigma_{\perp} = \sqrt{\sigma_{\mathrm{P}}^2 + \sigma_{\mathrm{H}}^2}. \tag{8.34}$$

The perpendicular conductivity is related to the Cowling conductivity defined in Eq. (2.119) by

$$\sigma_{\perp}^2 = \sigma_{\mathrm{C}}\sigma_{\mathrm{P}}. \tag{8.35}$$

In the case where the wave frequency is much less than the collision frequencies and the cyclotron frequencies, and the conductivities are given by Eqs. (8.26)-(8.28), the solution of the dispersion relation (8.31) is relatively simple; the two roots are (Wardle & Ng 1999)

$$\omega_1 = \left(\sqrt{1+b^2} + b\right) v_{\mathrm{A}} k \quad \text{and} \quad \omega_2 = \frac{v_{\mathrm{A}}^2 k^2}{\omega_1^*} \tag{8.36}$$

where ω_1^* is the complex conjugate of ω_1, and b is given by

$$b = \frac{v_{\mathrm{A}} k}{2\omega_{\mathrm{c}}} \exp(\pm i\theta). \tag{8.37}$$

This case also follows from Eq. (8.18) with Eq. (8.14), in the limit $\omega \ll \nu_{\mathrm{in}}, \nu_{\mathrm{dn}}$. A measure of the frequency of the wave at which the conductivity becomes important is provided by the frequency ω_{c} defined in Eq. (8.32); for $\omega \ll \omega_{\mathrm{c}}$, the LHP and RHP modes both have the usual Alfvén wave dispersion relation and are lightly damped, and a linearly polarized shear Alfvén wave can be formed as a combination of the two, while for $\omega \simeq \omega_{\mathrm{c}}$ the speeds of the two circularly polarized modes differ, and the waves are damped.

Following Wardle & Ng (1999), one can, in the case $\omega \ll \nu_{\mathrm{jn}}$, distinguish three conductivity regimes according to the magnitude of the typical Hall parameter β_{j}:

(i) $|\beta_{\mathrm{j}}| \gg 1$ for all the species, which gives $\sigma_{\parallel} \gg \sigma_{\mathrm{P}} \gg |\sigma_{\mathrm{H}}|$. In this case the charged particles make many cyclotron orbits between collisions, and so are effectively tied to the magnetic field. The ionized plasma drifts as a whole through the background neutrals, and the magnetic flux is frozen into the ionized component; this is called *ambipolar diffusion*.

(ii) $|\beta_{\mathrm{j}}| \ll 1$ for all the species, including the electrons, which leads to $\sigma_{\parallel} \simeq \sigma_{\mathrm{P}} \gg |\sigma_{\mathrm{H}}|$. There are many collisions in a cyclotron period, so the magnetic field has little effect and the conductivity is almost scalar.

(iii) $|\beta_{\mathrm{j}}| \simeq 1$ for any species, which defines the Hall regime.

The ambipolar diffusion approximation is generally valid for ions and electrons in interstellar clouds, however charged dust grains may have a value of β_{d} in the Hall regime. The effect of changing β_{d} on the dispersion relations given by Eqs. (8.36) is shown in Figure 8.1.

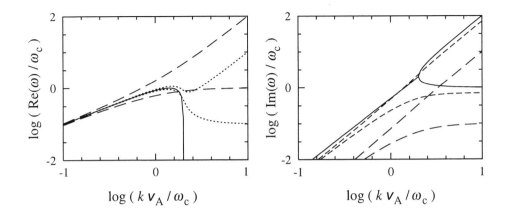

Figure 8.1: The dispersion relation for waves propagating parallel to the magnetic field in a weakly ionized, dusty interstellar plasma. The real and imaginary parts of the frequency (left and right panels respectively) are plotted against (real) wave number. The normalizing frequency ω_c is defined in Eq. (8.32). The curves are distinguished by the ratio of the Hall and Pedersen conductivities, $|\sigma_H|/\sigma_P$: *solid* - 0; *dotted* - 0.1; *short-dashed* - 1; *long-dashed* - 10. For clarity, the curves for the real part of ω for $|\sigma_H|/\sigma_P = 1$, and for the imaginary part for $|\sigma_H|/\sigma_P = 0.1$ have been omitted as they are similar to those for $|\sigma_H|/\sigma_P = 10$ and 0 respectively. (Reproduced from Wardle & Ng 1999.)

Noting that the relative sizes of the Hall and Pedersen conductivities sets the angle θ, we can see that, for $\omega \gtrsim \omega_c$, the phase speed and damping rate of the two modes depends strongly on the choice of $|\sigma_H|/\sigma_P$. In both the ambipolar regime and the scalar regime, $|\sigma_H|/\sigma_P \ll 1$, and the two modes are almost purely evanescent. In the Hall regime, $|\sigma_H|/\sigma_P \simeq 1$, and the two modes are weakly damped and have distinct phase speeds. The reason for the decrease in damping rate is that the Hall current is perpendicular to $\boldsymbol{E}'$ and so does not contribute to the energy dissipation in the plasma. In addition, the Hall conductivity introduces a handedness and asymmetry to the phase speeds of the LHP and RHP modes.

It is also instructive to use Eq. (8.18) to reproduce the results of Pilipp *et al.* (1987) for parallel propagation in an interstellar cloud of lower density, where the inertia of the charged particles is not neglected, that is, the wave frequency is not necessarily much less than the collision frequencies. As was done by Pilipp *et al.* (1987), a molecular interstellar cloud is considered, in which the density of molecular hydrogen $n_{n0} = 10^4$ cm^{-3}, $n_{i0}/n_{n0} = 10^{-7}$, the magnetic field is 10^{-4} G and the dominant ion species is HCO$^+$. It is assumed that 1% of the mass is contained in spherical dust grains of radius 10^{-5} cm composed of material with a mass density of 1 g cm^{-3}. We then have $\delta \simeq 10^{-4}$. The temperature of the cloud is assumed to be 20 K, so that the thermal pressure is approximately 10% of the magnetic pressure, and we are justified in neglecting the pressure gradient terms in Eqs. (8.1)-(8.4).

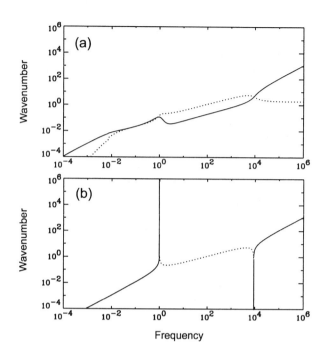

Figure 8.2: The wavenumber k plotted against frequency ω for propagation parallel to the magnetic field in an interstellar cloud, for the right–hand polarized mode. The frequency is normalized to Ω_d, and the wavenumber to Ω_d/v_{AT}. The dust parameter is $\delta = 0.8 \times 10^{-4}$. (a) Collisions between species are included. (b) Collisions are neglected. (Reproduced from Cramer & Vladimirov 1997.)

We choose to plot the real part of the wavenumber and the imaginary part of the wavenumber against the real frequency, as in Pilipp *et al.* (1987). We can thus readily determine from Eq. (8.17) or Eq. (8.18) the wavelength and damping length of the waves from a localized source oscillating at a given real frequency. The inverse determination of the real part of the frequency and the damping time of a wave of given real wavelength and angle of propagation is not so straightforward because the dispersion relation (8.17) is of eighth order in ω, rather than the dispersion relation quadratic in ω of Eq. (8.31). However, physically relevant solutions may be readily found numerically, and for lightly damped modes the ratio of imaginary to real frequency for real wavenumber is the same as the ratio of imaginary to real wavenumber for real frequency.

Figure 8.2(a) shows the plot of the real (solid curve) and imaginary parts (dotted curve) of k_z against frequency for the right–hand polarized mode (Cramer & Vladimirov 1997). The frequency is normalized to the dust cyclotron frequency Ω_d ($= 1.12 \times 10^{-2}$ yr^{-1}), and the wavenumber is normalized to Ω_d/v_{AT} ($= 7.6 \times 10^3$ pc^{-1}), where v_{AT} is the Alfvén

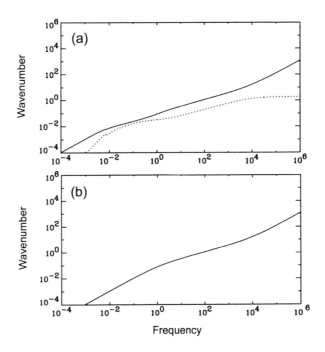

Figure 8.3: The wavenumber k plotted against frequency ω for propagation parallel to the magnetic field in an interstellar cloud, for the left–hand polarized mode. The frequency is normalized to Ω_d, and the wavenumber to Ω_d/v_{AT}. The dust parameter is $\delta = 0.8 \times 10^{-4}$. (a) Collisions between species are included. (b) Collisions are neglected. (Reproduced from Cramer & Vladimirov 1997.)

speed using the total mass density of the ions, neutral molecules and dust grains. We also have $\Omega_{m0}/\Omega_d = 0.88 \times 10^4$. The collision frequencies are as used by Pilipp *et al.* (1987). Figure 8.2(b) shows the corresponding result when all the collision frequencies are set to zero. The wave is now either purely propagating or purely evanescent, with the cutoff and resonance obtained from Eq. (7.16) evident. With collisions included, the wave is heavily damped between the resonance and cutoff frequencies. Figure 8.3 shows the corresponding dispersion relations for the left–hand polarized mode. This mode does not experience the dust–induced resonance and cutoff behaviour.

We turn now to the oblique propagation case: the wave propagates with the wavenumber k at an angle ϕ to the magnetic field, with $k_x = k \sin \phi$ and $k_z = k \cos \phi$. We have from Eq. (8.17) the following quadratic equation in k describing the two independent elliptically polarized modes:

$$k^4 \cos^2 \phi - k^2 \frac{(1 + \cos^2 \phi)}{2}(D_+ + D_-) + D_+ D_- = 0. \tag{8.38}$$

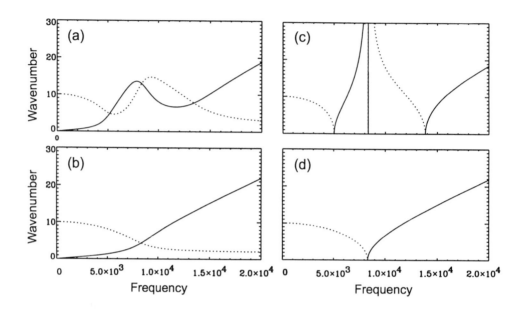

Figure 8.4: The wavenumber k_x perpendicular to the magnetic field plotted against frequency ω for fixed wavenumber parallel to the magnetic field $k_z = 10$, where the same normalizations as in Figure 8.2 are used. (a) Dust is present ($\delta = 0.8 \times 10^{-4}$) and collisions between species are included. (b) Dust is absent ($\delta = 0$) and collisions are included. (c) Dust is present, but there are no collisions. (d) Dust is not present and there are no collisions. (Reproduced from Cramer & Vladimirov 1997.)

With collisions included, it is found that the real and imaginary parts of k are comparable in size for frequencies between the resonance and cutoff, that is, the relative damping of the wave is less than for the parallel propagating wave (Cramer & Vladimirov 1997). The left–hand polarized mode at $\phi = 80°$ does not show much difference to the parallel propagating case.

If collisions are included, there is no longer a pure resonance in k_x at the Alfvén or dust-ion hybrid resonances defined by Eq. (8.19), but the real part of k_x attains a maximum there. This is shown in Figure 8.4(a), which is a plot of the real and imaginary parts of k_x against frequency for fixed $k_z = 10$. For our parameters, the Alfvén resonance frequency ω_r satisfies $\omega_r/\Omega_d = 0.83 \times 10^4$. For comparison, Figure 8.4(b) shows the corresponding plot for the case of no dust ($\delta = 0$). Also for comparison, Figure 8.4(c) shows the corresponding plots for no collisions between the species but with dust present, and Figure 8.4(d) shows the case for no collisions and no dust. With dust present, the cutoff-resonance-cutoff behaviour of the dispersion relation, which is absent for the dust–free plasma, is clearly shown. We see from Figure 8.4(a) that the presence of dust still radically alters the dispersion relation in the vicinity of the Alfvén resonance in the presence of collisions: the cutoff-resonance-cutoff triplet is still discernible and strong absorption occurs over a wide range of frequencies about the Alfvén resonant frequency.

Another view of the Alfvén resonance absorption mechanism is gained by considering a wave of fixed frequency and k_z propagating into a plasma of increasing ion density ρ_i in the x–direction. In the collisionless case, the wave will encounter the cutoff-resonance-cutoff triplet at successive spatial points in the density gradient, and wave energy will be absorbed at the resonance position where Eq. (8.19) is satisfied. The same will occur in the collisional case considered here, the only difference being that the resonance absorption occurs via the collisional damping processes.

For the interstellar cloud parameters considered here, the Alfvén resonant frequency is approximately 10^2 yr^{-1}, for a wavelength along the magnetic field of 10^{-4} pc ($\simeq 20$ Astronomical Units (AU)). At that frequency, the damping length is $\simeq 10$ times the wavelength for parallel propagation (and the damping time is $\simeq 10^{-1}$ yr for a real wavenumber), while for oblique propagation at the resonant frequency the wavelength and damping length are both $\simeq 20$ AU.

8.2.6   Nonlinear Waves

We turn now to some aspects of large–amplitude waves in astrophysical plasmas such as interstellar clouds. The important features of such an environment are the low degree of ionization and low temperature of the cloud, leading to strong collisional processes, as well as the presence of neutral and charged dust grains. Periodic density structures are sometimes observed, such as in the Orion A molecular cloud, oriented roughly perpendicular to the magnetic field, which may be due to nonlinear wave formation. It has been shown that the ion-neutral relative drift leading to ambipolar diffusion can lead to the steepening of the magnetic field profile and to the formation of singularities in the current density of hydromagnetic fluctuations and waves (Brandenburg & Zweibel 1994, 1995; Mac Low *et al.* 1995; Suzuki & Sakai 1997). The process may contribute to the production of the large–scale magnetic field in spiral galaxies, via dynamo generation of the fields and subsequent reconnection of the field at the small spatial scales produced by the ambipolar diffusion. The process may also aid the reconnection expected to occur during the star formation process, in neutral sheets formed during gravitational collapse. It may also occur in clumpy molecular clouds as the field lines are dragged about by the turbulent gas. The process of forming sharp magnetic field gradients is one of nonlinear diffusion, and stands in contrast to the expected smoothing out of the field produced by linear ambipolar diffusion.

Such steepening occurs in a nonlinear Alfvén wave that is linearly polarized, as was shown by Suzuki & Sakai (1997) by finding a steady–state solution for the wave, and also by using a one–dimensional simulation to show that a pulse–like perturbation of an Alfvén wave evolves to stable current sheet structures. A two–dimensional simulation showed that the current sheets that form from a large–amplitude Alfvén wave are unstable to the formation of sharp current filaments and associated density filaments. If charged dust is present, a nonlinear wave with sharp current features due to the ambipolar diffusion will involve a rotation of the wave magnetic field about the direction of propagation, and an oscillation of the field components, due to the dispersive effects of the dust. This is in contrast to the dust–free case, where the sharp reversal of the transverse magnetic field component occurs in a single plane.

Model and Wave Equations

As in Section 8.2.2, we consider waves in a uniform molecular cloud consisting of neutral atomic and molecular species, the ionized atomic and molecular species, the electrons, and negatively charged dust grains of a single size. A four-fluid model of the plasma is again used, differing from Eqs. (8.1)-(8.4) in that nonlinear terms in the equations of motion (1.5) are included, and the pressure gradients in the ions and neutrals are retained:

$$\rho_i \left(\frac{\partial \boldsymbol{v}_i}{\partial t} + \boldsymbol{v}_i \cdot \nabla \boldsymbol{v}_i \right) = -\nabla p_i + n_i e \left(\boldsymbol{E} + \boldsymbol{v}_i \times \boldsymbol{B} \right)$$
$$-\rho_i \nu_{in} \left(\boldsymbol{v}_i - \boldsymbol{v}_n \right) - \rho_i \nu_{id} \left(\boldsymbol{v}_i - \boldsymbol{v}_d \right) \tag{8.39}$$

$$\rho_n \left(\frac{\partial \boldsymbol{v}_n}{\partial t} + \boldsymbol{v}_n \cdot \nabla \boldsymbol{v}_n \right) = -\nabla p_n - \rho_n \nu_{ni} \left(\boldsymbol{v}_n - \boldsymbol{v}_i \right) - \rho_n \nu_{nd} \left(\boldsymbol{v}_n - \boldsymbol{v}_d \right) \tag{8.40}$$

$$\rho_d \left(\frac{\partial \boldsymbol{v}_d}{\partial t} + \boldsymbol{v}_d \cdot \nabla \boldsymbol{v}_d \right) = -Z_d n_d e \left(\boldsymbol{E} + \boldsymbol{v}_d \times \boldsymbol{B} \right)$$
$$-\rho_d \nu_{dn} \left(\boldsymbol{v}_d - \boldsymbol{v}_n \right) - \rho_d \nu_{di} \left(\boldsymbol{v}_d - \boldsymbol{v}_i \right) \tag{8.41}$$

$$0 = -n_e e \left(\boldsymbol{E} + \boldsymbol{v}_e \times \boldsymbol{B} \right) - \rho_e \nu_{en} \left(\boldsymbol{v}_e - \boldsymbol{v}_n \right) - \rho_e \nu_{ed} \left(\boldsymbol{v}_e - \boldsymbol{v}_d \right) - \rho_e \nu_{ei} \left(\boldsymbol{v}_e - \boldsymbol{v}_i \right) \tag{8.42}$$

where m_s is the species mass, ρ_s is the species mass density and $\boldsymbol{v}_s$ is the species velocity in the wave. p_i and p_n are the ion thermal and neutral thermal pressures, and ν_{st} is the collision frequency of a particle of species s with the particles of species t. We have neglected electron inertia in Eq. (8.42), momentum transfer to ions from electrons in Eq. (8.39) and to dust grains from electrons in Eq. (8.41), and the dust thermal pressure gradient in Eq. (8.41).

The fractional ionization n_i/n_n is assumed fixed, where n_n is the number density of neutrals. The charge on each dust grain is assumed constant, and for simplicity we also assume that $\delta = Z_d n_d / n_i$ is constant, even though n_n and n_i are variable. The neutral mass density obeys the continuity equation

$$\frac{\partial \rho_n}{\partial t} + \nabla \cdot \left(\rho_n \boldsymbol{v}_n \right) = 0. \tag{8.43}$$

The conduction current density is again given by Eq. (8.5), and equilibrium charge neutrality is expressed by Eq. (7.14).

Equations (8.42) and (8.5) lead to the following generalized Ohm's law:

$$\boldsymbol{E} + \left(\frac{\boldsymbol{v}_i}{1 - \delta} - \frac{\delta}{1 - \delta} \boldsymbol{v}_d \right) \times \boldsymbol{B} = \frac{\boldsymbol{J} \times \boldsymbol{B}}{n_e e} - \frac{m_e}{e} \nu_{en} \left(\boldsymbol{v}_e - \boldsymbol{v}_n \right)$$
$$- \frac{m_e}{e} \nu_{ed} \left(\boldsymbol{v}_e - \boldsymbol{v}_d \right) - \frac{m_e}{e} \nu_{ei} \left(\boldsymbol{v}_e - \boldsymbol{v}_i \right). \tag{8.44}$$

The expression for $\boldsymbol{E}$ obtained from Eq. (8.44) can now be substituted into the ion equation (8.39). We neglect the contribution of the electron collisional momentum transfer terms in Eq. (8.44), compared to the ion momentum transfer terms (i.e. we are neglecting resistivity), so that we may write

$$\boldsymbol{E} + \boldsymbol{v}_i \times \boldsymbol{B} = -\frac{\delta}{1 - \delta} \left(\boldsymbol{v}_i - \boldsymbol{v}_d \right) \times \boldsymbol{B} + \frac{\boldsymbol{J} \times \boldsymbol{B}}{n_e e}. \tag{8.45}$$

We can now simplify the discussion considerably by using the *strong ion coupling* approximation (Suzuki & Sakai 1996), whereby the ion inertia term (the left–hand side) and the ion
thermal pressure term are neglected in Eq. (8.39), leaving a balance between the remaining
terms. This is equivalent to assuming that $\omega \ll \nu_{in}$, as in the discussion of linear waves
in that limit in Section 8.2.2. At this point it is useful to normalize the magnetic field by a
reference field B_0, and define the Alfvén speed based on the field B_0 and the ion density:
$v_A = B_0/(\mu_0 \rho_i)^{1/2}$. Eq. (8.39) may then be written, using Eq. (8.45) and Faraday's law
neglecting the displacement current,

$$\frac{v_A^2}{\delta}(\nabla \times \boldsymbol{B}) \times \boldsymbol{B} = \Omega_{m0}(\boldsymbol{v}_i - \boldsymbol{v}_d) \times \boldsymbol{B} + \nu_{in}(\boldsymbol{v}_i - \boldsymbol{v}_n) + \nu_{id}(\boldsymbol{v}_i - \boldsymbol{v}_d) \qquad (8.46)$$

where $\Omega_{m0} = \Omega_i \delta/(1 - \delta)$ and Ω_i is the ion cyclotron frequency in the reference field B_0.
The presence of dust introduces the first and third terms on the right–hand side of Eq. (8.46).
In the absence of dust, with $\delta = 0$, $\Omega_{m0} = 0$ and $\nu_{id} = 0$, Eq. (8.46) may be solved to give
the relative drift velocity of ions and neutrals,

$$\boldsymbol{V}_i \equiv \boldsymbol{v}_i - \boldsymbol{v}_n = \frac{v_A^2}{\nu_{in}}(\nabla \times \boldsymbol{B}) \times \boldsymbol{B}. \qquad (8.47)$$

This is the strong coupling expression for the relative ion drift velocity in a dust–free partially
ionized plasma used by Suzuki & Sakai (1996).

The dust equation of motion (8.3) becomes

$$\frac{\partial \boldsymbol{v}_d}{\partial t} + \boldsymbol{v}_d \cdot \nabla \boldsymbol{v}_d = \frac{\Omega_d}{\Omega_i}\frac{v_A^2}{1 - \delta}(\nabla \times \boldsymbol{B}) \times \boldsymbol{B} + \frac{\Omega_d}{1 - \delta}(\boldsymbol{v}_i - \boldsymbol{v}_d) \times \boldsymbol{B}$$
$$-\nu_{dn}(\boldsymbol{v}_d - \boldsymbol{v}_n) + \nu_{di}(\boldsymbol{v}_i - \boldsymbol{v}_d). \qquad (8.48)$$

where Ω_d is the dust–grain cyclotron frequency using the reference field B_0. Inspection of
Eq. (8.48) shows that the acceleration of the dust due to the Lorentz force is proportional to
Ω_d/Ω_i and is thus very small compared to that of the ions.

To simplify the analysis we consider two limiting cases for the dust fluid velocity $\boldsymbol{v}_d$:
(a) the dust grains and neutral gas are strongly coupled, such that the inertia of the dust is
neglected, and (b) stationary dust grains, $\boldsymbol{v}_d = 0$. The first case holds approximately if the
neutral gas density is high enough that $\nu_{dn} > \omega$, where ω is the characteristic frequency of the
waves considered, which we assume here to be higher than the dust cyclotron frequency. For
a cloud with $n_n = 10^4 \text{ cm}^{-3}$, $\nu_{dn} \simeq 0.4\Omega_d$ (Pilipp *et al.* 1987), so that strong dust-neutral
coupling occurs for higher neutral densities. The second case occurs for high frequencies
and lower neutral densities, such that $\omega > (\Omega_d, \nu_{dn})$, so that the high dust inertia and small
dust collisional coupling implies that the dust is stationary on the time–scale of interest. We
proceed to discuss the two cases in some detail.

Strong Dust-Neutral Coupling

For strong coupling of the dust with the neutral gas, we neglect the inertia of the dust, just as
we neglect the ion inertia for strong ion coupling, so the LHS of Eq. (8.48) is zero. Defining

254 8 Astrophysical Plasmas

the dust fluid velocity relative to the neutrals as

$$\boldsymbol{V}_d \equiv \boldsymbol{v}_d - \boldsymbol{v}_n \tag{8.49}$$

Eqs. (8.46) and (8.48) can be written as (neglecting the ion-dust momentum transfer terms)

$$\frac{v_A^2}{1-\delta}(\nabla \times \boldsymbol{B}) \times \boldsymbol{B} = \Omega_m(\boldsymbol{V}_i - \boldsymbol{V}_d) \times \boldsymbol{B} + \nu_{in}\boldsymbol{V}_i \tag{8.50}$$

$$\frac{\Omega_d}{\Omega_i}\frac{v_A^2}{1-\delta}(\nabla \times \boldsymbol{B}) \times \boldsymbol{B} = -\frac{\Omega_d}{1-\delta}(\boldsymbol{V}_i - \boldsymbol{V}_d) \times \boldsymbol{B} + \nu_{dn}\boldsymbol{V}_d. \tag{8.51}$$

These equations can be solved for $\boldsymbol{V}_i$ and $\boldsymbol{V}_d$, yielding

$$\boldsymbol{V}_i = F\frac{v_A^2}{\nu_{in}(1-\delta)}\left[H(\nabla \times \boldsymbol{B}) \times \boldsymbol{B} - (R/\rho)((\nabla \times \boldsymbol{B}) \times \boldsymbol{B}) \times \boldsymbol{B}\right] \tag{8.52}$$

$$\boldsymbol{V}_d = \frac{\beta_d}{(1-\delta)\Omega_{m0}}\left[v_A^2(\nabla \times \boldsymbol{B}) \times \boldsymbol{B} - \nu_{in}\boldsymbol{V}_i\right] \tag{8.53}$$

where $R = \Omega_{m0}\rho_n/\nu_{in}\rho_0$, $B^2 = B_x^2 + B_y^2 + B_z^2$, $\beta_d = \Omega_d/\nu_{dn}$ is the dust Hall parameter defined in Eq. (8.29),

$$F = \left[1 + \frac{R^2 B^2}{\rho^2}\left(1 + \frac{\beta_d \rho}{R}\right)^2\right]^{-1} \tag{8.54}$$

and

$$H = \left[1 + \beta_d\frac{\nu_{dn}}{\nu_{in}}B^2\left(1 + \frac{\beta_d \rho}{R}\right)\right]. \tag{8.55}$$

We have normalized the density by ρ_0, defining $\rho = \rho_n/\rho_0$. Note that the limit $\beta_d \to 0$ corresponds to very strong coupling of the dust to the neutrals, such that $\boldsymbol{v}_d = \boldsymbol{v}_n$.

The inertia of the wave motion is provided predominantly by the neutrals. Summing the momentum conservation equations (8.1)-(8.3) of the ions, neutrals and dust, the equation of motion for the neutral velocity is obtained:

$$\rho_n\left(\frac{\partial \boldsymbol{v}_n}{\partial t} + \boldsymbol{v}_n \cdot \nabla \boldsymbol{v}_n\right) = -\nabla p_n + \frac{1}{\mu_0}(\nabla \times \boldsymbol{B}) \times \boldsymbol{B}. \tag{8.56}$$

The velocity is normalized by the Alfvén velocity based on ρ_0, $V_A = B_0/(\mu_0\rho_0)^{1/2}$, $\boldsymbol{v} = \boldsymbol{v}_n/V_A$, space and time are normalized by L_0 and $\tau_A = L_0/V_A$, and pressure is normalized by p_0. The result is, in the normalized variables,

$$\rho\left(\frac{\partial \boldsymbol{v}}{\partial t} + \boldsymbol{v} \cdot \nabla \boldsymbol{v}\right) = -\beta\nabla p + (\nabla \times \boldsymbol{B}) \times \boldsymbol{B}. \tag{8.57}$$

The magnetic induction equation gives, using Eq. (8.44) with the collisional electron momentum transfer terms neglected, and neglecting the Hall term,

$$\frac{\partial \boldsymbol{B}}{\partial t} = \nabla \times \left[\frac{1}{1-\delta}(\boldsymbol{v}_i - \delta\boldsymbol{v}_d) \times \boldsymbol{B}\right]. \tag{8.58}$$

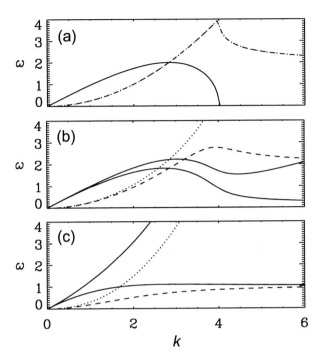

Figure 8.5: The dispersion relation for linear parallel propagating waves, and very strong coupling of dust. Frequencies and damping rates are plotted against wavenumber k (in normalized units). Here $A_\mathrm{D} = 0.5$, and (a) $R = 0$, (b) $R = 0.1$, and (c) $R = 1$. The upper (lower) solid lines are the frequencies of the left(right)–hand circularly polarized mode. The dotted (dashed) lines are the damping rates of the left(right)–hand polarized mode.

In the strong dust coupling limit, Eq. (8.58) becomes

$$\frac{\partial \boldsymbol{B}}{\partial t} = \nabla \times (\boldsymbol{v}_\mathrm{n} \times \boldsymbol{B}) + \nabla \times \left(\frac{1}{1 - \delta} (\boldsymbol{V}_\mathrm{i} - \delta \boldsymbol{V}_\mathrm{d}) \times \boldsymbol{B} \right). \tag{8.59}$$

Substituting Eq. (8.52), and using the normalizations defined above, we obtain

$$\frac{\partial \boldsymbol{B}}{\partial t} = \nabla \times (\boldsymbol{v} \times \boldsymbol{B})$$
$$+ A_\mathrm{D} \nabla \times \left[\frac{F}{\rho} \left(H \left((\nabla \times \boldsymbol{B}) \times \boldsymbol{B} \right) \times \boldsymbol{B} + (R/\rho) B^2 (\nabla \times \boldsymbol{B}) \times \boldsymbol{B} \right) \right] \tag{8.60}$$

where $B = |\boldsymbol{B}|$ and $A_\mathrm{D} = \rho_\mathrm{n}/\nu_\mathrm{in}\tau_\mathrm{A}\rho_\mathrm{i}$, and we have neglected terms of order $(\Omega_\mathrm{d}/\Omega_\mathrm{i})$.

We can obtain from the above nonlinear equations the linear dispersion relation for small–amplitude waves treated in Section 8.2.3. For example, consider the dispersion relation of linear waves for very strong dust coupling ($\beta_d \to 0$), with the waves propagating parallel to a steady magnetic field B_z in the z–direction, with frequency ω and wavenumber k. This can be derived directly from Eqs. (8.57) and (8.60), and also follows from the general dispersion relation for parallel propagating linear waves given by Eq. (8.18), in the limit $\nu_{dn} \gg \omega \gg \Omega_d$, $\omega > \nu_{ni}$ and $\Omega_{m0} < \nu_{in}$. The result is

$$\omega^2 = k^2 v_{An}^2 \left(1 - i\frac{\omega}{\nu_{ni}} \pm \frac{\omega \Omega_{m0}}{\nu_{ni}\nu_{in}} \right) = k^2 v_{An}^2 \left(1 - i A_D \omega \pm \frac{R}{\rho} A_D \omega \right) \tag{8.61}$$

where v_{An} is the Alfvén speed based on the magnetic field B_z and density ρ_n. The $+(-)$ sign corresponds to right (left) hand circular polarization of the wave magnetic field, which is transverse to the steady magnetic field.

The dispersion relation (8.61) is a generalization of the dispersion relation derived by Suzuki & Sakai (1996) to include dust, and shows the anisotropic effects of the dust in the different frequencies of the left and right–hand circularly polarized modes for a given wavenumber. The dispersion and anisotropy of the modes is shown in Figure 8.5, where the frequency and damping rate obtained from Eq. (8.61) is plotted against the wavenumber, for three values of the dust parameter R ($R = 0$, 0.1 and 1). The ambipolar diffusion parameter A_D is held fixed at 0.5 in each case. It is seen that a single frequency exists for the two modes of polarization in the absence of dust, and a linearly polarized Alfvén wave exists, which is dispersive at higher frequencies where the damping becomes appreciable. However in the presence of dust, the splitting of the two oppositely polarized modes, due to the effect of the dust on the ion and electron Hall currents, increases as R increases, and a linearly polarized wave can no longer be constructed. This splitting and dispersion due to the charged dust is not however as strong as results when all collisions are neglected and the plasma becomes uncoupled from the neutral gas (see Section 7.4). We find in the next sections that the anisotropic effect of the dust on the ion and electron Hall currents leads to a rotation of the magnetic field in nonlinear waves.

Stationary Dust

In the other limit of stationary dust ($v_d = 0$) the strong ion coupling equation (8.46) gives

$$v_i = \frac{F}{\nu_i} \left[\frac{v_A^2}{1-\delta} ((\nabla \times B) \times B - (R/\rho)((\nabla \times B) \times B) \times B)) \right.$$
$$\left. - \Omega_{m0} v_n \times B + \nu_i v_n + (\Omega_m^2/\nu_i)(B \cdot v_n)B \right] \tag{8.62}$$

which gives, on substitution into Eq. (8.40), the following normalized equation of motion for the neutrals:

$$\rho \left(\frac{\partial v}{\partial t} + v \cdot \nabla v \right) = -\beta \nabla \rho + F \left[(\nabla \times B) \times B - (R/\rho)((\nabla \times B) \times B) \times B \right]$$
$$- (RF/A_D) v \times B + (F\rho/A_D) \left[Gv + (R^2/\rho^2)(B \cdot v)B \right] \tag{8.63}$$

where $G = 1 - (1 + \nu_{nd}/\nu_{ni})/F$.

The magnetic induction equation becomes, using Eq. (8.44),

$$\frac{\partial \boldsymbol{B}}{\partial t} = \nabla \times \left[\frac{F}{1-\delta} \left(\boldsymbol{v} \times \boldsymbol{B} - (R/\rho)((\boldsymbol{B} \cdot \boldsymbol{v})\boldsymbol{B} - B^2 \boldsymbol{v}) \right) \right]$$

$$+ \frac{A_D}{(1-\delta)^2} \nabla \times \left[\frac{F}{\rho} \left(((\nabla \times \boldsymbol{B}) \times \boldsymbol{B}) \times \boldsymbol{B} + (R/\rho)B^2(\nabla \times \boldsymbol{B}) \times \boldsymbol{B} \right) \right]. \tag{8.64}$$

The greater complexity of Eqs. (8.63) and (8.64) compared to Eqs. (8.57) and (8.60) is found to lead to greater anisotropic and dispersive effects on the waves. However it is interesting to note that the linear dispersion relation, derived from Eqs. (8.63) and (8.64), or from the general dispersion relation (8.18), still with strong ion coupling but in the limit $\omega \gg \nu_{dn}$, is the same weakly dispersive relation, given by Eq. (8.61), as in the very strongly coupled dust case. Thus strong linear anisotropic splitting and dispersion depends on the removal of the strong ion coupling assumption, as in Section 8.2.3.

Steady–State Solutions

To illustrate some nonlinear solutions, we seek steady–state solutions propagating in the z–direction with the phase velocity V_0, and steady in the frame of reference with spatial coordinate $\xi = z - V_0 t$. For simplicity, we neglect the charged dust contribution, that is, $\delta = 0$ and $R = 0$. In the absence of dust the field variables v_x and B_x become uncoupled from v_y and B_y. We can therefore take the magnetic field to initially lie in the x-z plane.

Equations (8.43), (8.57) and (8.60) become

$$\rho(v_z - V_0) = D \tag{8.65}$$

$$\rho(v_z - V_0)\frac{dv_x}{d\xi} = B_z \frac{dB_x}{d\xi} \tag{8.66}$$

$$\rho(v_z - V_0)\frac{dv_z}{d\xi} = -\beta\frac{d\rho}{d\xi} - B_x \frac{dB_x}{d\xi} \tag{8.67}$$

$$(v_z - V_0)B_x = (A_D/\rho)(B_z^2 + B_x^2)\frac{dB_x}{d\xi} + C_x \tag{8.68}$$

where D and C_x are constants.

If we set $B_z = 0$ in Eqs. (8.66) and (8.68), the solution around $\xi = 0$ is then

$$B_x \sim \xi^{1/3}, \quad v_z \sim \xi^{2/3}. \tag{8.69}$$

This stationary solution was discussed by Brandenburg & Zweibel (1994). It leads to a singular current density, $J \sim \xi^{-2/3}$.

If $B_z \neq 0$, the solution of Eq. (8.68) in the vicinity of $\xi = 0$ is

$$B_x \sim \xi, \quad v_z \sim \xi^2 \tag{8.70}$$

so the current density is not singular. However the length–scale for variation of B_x can be small compared to the characteristic length, such as the wavelength of an Alfvén wave, with a relatively large associated current density.

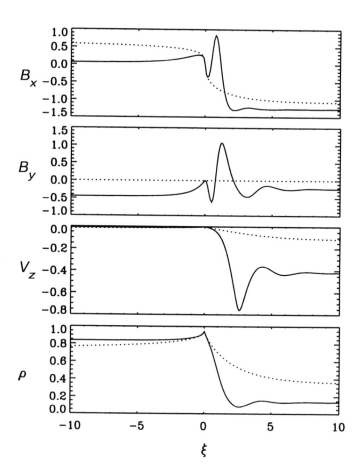

Figure 8.6: Profiles of a steady propagating nonlinear structure in a partially ionized, dusty plasma. The two transverse components of the magnetic field, B_x and B_y, the longitudinal velocity v_z and the density ρ are plotted as functions of position. The dotted curves correspond to no dust ($R = 0$), and the solid curves correspond to dust present ($R = 1$). The very strongly coupled dust model is used.

In the presence of dust, that is, $R \neq 0$, if $B_z = 0$ there is no coupling between B_x and B_y and there is no effect of the dust apart from a variation of the factor F in Eq. (8.54). The fields again have the variation given by Eq. (8.69) around $\xi = 0$. If $B_z \neq 0$, there is coupling between B_x and B_y, that is, a rotation of the magnetic field out of the x-z plane occurs. Solving the equations including dust for the parameter values $A_D = 0.5$, $V_0 = 0.07$, $\beta = 1$, $B_z = 0.1$, $D = -0.066$, $C_x = 0.04$ and $C_y = 0$, we obtain Figure 8.6. A shock transition is obtained, with uniform fields upstream and downstream of the transition. The boundary

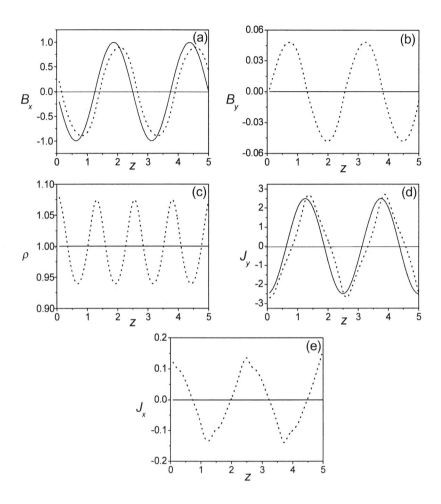

Figure 8.7: Field profiles of B_x, B_y, ρ, J_y and J_x against distance for an Alfvén wave that is initially linearly polarized in the x-z plane. The solid curves are the initial profiles, and the dotted curves are the profiles after the time $\tau_A / 2\pi$. The very strongly coupled dust model is used, with $R = 1$.

conditions chosen are $B_x = B_y = v_x = v_y = v_z = 0$ at $\xi = 0$. The figure shows the B_x, B_y, v_z and ρ profiles as functions of ξ for $R = 0$ (dotted curves, obtained by Suzuki & Sakai (1996)) and $R = 1$ (solid curves). Small oscillations occur in B_x, v_z and ρ, but an appreciable B_y is produced close to $\xi = 0$, with $B_y = 0$ upstream and downstream of the transition, that is, around $\xi = 0$ the magnetic field rotates out of the x-z plane. Otherwise, it is apparent that anisotropic effects are fairly weak in this case.

In the case of stationary dust, if $B_z = 0$, B_x and B_y are coupled together, that is, rotation of the magnetic field occurs, and B_x and B_y both have the variation $\sim \xi^{1/3}$, producing singular current densities in the x and y–directions. This is in contrast to the strongly coupled dust case, where the magnetic field remains in the x–direction. If $B_z \neq 0$, again the current density is nonsingular, but the anisotropic effect is found to be more pronounced.

Periodic Nonlinear Waves

It was shown by Suzuki & Sakai (1996) that, in the absence of dust, sharp current sheets can be generated by the ambipolar diffusion effect in periodic nonlinear Alfvén waves propagating in one dimension. The waves were assumed to be initially polarized in a single plane. The planar polarization was preserved in the time evolution of the sharp current sheets in the waves. With the inclusion of the charged dust, we expect that the magnetic field will rotate out of the initial plane of polarization. We show here the effect of that rotation on the formation of the current sheets. Figure 8.7 shows the time evolution of the fields in a large–amplitude periodic wave that is initially plane polarized with magnetic field in the x-z plane, for the very strongly coupled dust case. The magnetic field rotates out of the initial plane of polarization due to the presence of the dust. The nonlinear y component of the magnetic field so generated tends to have sharper gradients than the x component in each case, and so produces an appreciable current density in the x–direction, as shown in Figure 8.7(e). Parametric and two–dimensional filamentation instabilities of Alfvén waves can also be analysed for collisional astrophysical plasmas (Suzuki & Sakai 1997).

8.3   Pulsar Magnetospheres

Neutron stars are thought to be surrounded by a relativistic plasma of electrons and positrons, which is penetrated by beams of high–energy electrons and positrons travelling along the magnetic field. The plasma and beams are formed in the polar regions of the neutron star. In the very strong magnetic field of the star, synchrotron radiation losses mean that the particle kinetic energy perpendicular to the field is reduced to zero and the particle momenta are virtually one–dimensional along the field. The one–dimensional distribution of particles is unstable to the excitation of various types of waves: the maximum growth rate is found to be for longitudinal Langmuir (plasma) waves. The quasilinear interaction of these waves with the particles leads to a relaxed particle distribution with a long tail and no humps. This distribution in turn is unstable to the excitation of electromagnetic waves, and, in particular, Alfvén waves, via a cyclotron resonance (Lominadze, Machabeli & Mikhailovskii 1979). We proceed in this section to investigate the properties of Alfvén and magnetoacoustic waves under these extreme conditions of temperature and magnetic field.

In a very low–density plasma, or a strong magnetic field, the Alfvén speed may become comparable with the speed of light, in which case the MHD dispersion relations of Chapter 2 are altered by the inclusion of the displacement current in Maxwell's equations. The Alfvén, fast and slow dispersion relations (2.19)-(2.21) are easily generalized to

$$\omega_A = \frac{k v_A |\cos\theta|}{(1 + v_A^2/c^2)^{1/2}} \tag{8.71}$$

$$\omega_{\mathrm{F,S}} = k \left[\frac{\left(v_{\mathrm{A}}^2 + c_{\mathrm{s}}^2 \pm \left((v_{\mathrm{A}}^2 + c_{\mathrm{s}}^2)^2 - 4 v_{\mathrm{A}}^2 c_{\mathrm{s}}^2 \cos^2 \theta (1 + v_{\mathrm{A}}^2/c^2) \right)^{1/2} \right)}{2(1 + v_{\mathrm{A}}^2/c^2)} \right]^{1/2}. \tag{8.72}$$

These equations apply if the ion mass is very much greater than the electron mass. However, if the plasma is a pair plasma, that is, consists of electrons and positrons, and no other ions, the low–frequency modes are modified. It is known that electron-positron plasmas may exist in pulsar magnetospheres and in active galactic nuclei. Because the time and spatial scales associated with electrons and positrons are identical, pair plasmas will have dispersive properties quite different to electron-ion plasmas. However, an Alfvén mode may still be identified. We can gain insights into the properties of pair plasmas by first considering the case of a cold pair plasma.

8.3.1   Cold Pair Plasmas

We can investigate the transition from the usual cold plasma with normal ions to a pair plasma by letting the ion mass approach the electron mass. The plasma is neutral, so that $n_{\mathrm{i}} = n_{\mathrm{e}}$. The dispersion equation for waves propagating parallel to the magnetic field, obtained from the cold plasma expressions given by Eqs. (2.57) and (2.66), can be written as:

$$\alpha^2 = f^2 \left[v_{\mathrm{A}}^2/c^2 + \frac{1 + m_{\mathrm{e}}/m_{\mathrm{i}}}{(1 \pm f)(1 \mp (m_{\mathrm{e}}/m_{\mathrm{i}})f)} \right] \tag{8.73}$$

where $f = \omega/\Omega_{\mathrm{i}}$ and $\alpha = v_{\mathrm{A}} k_z/\Omega_{\mathrm{i}}$, and we have used the relation $\omega_{\mathrm{pi}}^2/\Omega_{\mathrm{i}}^2 = c^2/v_{\mathrm{A}}^2$, with v_{A} defined in terms of the ion mass density only. Figure 8.8 shows the resulting dispersion relations, for a sequence of three ion to electron mass ratios, and for two different values of v_{A}^2/c^2. A nonzero value of v_{A}^2/c^2 introduces extra high–frequency modes that are cut off: for $m_{\mathrm{e}} \ll m_{\mathrm{i}}$ the cutoff frequency in the LHP mode is $f = 1 + c^2/v_{\mathrm{A}}^2$, while for $m_{\mathrm{e}} = m_{\mathrm{i}}$ the cutoff is $f = (1 + 2c^2/v_{\mathrm{A}}^2)^{1/2}$ in both LHP and RHP modes. The LHP and RHP modes generally have different phase speeds, except in the limit of equal ion and electron masses (the positron-electron pair plasma), when the dispersion equation can be written

$$\frac{c^2 k_z^2}{\omega^2} = S = 1 - \frac{\omega_{\mathrm{p}}^2}{\omega^2 - \Omega^2} \tag{8.74}$$

where ω_{p} is the plasma frequency corresponding to the total electron and positron density, and to the electron mass, and Ω is the cyclotron frequency corresponding to the electron mass. S is the dielectric tensor element defined in Eq. (2.56). Eq. (8.74) may also be written in the notation of Eq. (8.73) as

$$\alpha^2 = f^2 \left[v_{\mathrm{A}}^2/c^2 + \frac{1}{1 - f^2} \right]. \tag{8.75}$$

A small equivalent ion mass may also arise in the form of the effective mass of a hole, in the electron and hole plasmas in solid state materials, and the resulting similar dispersion relations have been discussed by Baynham & Boardman (1971).

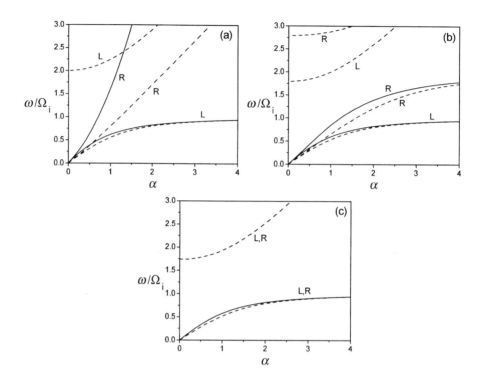

Figure 8.8: The normalized frequency plotted against the normalized wavenumber $\alpha = v_A k_z/\Omega_i$ for waves propagating parallel to the magnetic field in an electron-ion plasma of three values of the ion to electron mass ratio. The solid curves correspond to $v_A/c = 0$ and the dashed curves to $v_A/c = 1$. The left–hand (L) and right–hand (R) circularly polarized modes are indicated. (a) $m_i/m_e = 2000$, (b) $m_i/m_e = 2$, (c) $m_i/m_e = 1$.

For the charge-neutral pair plasma, the function D given by Eq. (2.56) entering the off–diagonal "gyrotropic" elements K_{12} and K_{21} of the dielectric tensor (Eq. 2.55) is zero, because the Hall currents of the positrons and electrons cancel, in contrast to the electron-ion plasma. As a result, the purely transverse LHP and RHP modes for parallel propagation have the same dispersion relation, as shown in Figure 8.8(c), and at frequencies $\omega \ll \Omega$ we have, for the lower solution branch,

$$\omega^2 = \frac{v_A^2 k_z^2}{1 + v_A^2/c^2} \tag{8.76}$$

where now v_A is the Alfvén speed using the total positron and electron mass density. If $v_A \ll c$, we obtain the usual Alfvén wave dispersion relation $\omega = v_A k_z$ for each mode. Thus we can form a linearly polarized Alfvén wave as a combination of the two oppositely circularly polarized modes. The electron whistler mode, that is, the RHP mode in the frequency range

$\Omega_i \ll \omega \ll \Omega_e$ for the case $m_i \gg m_e$, does not exist in the case $m_i = m_e$, because the RHP mode now has an identical (cyclotron) dispersion relation to the LHP mode.

Note that because the wave frequency of the Alfvén wave, in a very strongly magnetized, low–density pair plasma, can be comparable with the plasma frequency, the function P defined in Eq. (2.57) may be comparable in magnitude to S, in contrast to the situation in the electron-ion plasma. Here

$$P = 1 - \frac{\omega_p^2}{\omega^2} \tag{8.77}$$

so for parallel propagation, the longitudinal Langmuir wave is uncoupled from the transverse modes, and has the dispersion relation $\omega^2 = \omega_p^2$.

Let us now allow for different positron and electron densities, as occurs in the rotating pulsar magnetosphere (Arons & Barnard 1986), and define δ as one minus the ratio of electron to positron densities. The gyrotropic dielectric tensor terms K_{12} and K_{21} are no longer zero. The resulting dispersion equation for parallel propagation in a plasma with different ion and electron plasma frequencies, but with the same ion and electron cyclotron frequency, is

$$\frac{c^2 k_z^2}{\omega^2} = 1 - \frac{\omega_{pi}^2 + \omega_{pe}^2}{\omega^2 - \Omega^2} \pm \frac{\Omega(\omega_{pi}^2 - \omega_{pe}^2)}{\omega(\omega^2 - \Omega^2)} \tag{8.78}$$

and for $\omega \ll \Omega$ we have

$$\omega^2 = c^2 k_z^2 \left[1 + \frac{c^2(2 + \delta)}{2v_A^2} \pm \frac{\Omega}{\omega} \frac{c^2}{2v_A^2} \delta \right]^{-1} \tag{8.79}$$

where v_A is again defined using twice the mass density of the positrons. We now have low–frequency LHP and RHP modes of differing frequency; the result is reminiscent of the case of a dusty plasma discussed in Section 7.4, because again there is an imbalance in the electron and ion number densities.

The case of general angle of propagation in equal–mass plasmas has been considered by Stewart & Laing (1992). A fundamental difference to the electron-ion case occurs because, as noted above, $|P|$ must be considered finite. For a charge-neutral pair plasma, from Eq. (2.64) we obtain the dispersion equations

$$\frac{c^2 k^2}{\omega^2} = S \quad \text{and} \quad \frac{c^2 k^2}{\omega^2} = \frac{PS}{P\cos^2\theta + S\sin^2\theta}. \tag{8.80}$$

The first dispersion equation corresponds to the magnetoacoustic mode, which is linearly polarized in the direction purely transverse to B and k. This mode is also called the X (extraordinary) or t mode. The second dispersion equation in Eqs. (8.80) describes two modes that are polarized in the plane defined by B and k; the modes correspond to coupled Langmuir and Alfvén (A) modes. The two latter modes are pure Langmuir and Alfvén modes in the parallel propagation case, when the Alfvén mode becomes degenerate with the magneto-acoustic mode. The coupling of the Alfvén and Langmuir modes for oblique propagation is of the same character as the modification of the shear Alfvén wave by electron inertia or thermal effects in normal plasmas, leading to the IAW and KAW modes, as discussed in Section 2.8.

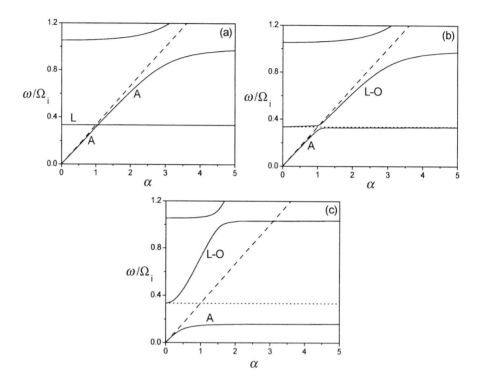

Figure 8.9: The normalized frequency plotted against the normalized wavenumber $\alpha = v_A k_z / \Omega_i$ (solid curves) for waves in the Langmuir-Alfvén modes propagating at three different angles θ to the magnetic field in an electron-positron plasma, with $v_A/c = 3$. The dotted lines indicate the parallel Langmuir wave frequency $\omega = \omega_p$, and the dashed lines are the light-lines, $\omega = ck$. The Langmuir (L), Alfvén (A) and Langmuir-Ordinary (L-O) modes are indicated. (a) $\theta = 0°$, (b) $\theta = 6°$, (c) $\theta = 60°$.

The dispersion equation for the coupled Langmuir and Alfvén modes may be written, in the notation of Eqs. (8.73) and (8.75), as

$$\alpha^2 = \frac{f^2(f^2 v_A^2/c^2 - 1)[v_A^2/c^2 + 1/(1 - f^2)]\cos^2\theta}{(f^2 v_A^2/c^2 - 1)\cos^2\theta + [v_A^2/c^2 + 1/(1 - f^2)]\sin^2\theta}. \tag{8.81}$$

Figure 8.9 shows the solutions of Eq. (8.81), indicating how the Alfvén and Langmuir wave dispersion relations are modified for oblique propagation. The Langmuir (L) mode reconnects with the upper (O or ordinary) section of the Alfvén (A) mode curve when $\theta \neq 0°$, to form the L-O mode, with a cutoff at the plasma frequency. The Alfvén mode extending to low frequencies has a resonance at $f \simeq \cos\theta/(v_A^2/c^2 + \sin^2\theta)^{1/2}$; this angle–dependent resonance is related to the resonances of the magnetoionic modes in a normal plasma (Melrose 1986). It is interesting to note that the Alfvén wave has its Poynting flux and group velocity along B at low frequency, even when the displacement current dominates the conduction current ($v_A > c$), and the phase and group velocities are almost equal to c, as is shown by Eq. (8.76) and Figure 8.9.

8.3.2 Relativistic Plasmas

Turning now to thermal effects on the dispersion relations of pair plasmas, we note that the case of greatest physical interest is that of highly relativistic plasmas in pulsar magnetospheres. The properties of the waves in this case have been discussed in the literature (Lominadze *et al.* 1987; Arons & Barnard 1986; Melrose & Gedalin 1999; Melrose *et al.* 1999; Melrose 2000, and references therein). As mentioned earlier, a good approximation is to assume the particle momenta are one–dimensional, along the magnetic field lines. The spread in particle energies in the pulsar plasma is thought to be extremely relativistic, so that a relativistic plasma dispersion function (RPDF) arises in the kinetic theory of the waves, replacing the nonrelativistic Maxwellian plasma dispersion function defined in Eq. (A.13):

$$W(z) = \int_{-\infty}^{\infty} \frac{\mathrm{d}p}{v - z} \frac{\mathrm{d}f(p)}{\mathrm{d}p} \quad \text{with} \quad \int_{-\infty}^{\infty} \mathrm{d}p\, f(p) = 1 \tag{8.82}$$

where v is the velocity normalized to c, the phase speed along the magnetic field is $z = \omega/k_{\parallel}$, and $f(p)$ is the one–dimensional distribution function expressed as a function of the particle momenta. Here $p = \gamma m v$ and $\gamma = (1 - v^2)^{-1/2}$. A Jüttner (relativistic thermal) distribution function, $f(p) \propto \exp(-\rho\gamma)$ with $\rho = mc^2/k_B T$, was used by Melrose & Gedalin (1999).

If Ω is the electron cyclotron frequency based on the rest mass of the electron, the relativistic cyclotron frequency is $\tilde{\Omega} = \Omega/\langle\gamma\rangle$, where angle brackets denote an average over the particle distribution function. The magnetic field is extremely strong close to the pulsar, so we assume here that $\omega \ll \tilde{\Omega}$, although cyclotron effects are important at greater distances from the star, where the problem of the escape of radiation from the pulsar arises. Also defining the plasma frequency ω_p in the rest frame of the plasma and with the total rest mass density, we have for the Alfvén speed (normalized by c), $v_A = \langle\gamma\rangle^{-1/2}\Omega/\omega_p$. The spread of particle velocities is measured by

$$\tilde{v}^2 = \frac{\langle\gamma v^2\rangle}{\langle\gamma\rangle}. \tag{8.83}$$

We now consider waves with a wavevector k lying in the x-z plane and at an angle θ to the magnetic field. The elements of the wave tensor Λ_{ij} defined in Eq. (1.39) may be written (Volokitin, Krasnosel'skikh & Machabeli 1985; Melrose *et al.* 1999)

$$\Lambda_{11} = a - \frac{b}{z^2}, \quad \Lambda_{22} = a - \frac{b - \sin^2\theta}{z^2} \tag{8.84}$$

$$\Lambda_{13} = \Lambda_{31} = \frac{b}{z^2}\tan\theta \tag{8.85}$$

$$\Lambda_{33} = 1 - \frac{\omega_p^2}{\omega^2}z^2 W(z) - \frac{b}{z^2}\tan^2\theta \tag{8.86}$$

where

$$a = 1 + \frac{1}{v_A^2}, \quad b = 1 - \frac{\tilde{v}^2}{v_A^2}. \tag{8.87}$$

We note that, as in the cold pair plasma case, the gyrotropic terms K_{12} and K_{21} of the dielectric tensor, and thence the elements Λ_{12} and Λ_{21}, are zero. This is no longer true if the electron and positron densities are unequal, or if the species move relative to each other, as occurs in the rotating pulsar, however such effects are expected to be small (Melrose *et al.* 1999). Nevertheless the gyrotropic terms, which are due to the nonbalance of the Hall currents and also arise in systems such as normal plasmas near the ion cyclotron frequency, and dusty plasmas (see Section 7.4), and which induce elliptic polarization rather than linear polarization of the waves, may be needed to explain observed circularly polarized radiation.

The dispersion equation then reduces to the two equations

$$\Lambda_{22} \;=\; 0 \tag{8.88a}$$
$$\Lambda_{11}\Lambda_{33} \;=\; \Lambda_{13}^2. \tag{8.88b}$$

The magnetoacoustic mode is described by Eq. (8.88a). As in the cold plasma case, it is polarized perpendicular to both B and k, and is independent of the two modes polarized in the plane of B and k, the Langmuir-Alfvén modes, which are described by Eq. (8.88b). The resulting dispersion relations for the Langmuir-Alfvén modes are shown in Figure 8.10 for the pulsar plasma with $v_A \simeq 3$ (Melrose *et al.* 1999). Comparing this figure with Figure 8.9, the cyclotron resonance is absent because cyclotron effects have been neglected for simplicity, and the lower angle–dependent resonance is washed out due to collisionless damping. For parallel propagation, the Alfvén mode is described by $\Lambda_{11} = 0$ and the Langmuir wave by $\Lambda_{33} = 0$. The parallel Langmuir and Alfvén modes cross near where the frequency of the Langmuir waves has a maximum.

For oblique propagation, these two modes are coupled by the term Λ_{13}. This induces a reconnection of the dispersion curves, with the former maximum frequency of the Langmuir mode now in the Alfvén mode, and the Langmuir mode joining on to the O mode branch of the former Alfvén mode curve, to form the L-O mode. The waves are now mixed electrostatic-electromagnetic modes, with an electric field component along the magnetic field. At low frequency the phase speed of the parallel Alfvén mode (and the degenerate magnetoacoustic mode) is

$$z_A = \frac{v_A}{(1 + v_A^2)^{1/2}} \simeq 1 - \frac{1}{2v_A^2} \tag{8.89}$$

as for the cold pair plasma. Thus the Alfvén mode is subluminous ($\omega/k < c$). The Langmuir mode, however, has a cutoff and is superluminous. However for oblique propagation, the L-O mode is subluminous at higher frequencies.

8.3.3 Instabilities

The mechanism of radio emission from pulsars is believed to be associated with an instability of the pair plasma produced in the polar cap regions of the rapidly rotating neutron star. Instabilities involving the Langmuir waves, and possibly the Alfvén waves, are widely favoured.

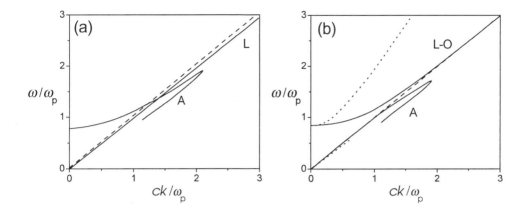

Figure 8.10: Dispersion curves for a mildly relativistic particle distribution in a pair plasma. The dashed diagonal line is the light–line. (a) For parallel propagation the solid line just to the right of the light–line corresponds to the parallel Alfvén wave, and the other curve is for the Langmuir mode, which has a maximum frequency, with a heavily damped branch or reverse dispersion to the right of the main dispersion curve. (b) As for (a) but for oblique propagation, with $\theta = 0.1$ rad for the solid curves and $\theta = 1$ rad for the dotted curves. (Reproduced from Melrose 2000.)

An energetic primary beam of particles in the magnetic field direction, with $\gamma_b \simeq 10^7$, generates a bulk plasma via a cascade process of generation of gamma rays and production of electron–positron pairs. The plasma so produced has a bulk velocity corresponding to $\gamma_p \simeq 10^3$-10^4 and a mean Lorentz factor $\langle \gamma \rangle \gg 1$ in the frame of the bulk plasma (Melrose 2000). A beam instability is most widely favoured, in which waves in the bulk plasma resonate with the energetic beam particles (see Chapter 5). Only subluminal waves with a phase speed less than the beam speed can grow in a beam instability, and the wave must have a component of electric field along the magnetic field. The latter requirement precludes the magnetoacoustic mode and the parallel Alfvén mode. The parallel electric field also leads to damping of the waves as they propagate through the magnetosphere.

If the beam excites Langmuir waves, the waves can, via nonlinear processes, produce transverse electromagnetic waves at the fundamental or second harmonic of the plasma frequency, a process known as relativistic plasma emission, in analogy to the process of plasma emission believed to be responsible for radio burst emission from the Sun. The beam instability of the Langmuir waves is reactive or hydrodynamic (see Section 5.2.2), but only occurs at a frequency where the wave becomes subluminal. Obliquely propagating L-O waves can be excited at a much higher frequency, and Alfvén waves can be excited at a much lower frequency than the frequency of the parallel Langmuir waves. However, none of these processes can account for the observed radio emission for plausible pulsar parameters (Melrose & Gedalin 1999).

The nonlinear interaction of Alfvén waves in pair plasmas has also been invoked in a form of relativistic plasma emission involving the Alfvén waves rather than the Langmuir waves, with the conversion of Alfvén waves into ordinary electromagnetic waves (Kaplan & Tsytovich 1973).

If there is an imbalance in the electron and positron number densities, we have seen that the parallel propagating modes are circularly polarized, which may account for observed circular polarization in pulsar radiation (Allen & Melrose 1982). A cyclotron instability exciting such modes, due to the anomalous Doppler effect (see Section 5.2.2), has been invoked by Kazbegi, Machabeli & Melikidze (1991). The cyclotron instability will operate near the light cylinder, and the waves will quasilinearly interact with the beam particles to cause diffusion of the particles in momentum space both along and across the magnetic field, resulting in further synchrotron losses and slowing of the beam. This synchrotron radiation from the beam particles has been postulated as the source of the gamma and X–radiation of young pulsars. As the pulsar evolves the region of the cyclotron instability approaches the light cylinder, and it disappears $\sim 10^4$ years after the pulsar forms.

8.4   Concluding Remarks

We have seen in this chapter that Alfvén waves play important roles in the most extreme physical environments of the universe, from tenuous interstellar clouds encompassing vast distances, where stars are born, to the extreme energies and magnetic fields of plasmas surrounding neutron stars. The same basic physics covers these waves as that governing the behaviour of Alfvén waves in the laboratory, and waves of wavelength relevant to the Earth, the Sun and interplanetary space, as we have found in this book. As the basic oscillation in a magnetized plasma, the Alfvén wave will continue to be a source of fascination for researchers, both those concerned with mundane questions of energy generation on the Earth, and those investigating the plasma universe.

Appendix: The Dielectric Tensor

The kinetic theory of collisionless plasmas uses the Vlasov equation for the particle momentum distribution function $f_\alpha(p)$, which satisfies the normalization condition

$$n_\alpha = \int d^3 p f_\alpha(p) \tag{A.1}$$

where n_α is the number density of the species α. The theory gives the following expression for the dielectric tensor (Melrose 1986):

$$K_{ij}(\omega, k) = \delta_{ij} + \sum_\alpha \frac{q_\alpha^2}{\epsilon_0 \omega^2} \int d^3 p \left[\frac{v_\parallel}{v_\perp} \left(v_\perp \frac{\partial}{\partial p_\parallel} - v_\parallel \frac{\partial}{\partial p_\perp} \right) f_\alpha(p) b_i b_j \right.$$
$$\left. + \sum_{s=-\infty}^{\infty} \frac{V_i(k, p; s) V_j^*(k, p; s)}{\omega - s\Omega_\alpha - k_\parallel v_\parallel} \left(\frac{\omega - k_\parallel v_\parallel}{v_\perp} \frac{\partial}{\partial p_\perp} + k_\parallel \frac{\partial}{\partial p_\parallel} \right) f_\alpha(p) \right] \tag{A.2}$$

with b a unit vector along the background magnetic field. The sum is over particle species α. Here

$$V(k, p; s) = \left(\frac{1}{2} v_\perp \{ e^{i\epsilon\psi} J_{s-1}(k_\perp R) + e^{-i\epsilon\psi} J_{s+1}(k_\perp R) \}, \right.$$
$$\left. -\frac{i\epsilon v_\perp}{2} \{ e^{i\epsilon\psi} J_{s-1}(k_\perp R) - e^{-i\epsilon\psi} J_{s+1}(k_\perp R) \}, v_\parallel J_s(k_\perp R) \right) \tag{A.3}$$

where

$$k = (k_\perp \cos\psi, k_\perp \sin\psi, k_\parallel), \tag{A.4}$$

$R = v_\perp/\Omega$ and J_s is an ordinary Bessel function. ϵ is the sign of the charge and Ω is the (positive) cyclotron frequency. The particle distribution function $f_\alpha(p)$ is assumed to depend only on the momenta $p_\perp$ and $p_\parallel$.

In the small gyroradius limit $k_\perp \rho = 0$, $J_s(k_\perp \rho)$ is zero except for $J_0(0) = 1$. Then Eq. (A.3) simplifies to

$$V(k, p; s) = \quad (0, i\epsilon v_\perp k_\perp R, v_\parallel) \quad s = 0 \tag{A.5}$$
$$\tfrac{1}{2} v_\perp (1, -is\epsilon, 0) \quad s = \pm 1 \tag{A.6}$$
$$0 \quad\quad |s| > 1. \tag{A.7}$$

For a bi–Maxwellian particle distribution with a drift velocity U along the magnetic field,

$$f(p) = \frac{n}{(2\pi)^{3/2} V_\perp^2 V_\parallel} \exp\left[-\frac{v_\perp^2}{2V_\perp^2} - \frac{(v_\parallel - U)^2}{2V_\parallel^2} \right] \tag{A.8}$$

with $V_\perp$ and $V_\parallel$ the thermal velocities corresponding to the perpendicular and parallel temperatures, given by

$$V_\perp^2 = k_B T_\perp / m, \quad V_\parallel^2 = k_B T_\parallel / m \tag{A.9}$$

where k_B is Boltzmann's constant.

The dielectric tensor with a bi–Maxwellian particle distribution function, defined by Eq. (A.8) for each species α may be written, setting $\psi = 0$, that is $k_y = 0$ and $k_x = k_\perp$,

$$
\begin{aligned}
K_{ij}(\omega, k) = \delta_{ij} + \sum_\alpha \frac{\omega_{p\alpha}^2}{\omega^2} &\left[A_\alpha a_{ij}(k) + \frac{\omega^2}{k_\parallel^2 V_{\alpha\parallel}^2} b_i b_j \right. \\
&\left. + \sum_{s=-\infty}^{\infty} z_s^\alpha Z(z_s^\alpha) \left\{ \frac{\omega - k_\parallel U_\alpha}{\omega - s\Omega_\alpha - k_\parallel U_\alpha} + A_\alpha \right\} N_{ij}^\alpha(\omega, k, s) \right]
\end{aligned}
\tag{A.10}
$$

where $A_\alpha = T_{\alpha\parallel} / T_{\alpha\perp} - 1$ and

$$z_s^\alpha = \frac{\omega - s\Omega_\alpha - k_\parallel U_\alpha}{\sqrt{2} k_\parallel V_{\alpha\parallel}}. \tag{A.11}$$

Also,

$$a_{ij} = \begin{bmatrix} 1 & 0 & -k_\perp/k \\ 0 & 1 & 0 \\ -k_\perp/k_\parallel & 0 & k_\perp^2/k_\parallel^2 \end{bmatrix}. \tag{A.12}$$

The function $Z(z)$ appearing in Eq. (A.10) is a form of the plasma dispersion function (Fried & Conte 1961),

$$Z(z) = \frac{1}{\sqrt{\pi}} \int_{-\infty}^{\infty} \frac{e^{-t^2} dt}{t - z}. \tag{A.13}$$

The approximate form of $Z(z)$ for $|z| \ll 1$ is

$$Z(z) \simeq -2z \left(1 - \frac{2}{3} z^2\right) + i\sqrt{\pi} e^{-z^2}, \tag{A.14}$$

while for $|z| \gg 1$ the asymptotic representation of Z is

$$Z(z) \simeq -\frac{1}{z} \left(1 + \frac{1}{2z^2}\right) + i\sigma\sqrt{\pi} e^{-z^2}, \tag{A.15}$$

with $\sigma = 0$, 1, 2 depending on whether $y > |x|$, $|y| < |x|$ or $y < -|x|$ respectively, where $z = x + iy$.

The N_{ij}^{α}'s appearing in Eq. (A.10) are given by (dropping the α index for simplicity)

$$N_{11} = e^{-\lambda} \frac{s^2}{\lambda} I_s \qquad (A.16)$$

$$N_{22} = e^{-\lambda} \left(\frac{s^2}{\lambda} + 2\lambda \right) I_s - 2\lambda I_s' \qquad (A.17)$$

$$N_{33} = e^{-\lambda} \frac{1}{\lambda} \frac{k_\perp^2}{k_\parallel^2} \left(\frac{\omega - s\Omega}{\Omega} \right)^2 I_s \qquad (A.18)$$

$$N_{12} = -N_{21} = e^{-\lambda} i\epsilon s (I_s' - I_s) \qquad (A.19)$$

$$N_{13} = N_{31} = e^{-\lambda} \frac{s}{\lambda} \frac{k_\perp}{k_\parallel} \left(\frac{\omega - s\Omega}{\Omega} \right) I_s \qquad (A.20)$$

$$N_{23} = -N_{32} = -i\epsilon \frac{k_\perp}{k_\parallel} \left(\frac{\omega - s\Omega}{\Omega} \right) (I_s' - I_s) \qquad (A.21)$$

where $I_s = I_s(\lambda)$ is a modified Bessel function of $\lambda = k_\perp^2 \rho^2$, where $\rho = V_\perp / \Omega$ is the particle Larmor radius. Two properties of the modified Bessel functions are that $I_{-s}(\lambda) = I_s(\lambda)$ and

$$\sum_{s=-\infty}^{\infty} e^{-\lambda} I_s(\lambda) = 0. \qquad (A.22)$$

Also, for $\lambda \ll 1$ we have the expansions

$$e^{-\lambda} I_0(\lambda) = 1 - \lambda + \frac{3}{4}\lambda^2 + \cdots \qquad (A.23)$$

$$e^{-\lambda} I_s(\lambda) = \frac{\lambda^s}{2^s s!} (1 - \lambda + \cdots) \quad (s \geq 1). \qquad (A.24)$$

The dielectric tensor for the bi–Maxwellian plasma and with $\psi \neq 0$ ($k_y \neq 0$), required for the description of surface waves by the kinetic theory in Section 4.8, may be obtained from Eqs. (A.2) and (A.3). It can also be easily obtained from the $k_y = 0$, $k_\perp = k_x$, tensor $\mathbf{K}^0$ given by Eq. (A.10), by a unitary rotation about the z–axis given by

$$\mathbf{K}(k_y \neq 0) = \mathbf{R}\mathbf{K}^0\mathbf{R}^{\mathrm{T}} \qquad (A.25)$$

where

$$\mathbf{R} = \begin{bmatrix} \cos\psi & -\sin\psi & 0 \\ \sin\psi & \cos\psi & 0 \\ 0 & 0 & 1 \end{bmatrix} \qquad (A.26)$$

is the unitary rotation matrix, $\mathbf{R}^{\mathrm{T}}$ is the transpose of $\mathbf{R}$ and $k_\perp = (k_x^2 + k_y^2)^{1/2}$. The new dielectric tensor components are then given by (Mikhailovskii 1967)

$$
\begin{aligned}
K_{11} &= \cos^2\psi\, K_{11}^0 + \sin^2\psi\, K_{22}^0 \\
K_{12} &= K_{12}^0 + \cos\psi\,\sin\psi\,(K_{11}^0 - K_{22}^0) \\
K_{13} &= \cos\psi\, K_{13}^0 - \sin\psi\, K_{23}^0 \\
K_{21} &= K_{21}^0 + \cos\psi\,\sin\psi\,(K_{11}^0 - K_{22}^0) \\
K_{22} &= \cos^2\psi\, K_{22}^0 + \sin^2\psi\, K_{11}^0 \\
K_{23} &= \cos\psi\, K_{23}^0 + \sin\psi\, K_{13}^0 \\
K_{31} &= \cos\psi\, K_{31}^0 - \sin\psi\, K_{32}^0 \\
K_{32} &= \cos\psi\, K_{32}^0 + \sin\psi\, K_{31}^0 \\
K_{33} &= K_{33}^0.
\end{aligned}
\tag{A.27}
$$

References

Achterberg, A. (1981), 'On the propagation of relativistic particles in a high β plasma', *Astron. Astrophys.* **98**, 161–172.

Achterberg, A. & Blandford, R. D. (1986), 'Transmission and damping of hydromagnetic waves behind a strong shock front: implications for cosmic ray acceleration', *Mon. Not. R. Astr. Soc.* **218**, 551–575.

Agranovich, V. M. & Chernyak, V. Y. (1982), 'Perturbation theory for weakly nonlinear p–polarized surface polaritons', *Solid State Communications* **44**, 1309–1311.

Akhiezer, A. I., Akhiezer, I. A., Polovin, R. V., Sitenko, A. G. & Stepanov, K. N. (1975), *Plasma Electrodynamics Vol I: Linear Theory*, Pergamon Press, Oxford.

Alexandrov, A. F., Bogdankevich, L. S. & Rukhadze, A. A. (1984), *Principles of Plasma Electrodynamics*, Springer, Berlin.

Alfvén, H. (1942), 'Existence of electromagnetic-hydrodynamic waves', *Nature* **150**, 405.

Allen, M. C. & Melrose, D. B. (1982), 'Elliptically polarized natural modes in pulsar magnetospheres', *Proc. Ast. Soc. Aust.* **4**, 365–370.

Amagishi, Y. & Tanaka, M. (1993), 'Ion-neutral collision effect on an Alfvén wave', *Phys. Rev. Lett.* **71**, 360–363.

Amagishi, Y. & Tsushima, A. (1984), 'Formation of spatial Alfvén resonance and cutoff fields in an inhomogeneous plasma', *Plasma Phys. Contr. Fusion* **26**, 1489–1503.

Amagishi, Y., Saeki, K. & Donnelly, I. J. (1989), 'Excitation of MHD surface waves propagating with shear Alfvén waves in an inhomogeneous cylindrical plasma', *Plasma Phys. Contr. Fusion* **31**, 675–691.

An, C.-H., Suess, S. T., Moore, R. L. & Musielak, Z. E. (1990), 'Reflection and trapping of Alfvén waves in a spherically symmetric stellar atmosphere', *Astrophys. J.* **350**, 309–323.

Appert, K., Balet, B., Gruber, R., Troyon, F., Tsunematsu, T. & Vaclavik, J. (1980), *in* 'Proceedings of 2nd joint Varenna-Grenoble international symposium on heating in toroidal plasmas', Como, p. 643.

Appert, K., Gruber, R. & Vaclavik, J. (1974), 'Continuous spectra of a cylindrical magnetohydrodynamic equilibrium', *Phys. Fluids* **17**, 1471–1472.

Appert, K., Gruber, R., Troyon, F. & Vaclavik, J. (1982), 'Excitation of global eigenmodes of the Alfvén wave in tokamaks', *Plasma Phys.* **24**, 1147–1159.

Appert, K., Vaclavik, J. & Villard, L. (1984), 'Spectrum of low–frequency, nonaxisymmetric oscillations in a cold, current-carrying plasma column', *Phys. Fluids* **27**, 432–437.

Arons, J. & Barnard, J. J. (1986), 'Wave propagation in pulsar magnetospheres: dispersion relations and normal modes of plasmas in superstrong magnetic fields', *Astrophys. J.* **302**, 120–137.

Arons, J. & Max, C. E. (1975), 'Hydromagnetic waves in molecular clouds', *Astrophys. J.* **196**, L77–L81.

Assis, A. S. & Busnardo-Neto, J. (1987), 'Cherenkov damping of surface waves', *Astrophys. J.* **323**, 399–405.

Balsara, D. (1996), 'Wave propagation in molecular clouds', *Astrophys. J.* **465**, 775–794.

Barnes, A. (1979), Hydromagnetic waves and turbulence in the solar wind, *in* C. F. Kennel, L. J. Lanzerotti & E. N. Parker, eds, 'Solar System Plasma Physics', Vol. 1, North Holland, Amsterdam.

Barr, H. C. & Boyd, T. J. M. (1972), 'Surface waves in hot plasmas', *J. Phys. A: Gen. Phys.* **A5**, 1108–1118.

Baynham, A. C. & Boardman, A. D. (1971), *Plasma Effects in Semiconductors: Helicons and Alfvén Waves*, Taylor and Francis, London.

Belcher, J. W. & Davis, L. (1971), 'Large-amplitude Alfvén waves in the interplanetary medium', *J. Geophys. Res.* **76**, 3534–3563.

Bellan, P. M. (1994), 'Alfvén 'resonance' reconsidered: exact equations for wave propagation across a cold inhomogeneous plasma', *Phys. Plasmas* **1**, 3523–3541.

Bellan, P. M. (1995), 'Response to Comments on "Alfvén resonance reconsidered: exact equations for wave propagation across a cold inhomogeneous plasma"', *Phys. Plasmas* **2**, 3552–3556.

Bellan, P. M. (1996a), 'Comment on "on the Alfvén resonance and its existence"', *Phys. Plasmas* **3**, 435–438.

Bellan, P. M. (1996b), 'Mode conversion into non–MHD waves at the Alfvén layer: the case against the field line resonance concept', *J. Geophys. Res.* **101**(A11), 24887–24898.

Bennett, K., Roberts, B. & Narain, U. (1999), 'Waves in twisted magnetic flux tubes', *Solar Phys.* **185**, 41–59.

Betti, R. & Freidberg, J. P. (1991), 'Ellipticity induced Alfvén eigenmodes', *Phys. Fluids* **B 3**, 1865–1870.

Biermann, L., Brosowski, B. & Schmidt, H. (1967), 'The interaction of the solar wind with a comet', *Solar. Phys.* **1**, 254–284.

Blandford, R. D. & Payne, D. G. (1982), 'Hydromagnetic flows from accretion discs and the production of radio jets', *Mon. Not. R. Ast. Soc.* **199**, 883–903.

Bliokh, P. V., Sinitsin, V. & Yaroshenko, V. (1995), *Dusty and Self-Gravitational Plasmas in Space*, Kluwer Academic Publishers, Dordrecht.

Borg, G. G. (1994), 'Shear Alfvén wave excitation by direct antenna coupling and fast-wave resonant-mode conversion', *Plasma Phys. Contr. Fusion* **36**, 1419–1438.

Borg, G. G. (1998), 'Direct antenna excitation of the shear Alfvén wave at finite frequency in the boundary of a fusion plasma', *J. Plasma Phys.* **59**, 151–168.

Brandenburg, A. & Zweibel, E. G. (1994), 'The formation of sharp structures by ambipolar diffusion', *Astrophys. J.* **427**, L91–L94.

Brandenburg, A. & Zweibel, E. G. (1995), 'Effects of pressure and resistivity on the ambipolar diffusion singularity: too little, too late', *Astrophys. J.* **448**, 734–741.

Braun, D. C., Duvall, T. L. & LaBonte, B. J. (1988), 'The absorption of high-degree p-mode oscillations in and around sunspots', *Astrophys. J.* **335**, 1015–1025.

Brinca, A. L., Borda de Auga, L. & Winske, D. (1993), 'On the stability of nongyrotropic ion populations: a first (analytic and simulation) assessment', *J. Geophys. Res.* **98**, 7549–7560.

Buchert, S. C. (1998), 'Magneto-optical Kerr effect for a dissipative plasma', *J. Plasma Phys.* **59**, 39–55.

Champeaux, S., Gazol, A., Passot, T. & Sulem, P.-L. (1997a), 'Plasma heating by Alfvén wave filamentation: a relevant mechanism in the solar corona and the interstellar medium', *Astrophys. J.* **486**, 477–483.

Champeaux, S., Passot, T. & Sulem, P. L. (1997b), 'On Alfvén wave filamentation', *J. Plasma Phys.* **58**, 665–690.

Chan, V. S., Miller, R. L. & Ohkawa, T. (1990a), 'Current drive by wave helicity injection', *Phys. Fluids* **B 2**, 1441–1445.

Chan, V. S., Miller, R. L. & Ohkawa, T. (1990b), 'Low–frequency current drive and helicity injection', *Phys. Fluids* **B 2**, 944–952.

Chance, M. S., Greene, J. M., Grimm, R. C. & Johnson, J. L. (1977), 'Study of the MHD spectrum of an elliptic plasma column', *Nucl. Fusion* **17**, 65–83.

Chen, L. & Hasegawa, A. (1974a), 'Plasma heating by spatial resonance of Alfvén waves', *Phys. Fluids* **17**, 1399–1403.

Chen, L. & Hasegawa, A. J. (1974b), 'A theory of long-period magnetic pulsations. 1. Steady state excitation of field line resonance', *J. Geophys. Res.* **79**, 1024–1032.

Cheng, C. Z. & Chance, M. S. (1986), 'Low–n shear Alfvén spectra in axisymmetric toroidal plasmas', *Phys. Fluids* **29**, 3695–3701.

Cheng, C. Z., Chen, L. & Chance, M. S. (1985), 'High–n ideal and resistive shear Alfvén waves in tokamaks', *Ann. Phys.* **161**, 21–47.

Chew, G. F., Goldberger, M. L. & Low, F. E. (1956), 'The Boltzmann equation and the one–fluid hydromagnetic equations in the absence of particle collisions', *Proc. R. Soc. Lond.* **A236**, 112.

Chian, A. C.-L. (1995), 'Nonlinear Alfvén wave phenomena in the planetary magnetosphere', *Phys. Scripta* **T60**, 36–43.

Chian, A. C.-L. & Oliveira, L. P. L. (1994), 'Magnetohydrodynamic parametric instabilities driven by a standing Alfvén wave in the planetary magnetosphere', *Astron. Astrophys.* **286**, L1–L4.

Choudhuri, A. R. (1998), *The Physics of Fluids and Plasmas*, Cambridge University Press, Cambridge.

Clancy, B. E. & Donnelly, I. J. (1986), 'The calculation of wave fields in a cylindrical plasma using a multiple shooting technique', *Comp. Phys. Commun.* **42**, 153–167.

Clemmow, P. C. & Dougherty, J. P. (1969), *Electrodynamics of Particles and Plasmas*, Addison–Wesley Publishing Company, Reading, Massachusetts.

Coleman, P. J. (1967), 'Wave–like phenomena in the interplanetary plasma: Mariner 2', *Planet. Space Sci.* **15**, 953–973.

Collins, G. A., Cramer, N. F. & Donnelly, I. J. (1984), 'MHD surface waves in a resistive cylindrical plasma at low to ion cyclotron frequency', *Plasma Phys. Contr. Fusion* **26**, 273–292.

Cramer, N. F. (1975), 'Parametric excitation of ion–cyclotron waves', *Plasma Phys.* **17**, 967–972.

Cramer, N. F. (1976), 'Parametric excitation of ion–cyclotron and fast waves by a nonuniform pump magnetic field', *Plasma Phys.* **18**, 749–760.

Cramer, N. F. (1977), 'Parametric excitation of magnetoacoustic waves by a pump magnetic field in a high β plasma', *J. Plasma Phys.* **17**, 93–103.

Cramer, N. F. (1991), 'Nonlinear surface Alfvén waves', *J. Plasma Phys.* **46**, 15–27.

Cramer, N. F. (1994), 'Magnetoaoustic surface waves on current sheets', *J. Plasma Phys.* **51**, 221–232.

Cramer, N. F. (1995), 'The theory of Alfvén surface waves', *Phys. Scripta* **T60**, 185–197.

Cramer, N. F. & Donnelly, I. J. (1981), 'The parametric excitation of kinetic Alfvén waves by a magnetic pump', *J. Plasma Phys.* **26**, 253–266.

Cramer, N. F. & Donnelly, I. J. (1983), 'Fast and ion–cyclotron surface waves at a plasma-vacuum interface', *Plasma Phys.* **25**, 703–712.

Cramer, N. F. & Donnelly, I. J. (1984), 'Surface and discrete Alfvén waves in a current carrying plasma', *Plasma Phys. Contr. Fusion* **26**, 1285–1298.

Cramer, N. F. & Donnelly, I. J. (1985), 'Perturbations of a twisted solar coronal loop: the relation between surface waves and instabilities', *Solar Phys.* **99**, 119–132.

Cramer, N. F. & Donnelly, I. J. (1992), Magnetohydrodynamic surface waves, *in* J. Halevi, ed., 'Spatial Dispersion in Solids and Plasmas', Elsevier, Amsterdam, pp. 521–555.

Cramer, N. F. & Sy, W. N.-C. (1979), 'Parametric decay of magnetoacoustic oscillations in a cylindrical plasma', *J. Plasma Phys.* **22**, 549–562.

Cramer, N. F. & Vladimirov, S. V. (1996), 'The Alfvén resonance in a magnetized dusty plasma', *Phys. Scripta* **53**, 586–590.

Cramer, N. F. & Vladimirov, S. V. (1996b), 'Alfvén surface waves in a magnetized dusty plasma', *Phys. Plasmas* **3**, 4740–4747.

Cramer, N. F. & Vladimirov, S. V. (1997), 'Alfvén waves in dusty interstellar clouds', *Publ. Astron. Soc. Aust.* **14**, 170–178.

Cramer, N. F. & Vladimirov, S. V. (1998), 'The resonance absorption of wave energy in a dusty plasma', *Phys. Scripta* **T75**, 213–215.

Cramer, N. F. & Yung, C.-M. (1986), 'Surface waves in a two–ion–species magnetized plasma', *Plasma Phys. Contr. Fusion* **28**, 1043–1054.

Cramer, N. F., Verheest, F. & Vladimirov, S. V. (1999), 'Instabilities of Alfvén and magnetosonic waves in dusty cometary plasmas with an ion ring beam', *Phys. Plasmas* **6**, 36–43.

Cramer, N. F., Yeung, L. K. & Vladimirov, S. V. (1998), 'Surface waves in a magnetized plasma with mobile dust grains', *Phys. Plasmas* **5**, 3126–3134.

Cranmer, S. R., Field, G. B. & Kohl, J. L. (1999), 'Spectroscopic constraints of ion cyclotron resonance heating in the polar solar corona and high–speed solar wind', *Astrophys. J.* **518**, 937–947.

Cross, R. C. (1983), 'Experimental observations of localized Alfvén and ion acoustic waves in a plasma', *Plasma Phys.* **25**, 1377–1387.

Cross, R. C. (1988), *An Introduction to Alfvén Waves*, Adam Hilger, Bristol.

Cross, R. C. & Lehane, J. A. (1968), 'Hydromagnetic waves in a nonuniform hydrogen plasma', *Phys. Fluids* **11**, 2621–2626.

Cross, R. C. & Murphy, A. B. (1986), 'Alfvén wave modes in a cylindrical plasma with finite edge density', *Plasma Phys. Contr. Fusion* **28**, 597–612.

Davidson, R. C. & Ogden, J. M. (1975), 'Electromagnetic ion cyclotron instability driven by ion energy anisotropy in high–beta plasmas', *Phys. Fluids* **18**, 1045–1050.

Davila, J. M. & Scott, J. S. (1984), 'The scattering of energetic particles by waves in a finite β plasma', *Astrophys. J.* **280**, 334–338.

Davis, C. L. & Donnelly, I. J. (1986), 'Discrete Alfvén waves in planar flux tubes', *Astrophys. Sp. Sci.* **127**, 265–284.

de Assis, A. S. & Tsui, K. H. (1991), 'Coronal loop heating by wave-particle interactions', *Astrophys. J.* **366**, 324–327.

de Chambrier, A., Heym, A., Hoffmann, F., Joye, B., Keller, R., Liette, A., Lister, J. B., Morgan, P. D., Peacock, N. J., Pochelon, A. & Stamp, M. F. (1983), 'Measurements of electron and ion heating by Alfvén waves in the TCA tokamak', *Plasma Phys.* **25**, 1021–1035.

Derby, N. F. (1978), 'Modulational instability of finite–amplitude, circularly polarized Alfvén waves', *Astrophys. J.* **224**, 1013–1016.

Dewar, R. L. & Davies, B. J. (1984), 'Bifurcation of the resistive Alfvén wave spectrum', *J. Plasma Phys.* **32**, 443–461.

Dewar, R. L., Grimm, R. C., Johnson, J. L., Frieman, E. A., Greene, J. M. & Rutherford, P. H. (1974), 'Long–wavelength kink instabilities in low–pressure, uniform axial current, cylindrical plasmas with elliptic cross sections', *Phys. Fluids* **17**, 930–938.

D'Ippolito, D. A. & Goedbloed, J. P. (1980), 'Mode coupling in a toroidal, sharp–boundary plasma. I. Weak- coupling limit', *Plasma Phys.* **22**, 1091–1107.

Dolgopolov, V. V. & Stepanov, K. N. (1966), 'Resonance absorption of low–frequency wave energy in a cold inhomogeneous plasma', *Sov. Phys. Tech. Phys.* **11**, 741–743.

Donnelly, I. J. & Cramer, N. F. (1984), 'MHD surface waves in a cylindrical plasma with a spatial density variation', *Plasma Phys. Contr. Fusion* **26**, 769–787.

Donnelly, I. J., Clancy, B. E. & Cramer, N. F. (1985), 'Alfvén wave heating of a cylindrical plasma using axisymmetric waves. Part 1. MHD theory.', *J. Plasma Phys.* **34**, 227–246.

Donnelly, I. J., Clancy, B. E. & Cramer, N. F. (1986), 'Alfvén wave heating of a cylindrical plasma using axisymmetric waves. Part 2. Kinetic theory', *J. Plasma Phys.* **35**, 75–106.

Dungey, J. W. (1954), Electrodynamics of the outer atmosphere, Report 69, Pennsylvania State University Ionosphere Research Laboratory.

Durrer, R., Kashnioshvili, T. & Yates, A. (1998), 'Microwave background anisotropies from Alfvén waves', *Phys. Rev.* **D 58**, 123004/1–123004/9.

Edwin, P. M. & Roberts, B. (1983), 'Wave propagation in a magnetic cylinder', *Solar Phys.* **88**, 179–191.

Einaudi, G. & van Hoven, G. (1983), 'The stability of coronal loops: finite–length and pressure–profile limits', *Solar Phys.* **88**, 163–177.

Elfimov, A. G. & Nekrasov, F. M. (1973), 'Decay instability of magnetohydrodynamic oscillations in a plasma cylinder in the field of a direct magneto–sonic wave', *Nucl. Fusion* **13**, 653–659.

Elfimov, A. G., Petrzilka, V. & Tataronis, J. A. (1994a), 'Radial plasma transport and toroidal current driven by nonresonant ponderomotive forces', *Phys. Plasmas* **1**, 2882–2889.

Elfimov, A. G., Tataronis, J. A. & Hershkowitz, N. (1994b), 'The influence of multiple ion species on Alfvén wave dispersion and Alfvén wave plasma heating', *Phys. Plasmas* **1**, 2637–2644.

Faria, R. T., Mirza, A. M., Shukla, P. K. & Pokhotelov, O. A. (1998), 'Linear and nonlinear dispersive Alfvén waves in two–ion plasmas', *Phys. Plasmas* **5**, 2947–2951.

Fiedler, R. (1986), 'Magnetohydrodynamic solitons and radio knots in jets', *Astrophys. J.* **305**, 100–108.

Fisch, N. J. (1987), 'Theory of current drive in plasmas', *Rev. Mod. Phys.* **59**, 175–234.

Foote, E. A. & Kulsrud, R. M. (1979), 'Hydromagnetic waves in high β plasmas', *Astrophys. J.* **233**, 302–316.

Fried, B. D. & Conte, S. D. (1961), *The Plasma Dispersion Function*, Academic Press, New York.

Fu, G. Y. & Van Dam, J. W. (1989), 'Excitation of the toroidicity–induced shear Alfvén eigenmode by fusion alpha particles in an ignited tokamak', *Phys. Fluids* **B 1**, 1949–1952.

Galeev, A. A. (1991), *in* R. L. Newburn, ed., 'Comets in the Post–Halley Era', Kluwer, Dordrecht.

Galeev, A. A. & Oraevskii, V. N. (1963), 'The stability of Alfvén waves', *Soviet Phys. Dokl.* **7**, 988.

Gary, S. P. (1986), 'Low-frequency waves in a high-beta collisionless plasma: polarization, compressibility and helicity', *J. Plasma Phys.* **35**, 431–447.

Gary, S. P. (1991), 'Electromagnetic ion/ion instabilities and their consequences in space plasmas: a review', *Space Sci. Revs.* **56**, 373–415.

Gary, S. P. (1993), *Theory of Space Plasma Microinstabilities*, Cambridge Atmospheric and Space Science Series, Cambridge University Press, Cambridge.

Gary, S. P., Montgomery, M. D., Feldman, W. C. & Forslund, D. W. (1976), 'Proton temperature anisotropy instabilities in the solar wind', *J. Geophys. Res.* **81**, 1241–1246.

Gekelman, W. (1999), 'Review of laboratory experiments on Alfvén waves and their relationship to space observations', *J. Geophys. Res.* **104**, 14417–14435.

Gekelman, W., Leneman, D., Maggs, J. & Vincena, S. (1994), 'Experimental observation of Alfvén wave cones', *Phys. Plasmas* **1**, 3775–3783.

Gekelman, W., Vincena, S., Leneman, D. & Maggs, J. (1997), 'Laboratory experiments on shear Alfvén waves and their relationship to space plasmas', *J. Geophys. Res.* **102**, 7225–7235.

Ghosh, S. & Papadopoulos, K. (1987), 'The onset of Alfvénic turbulence', *Phys. Fluids* **30**, 1371–1387.

Ginzburg, V. L. (1964), *The Propagation of Electromagnetic Waves in Plasmas*, Pergamon Press, Oxford.

Goedbloed, J. P. (1975), 'Spectrum of ideal magnetohydrodynamics of axisymmetric toroidal systems', *Phys. Fluids* **18**, 1258–1268.

Goedbloed, J. P. (1998), 'Once more: the continuous spectrum of ideal magnetohydrodynamics', *Phys. Plasmas* **5**, 3143–3154.

Goedbloed, J. P. & Hagebeuk, H. J. L. (1972), 'Growth rates of instabilities of a diffuse linear pinch', *Phys. Fluids* **15**, 1090–1101.

Goedbloed, J. P. & Lifschitz, A. (1995), 'Comment on "Alfvén resonance reconsidered: exact equations for wave propagation across a cold inhomogeneous plasma"', *Phys. Plasmas* **2**, 3550–3551.

Goertz, C. K. (1984), 'Kinetic Alfvén waves on auroral field lines', *Planet. Space Sci.* **32**, 1387–1392.

Goertz, C. K. (1989), 'Dusty plasmas in the solar system', *Rev. Geophys.* **27**, 271.

Goldstein, M. L. (1978), 'An instability of finite amplitude circularly polarized Alfvén waves', *Astrophys. J.* **219**, 700–704.

Goossens, M. & Hollweg, J. V. (1993), 'Resonant behaviour of MHD waves on magnetic flux tubes. IV. Total resonant absorption and MHD radiating eigenmodes', *Solar Phys.* **145**, 19–44.

Goossens, M. & Poedts, S. (1992), 'Linear resistive magnetohydrodynamic computations of resonant absorption of acoustic oscillations in sunspots', *Astrophys. J.* **384**, 348–360.

Goossens, M., Hollweg, J. & Sakurai, T. (1992), 'Resonant behaviour of MHD waves on magnetic flux tubes. II. Effect of equilibrium flow', *Solar Phys.* **138**, 233–255.

Goossens, M., Ruderman, M. S. & Hollweg, J. V. (1995), 'Dissipative MHD solutions for resonant Alfvén waves in 1-dimensional magnetic flux tubes', *Solar Phys.* **157**, 75–102.

Gordon, B. E. & Hollweg, J. V. (1983), 'Collisional damping of surface waves in the solar corona', *Astrophys. J.* **266**, 373–382.

Griffths, D. J. (1989), *Introduction to Electrodynamics*, 2nd edn, Prentice Hall, Englewood Cliffs, New Jersey.

Grozev, D., Shivarova, A. & Boardman, A. D. (1987), 'Envelope solitons of surface waves in a plasma column', *J. Plasma Phys.* **38**, 427–437.

Guernsey, R. L. (1969), 'Surface waves in hot plasmas', *Phys. Fluids* **12**, 1852–1857.

Gurnett, D. A. & Goertz, C. K. (1981), 'Multiple Alfvén wave reflections excited by Io: origin of the Jovian decametric arcs', *J. Geophys. Res.* **86**, 717–722.

Hain, K. & Lüst, R. (1958), 'Zur Stabilität zylindersymmetrischer Plasmakonfigurationen mit Volumenströmen', *Z. Naturforschung* **13a**, 936.

Hamabata, H. (1993), 'Parametric instabilities of circularly polarized Alfvén waves in high-β plasmas', *Astrophys. J.* **406**, 563–568.

Harris, E. (1962), 'On a plasma sheath separating regions of oppositely directed magnetic field', *Il Nuovo Cim.* **23**, 115–121.

Hasegawa, A. . (1976), 'Particle acceleration by MHD surface wave and formation of aurora', *J. Geophys. Res.* **81**, 5083–5090.

Hasegawa, A. & Chen, L. (1974), 'Theory of magnetic pulsations', *Space Sci. Rev.* **16**, 347–359.

Hasegawa, A. & Chen, L. (1976), 'Kinetic processes in plasma heating by resonant mode conversion of Alfvén waves', *Phys. Fluids* **19**, 1924–1934.

Hasegawa, A. & Mima, K. (1976), 'Exact solitary Alfvén wave', *Phys. Rev. Lett.* **37**, 690–693.

Hasegawa, A. & Uberoi, C. (1982), *The Alfvén Wave*, Technical Information Center, U.S. Dept of Energy, Oak Ridge, Tennessee.

Havnes, O., Hartquist, T. W. & Pilipp, W. (1989), 'Wave propagation in dusty cool stellar envelopes', *Astron. Astrophys.* **217**, L13–L15.

Herlofsen, N. (1950), 'Magnetohydrodynamic waves in a compressible fluid conductor', *Nature* **165**, 1020.

Heyvaerts, J. & Priest, E. R. (1983), 'Coronal heating by phase–mixed Alfvén waves', *Astron. Astrophys.* **117**, 220–234.

Hollweg, J. V. (1971), 'Density fluctuations driven by Alfvén waves', *J. Geophys. Res.* **76**, 5155–5161.

Hollweg, J. V. (1982), 'Surface waves on solar wind tangential discontinuities', *J. Geophys. Res.* **87**, 8065–8076.

Hollweg, J. V. (1988), 'Resonance absorption of solar p-modes by sunspots', *Astrophys. J.* **335**, 1005–1014.

Hollweg, J. V. (1999), 'Kinetic Alfvén wave revisited', *J. Geophys. Res.* **104**, 14811–14819.

Hollweg, J. V. & Roberts, B. (1984), 'Surface solitary waves and solitons', *J. Geophys. Res.* **89**, 9703–9710.

Hood, A. W. & Priest, E. R. (1979), 'Kink instability of solar coronal loops as the cause of solar flares', *Solar Phys.* **64**, 303–321.

Hopcraft, K. I. & Smith, P. R. (1986), 'Magnetohydrodynamic waves in a neutral sheet', *Planet. Space Sci.* **34**, 1253–1257.

Huba, J. D. (1996), 'Impulsive plasmoid penetration of a tangential discontinuity: two–dimensional ideal and Hall magnetohydrodynamics', *J. Geophys. Res.* **101**, 24855–24868.

Huddleston, D. E., Coates, A. J., Johnston, A. D. & Neubauer, F. M. (1993), 'Mass loading and velocity diffusion models for heavy pickup ions at comet Grigg-Skjellerup', *J. Geophys. Res.* **98**, 20995–21002.

Hughes, W. J. (1995), *in* M. G. Kivelson & C. T. Russell, eds, 'Introduction to Space Physics', Cambridge University Press, chapter 9, p. 227.

Hung, N. T. (1974), 'Parametric excitation of Alfvén and acoustic waves', *J. Plasma Phys.* **12**, 445–453.

Ionson, J. A. (1978), 'Resonant absorption of Alfvénic surface waves and the heating of solar coronal loops', *Astrophys. J.* **226**, 650–673.

Jayanti, V. & Hollweg, J. V. (1993), 'Parametric instabilities of parallel–propagating Alfvén waves: some analytical results', *J. Geophys. Res.* **98**, 19049–19063.

Kalita, M. K. & Kalita, B. C. (1986), 'Finite–amplitude solitary Alfvén waves in a low–beta plasma', *J. Plasma Phys.* **35**, 267–272.

Kaplan, S. A. & Tsytovich, V. N. (1973), *Plasma Astrophysics*, Pergamon Press, Oxford.

Karney, C. F. F., Perkins, F. W. & Sun, Y.-C. (1979), 'Alfvén resonance effects on magnetosonic modes in large tokamaks', *Phys. Rev. Lett.* **42**, 1621–1624.

Kazbegi, A. Z., Machabeli, G. & Melikidze, G. I. (1991), 'On the circular polarization in pulsar emission', *Mon. Not. R. Ast. Soc.* **253**, 377–387.

Kennel, C. F., Buti, B., Hada, T. & Pellat, R. (1988), 'Nonlinear, dispersive, elliptically polarized Alfvén waves', *Phys. Fluids* **31**, 1949–1461.

Kirk, J. G. (1994), Particle acceleration, *in* A.O.Benz & T. J.-L. Courvoisier, eds, 'Plasma Astrophysics', Saas-Fee Advanced Course 24, Lecture Notes 1994, Swiss Society for Astrophysics and Astronomy, Springer-Verlag, Berlin, chapter 3, p. 225.

Kivelson, M. G. (1995a), *in* M. G. Kivelson & C. T. Russell, eds, 'Introduction to Space Physics', Cambridge University Press, chapter 2, p. 27.

Kivelson, M. G. (1995b), *in* M. G. Kivelson & C. T. Russell, eds, 'Introduction to Space Physics', Cambridge University Press, chapter 11, p. 330.

Klima, R., Longinov, A. V. & Stepanov, K. N. (1975), 'High frequency heating of plasma with two ion species', *Nucl. Fusion* **15**, 1157–1171.

Krauss-Varban, D., Omidi, N. & Quest, K. B. (1994), 'Mode properties of low–frequency waves: kinetic theory versus Hall-MHD', *J. Geophys. Res.* **99**, 5987–6009.

Kuo, S. P., Whang, M. H. & Lee, M. C. (1988), 'Filamentation instability of large–amplitude Alfvén waves', *J. Geophys. Res.* **93**, 9621–9627.

Kuo, S. P., Whang, M. H. & Schmidt, G. (1989), 'Convective filamentation instability of circularly polarized Alfvén waves', *Phys. Fluids* **B1**, 734–740.

Kuvshinov, B. N., Pegoraro, F., Rem, J. & Schep, T. J. (1999), 'Drift-Alfvén vortices with finite ion gyroradius and electron inertia effects', *Phys. Plasmas* **6**, 713–728.

Lashmore-Davies, C. N. (1976), 'Modulated instability of a finite amplitude Alfvén wave', *Phys. Fluids* **19**, 857–859.

Lashmore-Davies, C. N. & Ong, R. S. B. (1974), 'Parametric excitation of Alfvén and ion acoustic waves', *Phys. Rev. Lett.* **32**, 1172–1175.

Lee, L. C., Albano, R. K. & Kan, J. R. (1981), 'Kelvin-Helmholtz instability in the magnetopause-boundary layer region', *J. Geophys. Res.* **86**, 54–58.

Lee, M. A. & Ip, W. H. (1987), 'Hydromagnetic wave excitation by ionized interstellar hydrogen and helium in the solar wind', *J. Geophys. Res.* **92**, 11041–11052.

Lee, M. A. & Völk, H. J. (1973), 'Damping and nonlinear wave-particle interactions of Alfvén waves in the solar wind', *Astrophys. Sp. Sci.* **24**, 31–49.

Lehane, J. A. & Paoloni, F. J. (1972a), 'Parametric amplification of Alfvén waves', *Plasma Phys.* **14**, 461–471.

Lehane, J. A. & Paoloni, F. J. (1972b), 'The propagation of non–axisymmetric Alfvén waves in an argon plasma', *Plasma Phys.* **14**, 701–711.

Leonovich, A. S., Mazur, V. A. & Senatorov, V. N. (1983), 'Alfven waveguide', *Soviet Phys. JETP* **58**, 83–85.

Litwin, C. & Prager, S. C. (1998), 'Alpha effect of Alfvén waves and current drive in reversed–field pinches', *Phys. Plasmas* **5**, 553–555.

Lominadze, D. G., Machabeli, G. Z. & Mikhailovskii, A. B. (1979), 'Influence of magneto-bremsstrahlung on quasilinear relativistic plasma relaxation in a strong magnetic field', *Sov. J. Plasma Phys.* **5**, 748–752.

Lominadze, D. G., Machabeli, G. Z., Melikidze, G. I. & Pataraya, A. D. (1986), 'Magneto-spheric plasma of a pulsar', *Sov. J. Plasma Phys.* **12**, 712–721.

Longtin, M. & Sonnerup, B. U. O. (1986), 'Modulational instability of circularly polarized Alfvén waves', *J. Geophys. Res.* **91**, 6816–6824.

Lottermoser, R.-F. & Scholer, M. (1997), 'Undriven magnetic reconnection in magnetohydro-dynamics and Hall magnetohydrodynamics', *J. Geophys. Res.* **102**, 4875–4892.

Lou, Y.-Q. (1990), 'Viscous magnetohydrodynamic modes and p–mode absorption by sunspots', *Astrophys. J.* **350**, 452–462.

Louarn, P., Wahlund, J. E., Chust, T., de Feraudy, H., Roux, A., Holback, B., Dovner, P. O., Eriksson, A. I. & Holmgren, G. (1994), 'Observation of kinetic Alfvén waves by the Freja spacecraft', *Geophys. Res. Lett.* **21**, 1847–1850.

Luhmann, J. G. (1995), *in* M. G. Kivelson & C. T. Russell, eds, 'Introduction to Space Physics', Cambridge University Press, chapter 7, p. 183.

Lundberg, J. (1994), 'Nonlinear wave propagation along a magnetic flux tube', *Solar Phys.* **154**, 215–230.

Lundquist, S. (1949), 'Experimental investigations of magneto–hydrodynamic waves', *Phys. Rev.* **76**, 1805.

Lysak, R. L. & Lotko, W. (1996), 'On the kinetic dispersion relation for shear Alfvén waves', *J. Geophys. Res.* **101**, 5085–5094.

Mac Low, M.-M., Norman, M. L., Königl, A. & Wardle, M. (1995), 'Incorporation of ambipo-lar diffusion into the Zeus magnetohydrodynamics code', *Astrophys. J.* **442**, 726–735.

Marsch, E. & Tu, C. Y. (1993), 'Modeling results on spatial transport and spectral transfer of solar wind alfvénic turbulence', *J. Geophys. Res.* **98**, 21045–21059.

Mathis, J. S., Rumpl, W. & Nordsiek, K. H. (1977), 'The size distribution of interstellar grains', *Astrophys. J.* **217**, 425.

McKean, M. E., Winske, D. & Gary, S. P. (1992), 'Mirror and ion cyclotron anisotropy insta-bilities in the magnetosheath', *J. Geophys. Res.* **97**, 19421–19432.

Melrose, D. (1986), *Instabilities in Space and Laboratory Plasmas*, Cambridge University Press, Cambridge.

Melrose, D. & Gedalin, M. E. (1999), 'Relativistic plasma emission and pulsar radio emission: a critique', *Astrophys. J.* **521**, 351–361.

Melrose, D. & McPhedran, R. (1991), *Electromagnetic Processes in Dispersive Media*, Cam-bridge University Press, Cambridge.

Melrose, D. B. (2000), 'Wave dispersion in highly relativistic plasma', *Phys. Scripta* **T84**, 7–11.

Melrose, D. B., Gedalin, M. E., Kennet, M. P. & Fletcher, C. S. (1999), 'Dispersion in an intrinsically relativistic, one–dimensional, strongly magnetized pair plasma', *J. Plasma Phys.* **62**, 233–248.

Mendis, D. A. & Rosenberg, M. (1994), 'Cosmic dusty plasma', *Ann. Rev. Astron. Astrophys.* **32**, 419–463.

Mestel, L. (1968), 'Magnetic braking by a stellar wind. I.', *Mon. Not. R. Ast. Soc.* **138**, 359–391.

Mestel, L. & Spitzer, L. (1956), 'Star formation in magnetic dust clouds', *Mon. Not. R. Ast. Soc.* **116**, 503–514.

Mett, R. R. & Tataronis, J. A. (1989), 'Current drive via magnetohydrodynamic helicity waves', *Phys. Rev. Lett.* **63**, 1380–1383.

Mikhailovskii, A. B. (1967), Oscillations of an inhomogeneous plasma, *in* M. A. Leontovich, ed., 'Reviews of Plasma Physics', Vol. 3, Consultants Bureau, New York, p. 159.

Mio, K., Ogino, T., Minami, K. & Takeda, S. (1976), 'Modified nonlinear Schrödinger equation for Alfvén waves propagating along the magnetic field in cold plasmas', *J. Phys. Soc. Japan* **41**, 265–271.

Mjølhus, E. (1976), 'On the modulational instability of hydromagnetic waves parallel to the magnetic field', *J. Plasma Phys.* **16**, 321–334.

Mjølhus, E. (1989), 'Nonlinear Alfvén waves and the DNLS equation: oblique aspects', *Phys. Scripta* **40**, 227–237.

Mjølhus, E. & Wyller, J. (1986), 'Alfvén solitons', *Phys. Scripta* **33**, 442–451.

Montgomery, D. & Harding, R. (1966), 'Parametric excitation of Alfvén waves', *Phys. Lett.* **23**, 670.

Morales, G. J. & Maggs, J. E. (1997), 'Structure of kinetic Alfvén waves with small transverse scale length', *Phys. Plasmas* **4**, 4118–4125.

Morales, G. J., Loritsch, R. S. & Maggs, J. E. (1994), 'Structure of Alfvén waves at the skin–depth scale', *Phys. Plasmas* **1**, 3765–3774.

Musielak, Z. E. & Suess, S. T. (1989), 'MHD surface waves in high and low beta plasmas. Part 1. Normal–mode solutions', *J. Plasma Phys.* **42**, 75–89.

Nishikawa, K. (1968), 'Parametric excitation of coupled waves. 1. General formulation', *J. Phys. Soc. Japan* **24**, 916–922.

Osterbrock, D. E. (1961), 'The heating of the solar chromosphere, plages, and corona by magnetohydrodynamic waves', *Astrophys. J.* **134**, 347–387.

Ovenden, C. R., Shah, H. A. & Schwartz, S. J. (1983), 'Alfvén solitons in the solar wind', *J. Geophys. Res.* **88**, 6095–6101.

Paoloni, F. J. (1978), 'Coupling to fast eigenmodes in a non–uniform plasma', *Nucl. Fusion* **18**, 359–366.

Parker, E. N. (1994), *Spontaneous Discontinuities in Magnetic Fields*, Oxford University Press, Oxford.

Perkins, F. W. (1977), 'Heating tokamaks via the ion cyclotron and ion-ion hybrid resonances', *Nucl. Fusion* **17**, 1197–1224.

Pilipp, W. & Hartquist, T. W. (1994), 'Grains in shocks in weakly ionized clouds. II. Magnetic field rotation in oblique shocks', *Mon. Not. R. Astron. Soc.* **267**, 801–810.

Pilipp, W., Hartquist, T. W., Havnes, O. & Morfill, G. E. (1987), 'The effects of dust on the propagation and dissipation of Alfvén waves in interstellar clouds', *Astrophys. J.* **314**, 341–351.

Pippard, A. B. (1978), *The Physics of Vibration*, Vol. 1, Cambridge University Press, Cambridge.

Pneuman, G. W. (1965), 'Hydromagnetic waves in a nonuniform current carrying plasma column', *Phys. Fluids* **8**, 507–516.

Poedts, S., Goossens, M. & Kerner, W. (1989), 'Numerical simulation of coronal heating by resonant absorption of Alfvén waves', *Solar Phys.* **123**, 83–115.

Pridmore-Brown, D. C. (1966), 'Alfvén waves in a stratified incompressible fluid', *Phys. Fluids* **9**, 1290–1292.

Pu, Z.-Y. & Kivelson, M. G. (1983), 'Kelvin-Helmholtz instability at the magnetopause: solution for compressible plasmas', *J. Geophys. Res.* **88**(A2), 841–852.

Rauf, S. & Tataronis, J. A. (1995), 'On the Alfvén resonance and its existence', *Phys. Plasmas* **2**, 340–342.

Roberts, B. (1981), 'Wave propagation in a magnetically structured atmosphere. I. Surface waves at a magnetic interface', *Solar Phys.* **69**, 27–38.

Roberts, B. (1985), 'Solitary waves in a magnetic flux tube', *Phys. Fluids* **28**, 3280–3286.

Roberts, B. & Mangeney, A. (1982), 'Solitons in solar magnetic flux tubes', *Mon. Not. R. Astron. Soc.* **198**, 7P–11P.

Roberts, D. A., Goldstein, M. L., Matthaeus, W. H. & Ghosh, S. (1992), 'Velocity shear generation of solar wind turbulence', *J. Geophys. Res.* **97**, 17115.

Ross, D. W., Chen, G. L. & Mahajan, S. M. (1982), 'Kinetic description of Alfvén wave heating', *Phys. Fluids* **25**, 652–667.

Rowe, G. W. (1991), 'General dispersion relation for surface waves on a plasma-vacuum interface: an image approach', *J. Plasma Phys.* **46**, 495–511.

Rowe, G. W. (1992), 'General dispersion relation for surface waves on a plasma-vacuum interface: application to magnetised plasmas', *Aust. J. Phys.* **45**, 55–74.

Rowe, G. W. (1993), 'Collisionless damping of fast and ion cyclotron surface waves', *Aust. J. Phys.* **46**, 271–304.

Ruderman, M. S. (1988), 'Longitudinal propagation of nonlinear surface Alfvén waves at a magnetic interface in a compressibleatmosphere', *Plasma Phys. Contr. Fusion* **30**, 1117–1125.

Ruderman, M. S. & Goossens, M. (1993), 'Nonlinearity effects on resonant absorption of surface Alfvén waves in incompressibleplasmas', *Solar Phys.* **143**, 69–88.

Ruderman, M. S., Goossens, M. & Zhelyazkov, I. (1995), 'Comment on "Alfvén resonance reconsidered: exact equations for wave propagation across a cold inhomogeneous plasma"', *Phys. Plasmas* **2**, 3547–3549.

Ryutova, M. P. (1988), 'Negative–energy waves in a plasma with structured magnetic fields', *Sov. Phys. JETP* **67**, 1594–1601.

Sagdeev, R. Z. & Galeev, A. A. (1969), *Nonlinear Plasma Theory*, W. A. Benjamin, Inc., New York.

Sahyouni, W., Zhelyazkov, I. & Nenovski, P. (1988), 'Dark envelope solitons of fast magnetosonic surface waves in solar flux tubes', *Solar Phys.* **115**, 17–32.

Sakai, J.-I. & Sonnerup, B. U. Ö. (1983), 'Modulational instability of finite amplitude dispersive Alfvén waves', *J. Geophys. Res.* **88**, 9069–9079.

Sakai, J.-I., Kawata, T., Yoshida, K., Furusawa, K. & Cramer, N. F. (2000a), 'Simulation of a collision between shock waves and a magnetic flux tube: excitation of surface Alfvén waves and body Alfvén waves', *Astrophys. J.* **537**, 1063–10072.

Sakai, J.-I., Mizuhata, Y., Kawata, T. & Cramer, N. F. (2000b), 'Simulation of nonlinear waves in a magnetic flux tube near the quiet solar photospheric network', *Astrophys. J.* **544**, 1108–1121.

Sakurai, T., Goosens, M. & Hollweg, J. V. (1991a), 'Resonant behaviour of MHD waves on magnetic flux tubes. I: Connection formulae at the resonant surfaces', *Solar Phys.* **133**, 227–245.

Sakurai, T., Goossens, M. & Hollweg, J. V. (1991b), 'Resonant behaviour of MHD waves on magnetic flux tubes. II: Absorption of sound waves by sunspots', *Solar Phys.* **133**, 247–262.

Schmidt, G. (1979), *Physics of High Temperature Plasmas*, 2nd edn, Academic Press, New York.

Seboldt, W. (1990), 'Nonlocal analysis of low–frequency waves in the plasma sheet', *J. Geophys. Res.* **95**, 10471–10479.

Sedlacek, Z. (1971), 'Electrostatic oscillations in cold inhomogeneous plasma. I. Differential equation approach', *J. Plasma Phys.* **5**, 239–263.

Seyler, C. E., Clark, A. E., Bonnell, J. & Wahlund, J.-E. (1998), 'Electrostatic broadband ELF wave emission by Alfvén wave breaking', *J. Geophys. Res.* **103**, 7027–7041.

Seyler, C. E., Wahlund, J. E. & Holback, B. (1995), 'Theory and simulation of low–frequency plasma waves and comparison to Freja satellite observations', *J. Geophys. Res.* **100**, 21453–21472.

Sharma, A. S., Cargill, P. J. & Papadopoulos, K. (1988), 'Resonance absorption of Alfvén waves at comet-solar wind interaction regions', *Geophys. Res. Lett.* **15**, 740–743.

Sharma, O. P. & Patel, V. L. (1986), 'Low–frequency electromagnetic waves driven by gyrotropic gyrating ion beams', *J. Geophys. Res.* **91**, 1529–1534.

Shercliff, J. A. (1965), *A Textbook of Magnetohydrodynamics*, Pergamon Press, Oxford.

Shivarova, A. & Zhelyazkov, I. (1982), Surface waves in gas–discharge plasmas, *in* A. D. Boardman, ed., 'Electromagnetic Surface Modes', Wiley, Chichester, p. 465.

Shukla, P. K., Rahman, H. U. & Sharma, R. P. (1982), 'Alfvén soliton in a low–beta plasma', *J. Plasma Phys.* **28**, 125–131.

Smith, J. M., Roberts, B. & Oliver, R. (1997), 'Magnetoacoustic wave propagation in current sheets', *Astron. Astrophys.* **327**, 377–387.

Southwood, D. J. & Hughes, W. J. (1983), 'Theory of hydromagnetic waves in the magnetosphere', *Space Sci. Revs.* **35**, 301–366.

Spangler, S. R. & Plapp, B. B. (1992), 'Characteristics of obliquely propagating, nonlinear Alfvén waves', *Phys. Fluids* **B 4**, 3356–3370.

Spangler, S. R. & Sheerin, J. P. (1982), 'Properties of Alfvén solitons in a finite–beta plasma', *J. Plasma Phys.* **27**, 193–198. Corrigendum **32**, 347 (1985).

Spangler, S. R., Sheerin, J. P. & Payne, G. L. (1985), 'A numerical study of nonlinear Alfvén waves and solitons', *Phys. Fluids* **28**, 104–109.

Spies, G. O. & Li, J. (1990), 'On the kinetic Alfvén wave', *Phys. Fluids* **B 2**, 2287–2293.

Spitzer, L. (1978), *Physical Processes in the Interstellar Medium*, Wiley, New York.

Stasiewicz, K., Gustafsson, G., Marklund, G., Lindqvist, P.-A., Clemmons, J. & Zanetti, L. (1997), 'Cavity resonators and Alfvén resonance cones observed on Freja', *J. Geophys. Res.* **102**, 2565–2575.

Stefant, R. J. (1970), 'Alfvén wave damping from finite gyroradius coupling to the ion acoustic mode', *Phys. Fluids* **13**, 440–450.

Steinolfson, R. S., Priest, E. R., Poedts, S., Nocera, L. & Goosens, M. (1986), 'Viscous normal modes on coronal inhomogeneities and their role as a heating mechanism', *Astrophys. J.* **304**, 526–531.

Stenuit, H., Poedts, S. & Goossens, M. (1993), 'Total resonant absorption of acoustic oscillations in sunspots', *Solar Phys.* **147**, 13–28.

Stepanov, K. N. (1965), 'Concerning the effect of plasma resonance on surface wave propagation in a nonuniform plasma', *Soviet Phys. Tech. Phys.* **10**, 773.

Stewart, G. A. & Laing, E. W. (1992), 'Wave propagation in equal–mass plasmas', *J. Plasma Phys.* **47**, 295–315.

Stix, T. H. (1980), *in* 'Proceedings of 2nd joint Varenna-Grenoble International Symposium on Heating in Toroidal Plasmas', Como, p. 631.

Stix, T. H. (1992), *Waves in Plasmas*, American Institute of Physics, New York.

Storer, R. G. & Schellhase, A. R. (1995), 'Stable resistive Alfvén spectrum', *Phys. Scripta* **T60**, 57–64.

Stringer, T. E. (1963), 'Low–frequency waves in an unbounded plasma', *Plasma Phys. (J. Nuc. Energy Part C)* **5**, 89–107.

Suzuki, M. & Sakai, J.-I. (1996), 'Nonlinear Alfvén waves associated with filament currents in weakly ionized plasmas', *Astrophys. J.* **465**, 393–405.

Suzuki, M. & Sakai, J.-I. (1997), 'Filamentation instability of nonlinear Alfvén waves in weakly ionized plasmas', *Astrophys. J.* **487**, 921–929.

Swanson, D. G. (1975), 'Mode conversion of toroidal Alfvén waves', *Phys. Fluids* **18**, 1269–1276.

Sy, W. N.-C. (1978), 'Spectrum of axisymmetric torsional Alfvén waves', *Phys. Fluids* **21**, 702–704.

Sy, W. N.-C. (1984), 'Influence of resistivity on localized shear Alfvén wave propagation in a nonuniform plasma', *Plasma Phys. Contr. Fusion* **26**, 915–920.

Sy, W. N.-C. (1985), 'High frequency spectrum of a diffuse linear pinch with multiple ion species', *Aust. J. Phys.* **38**, 143–155.

Tanaka, M., Sato, T. & Hasegawa, A. (1987), 'Macroscale particle simulation of kinetic Alfvén waves', *Geophys. Res. Lett.* **14**, 868–871.

Tanenbaum, B. S. (1967), *Plasma Physics*, McGraw Hill, New York.

Tataronis, J. A. (1975), 'Energy absorption in the continuous spectrum of ideal MHD', *J. Plasma Phys.* **13**, 87–105.

Tataronis, J. A. & Grossman, W. (1976), 'On Alfvén wave heating and transit time magnetic pumping in the guiding–centre model of a plasma', *Nucl. Fusion* **16**, 667–678.

Tataronis, J. A. & Petrzilka, V. (1996), 'Magnetohydrodynamic ponderomotive forces generated about Alfvén resonance layers', *Phys. Plasmas* **3**, 4434–4439.

Taylor, J. B. (1989), 'Current drive by plasma waves and helicity conservation', *Phys. Rev. Lett.* **63**, 1384–1385.

Terasawa, T., Hoshino, M., Sakai, J.-I. & Hada, T. (1986), 'Decay instability of finite–amplitude circularly polarized Alfvén waves: a numerical simulation of stimulated brillouin scattering', *J. Geophys. Res.* **91**, 4171–4187.

Thompson, W. B. (1961), 'The dynamics of high temperature plasmas', *Rep. Prog. Phys.* **24**, 363–424.

Thorne, R. M. & Tsurutani, B. T. (1987), 'Resonant interactions between cometary ions and low frequency electromagnetic waves', *Planet. Sp. Sci.* **35**, 1501–1511.

Tripathi, K. D. & Sharma, S. K. (1996), 'Dispersion properties of low–frequency waves in magnetized dusty plasmas with dust size distribution', *Phys. Plasmas* **3**, 4380–4385.

Tsurutani, B. T. & Smith, E. J. (1986), 'Hydromagnetic waves and instabilities associated with cometary ion pickup: ICE observations', *Geophys. Res. Lett.* **13**, 263–266.

Tsurutani, B. T., Richardson, I. G., Thorne, R. M., Butler, W., Smith, E. J., Cowley, S. W. H., Gary, S. P., Akasofu, S. I. & Zwickl, R. D. (1985), 'Observations of the right–hand resonant ion beam instability in the distant plasma sheet boundary layer', *J. Geophys. Res.* **90**, 12159–12172.

Tsurutani, B. T., Thorne, R. M., Smith, E. J., Gosling, J. T. & Matsumoto, H. (1987), 'Steepened magnetosonic waves at comet Giacobini-Zinner', *J. Geophys. Res.* **92**, 11074–11082.

Tsytovich, V. N. & Havnes, O. (1993), 'Charging processes, dispersion properties and anomalous transport in dusty plasma', *Comm. Plasma Phys. Contr. Fusion* **15**, 267–280.

Uberoi, C. (1972), 'Alfvén waves in inhomogeneous magnetic fields', *Phys. Fluids* **15**, 1673–1675.

Uberoi, C. (1994), 'Resonant absorption of Alfvén waves near a neutral point', *J. Plasma Phys.* **52**, 215–221.

Uberoi, C. & Datta, A. (1998), 'Ion-neutral collision effects on Alfvén surface waves', *Phys. Plasmas* **5**, 4149–4155.

Uberoi, C. & Narayan, A. S. (1986), 'Effect of variation of magnetic field direction of hydromagnetic surface waves', *Plasma Phys. Contr. Fusion* **28**, 1635–1643.

Uberoi, C. & Somasundaram, K. (1982), 'Two–mode structure of Alfvén surface waves', *Phys. Rev Lett.* **49**, 39–42.

Uberoi, C., Lanzerotti, L. J. & Wolfe, A. (1996), 'Surface waves and magnetic reconnection at a magnetopause', *J. Geophys. Res.* **101**, 24979–24983.

Vahala, G. & Montgomery, D. (1971), 'Parametric amplification of Alfvén waves', *Phys. Fluids* **14**, 1137–1140.

Verheest, F. (2000), *Waves in Dusty Space Plasmas*, Kluwer Academic Publishers, Dordrecht.

Verheest, F. & Meuris, P. (1995), 'Whistler–like instabilities due to charge fluctuations in dusty plasmas', *Phys. Lett.* **A198**, 228–232.

Verheest, F. & Meuris, P. (1998), 'Interaction of the solar wind with dusty cometary plasmas', *Phys. Scripta* **T75**, 84–87.

Verheest, F. & Shukla, P. K. (1995), 'Linear and quasilinear Alfvén waves in dusty plasmas with anisotropic pressures', *Phys. Scripta* **T60**, 136–139.

Villard, L., Brunner, S., Jaun, A. & Vaclavik, J. (1995), 'Alfvén wave heating and stability', *Phys. Scripta* **T60**, 44–56.

Vladimirov, S. V. (1994a), 'Amplification of electromagnetic waves in dusty nonstationary plasmas', *Phys. Rev.* **E 49**, R997–R999.

Vladimirov, S. V. (1994b), 'Propagation of waves in dusty plasmas with variable charges on dust particles', *Phys. Plasmas* **1**, 2762–2768.

Vladimirov, S. V. & Cramer, N. F. (1996), 'Nonlinear Alfvén waves in magnetised plasmas with impurities or dust', *Phys. Rev* **E 54**, 6762–6768.

Vladimirov, S. V., Tsytovich, V. N., Popel, S. I. & Khakimov, F. K. (1995), *Modulational Interactions in Plasmas*, Kluwer Academic Publishers, Dordrecht.

Vladimirov, S. V., Yu, M. Y. & Tsytovich, V. N. (1994), 'Recent advances in the theory of nonlinear surface waves', *Phys. Rep.* **241**, 1–63.

Völk, H. J. & Cesarsky, C. J. (1982), 'Nonlinear Landau damping of Alfvén waves in a high β plasma', *Z. Naturforsch.* **37a**, 809–815.

Volokitin, A. S., Krasnosel'skikh, V. V. & Machabeli, G. Z. (1985), 'Waves in the relativistic electron-positron plasma of a pulsar', *Sov. J. Plasma Phys.* **11**, 310–314.

Wardle, M. (1998), 'Dust grains and the structure of steady C-type magnetohydrodynamic shock waves in molecular clouds', *Mon. Not. R. Ast. Soc.* **298**, 507–524.

Wardle, M. & Ng, C. (1999), 'The conductivity of dense molecular gas', *Mon. Not. R. Ast. Soc.* **303**, 239–246.

Weber, E. J. & Davis, L. (1967), 'The angular momentum of the solar wind', *Astrophys. J.* **148**, 217–227.

Wentzel, D. G. (1979a), 'The dissipation of hydromagnetic surface waves', *Astrophys. J.* **233**, 756–764.

Wentzel, D. G. (1979b), 'Hydromagnetic surface waves', *Astrophys. J.* **227**, 319–322.

Wessen, K. P. & Cramer, N. F. (1991), 'Finite–frequency surface waves on current sheets', *J. Plasma Phys.* **45**, 389–406.

Whang, Y. C. (1997), 'Attenuation of magnetohydrodynamic waves', *Astrophys. J.* **485**, 389–397.

Winglee, R. M. (1984), 'Alfvén wave heating in a multiple ion–component plasma', *Plasma Phys. Contr. Fusion* **26**, 511–524.

Winglee, R. M. (1994), 'Non–MHD influences on the magnetospheric current system', *J. Geophys. Res.* **99**, 13437–13454.

Winske, D., Wu, C. S., Li, Y. Y., Mou, Z. Z. & Guo, S. Y. (1985), 'Coupling of newborn ions to the solar wind by electromagnetic instabilities and their interaction with the bow shock', *J. Geophys. Res.* **90**, 2713–2726.

Wong, H. K. & Goldstein, M. L. (1986), 'Parametric instabilities of circularly polarized Alfvén waves including dispersion', *J. Geophys. Res.* **91**, 5617–5628.

Woods, L. C. (1962), 'Hydromagnetic waves in a cylindrical plasma', *J. Fluid Mechanics* **13**, 570–586.

Woods, L. C. (1987), *Principles of Magnetoplasma Dynamics*, Clarendon Press, Oxford.

Wort, D. J. H. (1971), 'The peristaltic tokamak', *Plasma Phys.* **13**, 258–262.

Wu, C. C. & Kennel, C. F. (1992), 'Evolution of small–amplitude intermediate shocks in a dissipative and dispersive system', *J. Plasma Phys.* **47**, 85–109.

Wu, C. S. (1995), 'Alfvén waves in the solar wind', *Phys. Scripta* **T60**, 91–96.

Wu, C. S. & Davidson, R. C. (1972), 'Electromagnetic instabilities produced by neutral-particle ionization by interplanetary space', *J. Geophys. Res.* **77**, 5399–5406.

Wu, D.-J. & Wang, D.-Y. (1996), 'Solitary kinetic Alfvén waves on the ion–acoustic velocity branch in a low–beta plasma', *Phys. Plasmas* **3**, 4304–4306.

Wu, D.-J., Huang, G.-L., Wang, D.-Y. & Fälthammar, C.-G. (1996), 'Solitary kinetic Alfvén waves in the two–fluid model', *Phys. Plasmas* **3**, 2879–2884.

Yamauchi, M. & Lui, A. T. Y. (1997), 'Modified magnetohydrodynamic waves in a current sheet in space', *Phys. Plasmas* **4**, 4382–4387.

Yu, M. Y. & Shukla, P. K. (1978), 'Finite amplitude solitary Alfvén waves', *Phys. Fluids* **21**, 1457–1458.

Zhelyazkov, I., Murawski, K. & Goossens, M. (1996), 'MHD surface waves in a complex (longitudinal + sheared) magnetic field', *Solar Phys.* **165**, 99–114.

Index